FUNDAMENTALS of NATURAL GAS PROCESSING

MECHANICAL ENGINEERING
A Series of Textbooks and Reference Books

Founding Editor

L. L. Faulkner

*Columbus Division, Battelle Memorial Institute
and Department of Mechanical Engineering
The Ohio State University
Columbus, Ohio*

FUNDAMENTALS OF NATURAL GAS PROCESSING

Arthur J. Kidnay

William R. Parrish

Taylor & Francis
Taylor & Francis Group
Boca Raton London New York

CRC is an imprint of the Taylor & Francis Group,
an informa business

CRC Press
Taylor & Francis Group
6000 Broken Sound Parkway NW, Suite 300
Boca Raton, FL 33487-2742

International Standard Book Number-10: 0-8493-3406-3 (Hardcover)
International Standard Book Number-13: 978-0-8493-3406-1 (Hardcover)
Library of Congress Card Number 2005036359

Library of Congress Cataloging-in-Publication Data

Kidnay, A. J.
    Fundamentals of natural gas processing / Arthur J. Kidnay and William Parrish.
        p. cm. -- (Mechanical engineering)
    Includes bibliographical references and index.
    ISBN-13: 978-0-8493-3406-1 (acid-free paper)
    ISBN-10: 0-8493-3406-3 (acid-free paper)
    1. Gas industry. I. Parrish, William, 1914- II. Title. III. Mechanical engineering
series (Boca Raton, Fla.)

TP751.K54 2006
665.7'3--dc22   2005036359

Visit the Taylor & Francis Web site at
http://www.taylorandfrancis.com

and the CRC Press Web site at
http://www.crcpress.com

Dedication

To our wives, Joan and Joan, for their enduring support and patience throughout the preparation of this book.

Preface

The natural gas industry began in the early 1900s in the United States and is still evolving. This high-quality fuel and chemical feedstock plays an important role in the industrial world and is becoming an important export for other countries. Several high-quality books* provide guidance to those experienced in natural gas processing. This book introduces the natural gas industry to a reader entering the field. It also helps those providing a service to the industry in a narrow application to better understand how their products and services fit into the overall process.

To help the reader understand the need of each processing step, the book follows the gas stream from the wellhead to the market place. The book focuses primarily on the gas plant processes. Wherever possible, the advantages, limitations, and ranges of applicability of the processes are discussed so that their selection and integration into the overall gas plant can be fully understood and appreciated.

The book compiles information from other books, open literature, and meeting proceedings** to hopefully give an accurate picture of where the gas processing technology stands today, as well as indicate some relatively new technologies that could become important in the future. An invaluable contribution to the book is the insight provided to the authors by experts in certain applications.

---

* For example, *GPSA Engineering Data Book* (Gas Processors Supply Association, Tulsa, OK, 12th Edition, 2004), and the fifth edition of Kohl and Nielsen, *Gas Purification* (Gulf Publishing, Houston, TX, 1997).

** The two most important meetings involving natural gas processing in the United States are the annual meeting of the Gas Processors Association and the Laurance Reid Gas Conditioning Conference.

Acknowledgments

The authors communicated with numerous people in preparing this book. It could not have been written without the aid of the Gas Processors Association (GPA). Ron Brunner graciously supplied requested material from the vast literature on gas processing available through the GPA. Dan McCartney provided valuable insight and comments while he generously took time to review the manuscript. Carter Tannehill kindly provided us with the cost data provided in Chapter 14. In most cases, the private communications referenced in this book involved numerous letters and conversations. Phil Richman and John Peranteaux willingly provided both technical input and editorial comment. Others who provided valuable input include Joe Kuchinski, Charles Wallace, Ed Wichert, Dendy Sloan, Veet Kruka, and Dale Embry. A number of companies graciously provided us with drawings and photographs. One company generously supplied a modified drawing that replaced their product names with generic names so that the figure could be used. Finally, we appreciate the patience and assistance of the editorial staff at Taylor and Francis.

AUTHORS

Arthur Kidnay, Ph.D., P. E., is professor emeritus, Chemical Engineering Department, Colorado School of Mines (CSM). He was a research engineer with the National Institute of Standards and Technology (NIST) for 9 years before joining the faculty of CSM. He has taught and conducted extensive research in the fields of vapor–liquid equilibria, physical adsorption, and heat transfer. Dr. Kidnay is the author of 69 technical papers and has advised 42 M.S. and Ph.D. students. He remains very active in professional activities at CSM and presently teaches a senior course in natural gas processing. For 26 years, Dr. Kidnay and four colleagues have taught a continuing education course in gas processing to engineers and scientists from the natural gas industry.

In recognition of his services to the engineering profession, he was elected a Fellow of the American Institute of Chemical Engineers, in 1987 and was appointed by the governor of Colorado to two terms (1984–1992) on the Board of Registration for Professional Engineers. He served on the Cryogenic Conference Executive Board from 1969 through 1972 and received the Russell B. Scott Memorial Award for the outstanding technical paper presented at the 1966 Cryogenic Engineering Conference. Professor Kidnay was NATO Senior Science Fellow at Oxford University in the summer of 1972.

William R Parrish, Ph.D., P.E., is a retired senior research associate. He spent 25 years in research and development at ConocoPhillips (formerly Phillips Petroleum Company) where he obtained physical properties data needed for new processes and for resolving operation problems. He provided company-wide technical expertise on matters involving physical properties and gas hydrates. He also participated on six gas plant optimization teams. His work has appeared in 49 technical publications and he holds two patents. He presently teaches a continuing education course in gas processing for engineers and scientists from industry.

Dr. Parrish represented his company on various committees including the Gas Processors Association's Enthalpy Committee of Section F. He also participated on Department of Energy peer review committees. He is a Fellow of the American Institute of Chemical Engineers and is actively involved in professional engineer examination development.

Table of Contents

1 Overview of the Natural Gas Industry

1.1 INTRODUCTION

The Chinese are reputed to have been the first to use natural gas commercially, some 2,400 years ago. The gas was obtained from shallow wells, transported in bamboo pipes, and used to produce salt from brine in gas-fired evaporators. Manufactured, or town, gas (gas manufactured from coal) was used in both Britain and the United States in the late 17th and early 18th centuries for streetlights and house lighting The next recorded commercial use of natural gas occurred in 1821. William Hart drilled a shallow 30-foot (9-meter) well in Fredonia, New York, and, by use of wooden pipes, transported the gas to local houses and stores (Natural Gas Suppliers Association, 2004).

During the following years, a number of small, local programs involved natural gas, but large-scale activity began in the early years of the 20th century. The major boom in gas usage occurred after World War II, when engineering advances allowed the construction of safe, reliable, long-distance pipelines for gas transportation. At the end of 2004, the United States had more than 297,000 miles (479,000 kilometers) of gas pipelines, both interstate and intrastate. In 2004 the U.S. was the world's second largest producer of natural gas (19.2 trillion cubic feet [Tcf]*, 543 BSm³) and the leading world consumer (22.9 Tcf, 647 BSm³). (Energy Information Administration, 2005h and BP Statistical Review of World Energy, 2005)

Although the primary use of natural gas is as a fuel, it is also a source of hydrocarbons for petrochemical feedstocks and a major source of elemental sulfur, an important industrial chemical. Its popularity as an energy source is expected to grow substantially in the future because natural gas presents many environmental advantages over petroleum and coal, as shown in Table 1.1. Carbon dioxide, a greenhouse gas linked to global warming, is produced from oil and coal at a rate approximately 1.4 to 1.75 times higher than production from natural gas.

Both atmospheric nitrogen and nitrogen in fuel are sources of nitrogen oxides (NO_X), which are greenhouse gases and a source of acid rain. Because both oil and coal contain nitrogen compounds not present in natural gas, the nitrogen oxides formed from burning natural gas are approximately 20% of those produced

---

* Gas volumes are normally reported in terms of standard cubic feet (scf) at standard conditions of 60°F and 14.7 psia. In metric units, the volumes are given in either normal cubic meters, Nm³, where standard conditions are 0°C, 1 bar, or standard cubic meters, Sm³, where the standard conditions are 15°C, 1 bar. In the U.S. gas industry, prefix M represents 10^3, and MM, B, and T represent 10^6, 10^9, and 10^{12}, respectively. We use this convention for both engineering and SI units.

TABLE 1.1
Pounds of Air Pollutants Produced per Billion Btu of Energy

Pollutant	Natural Gas[a]	Oil[b]	Coal[c]
Carbon dioxide	117,000	164,000	208,000
Carbon monoxide	40	33	208
Nitrogen oxides	92	448	457
Sulfur dioxide	0.6	1,122	2,591
Particulates	7.0	84	2,744
Formaldehyde	0.750	0.220	0.221
Mercury	0.000	0.007	0.016

[a] Natural gas burned in uncontrolled residential gas burners.
[b] Oil is # 6 fuel oil at 6.287 million Btu per barrel and 1.03% sulfur with no postcombustion removal of pollutants.
[c] Bituminous coal at 12,027 Btu per pound and 1.64% sulfur with no postcombustion removal of pollutants.

Source: Energy Information Administration (1998).

when oil or coal is burned. Particulate formation is significantly less in gas compared with coal and oil, an important environmental consideration because in addition to degrading air quality, high levels of particulates may pose significant health problems.

The values reported in Table 1.1 for sulfur dioxide can be misleading. Many natural gases contain considerable quantities of sulfur at the wellhead, but specifications for pipeline-quality gas require almost total sulfur removal before pipelining and sale. Consequently, the tabular values for natural gas represent combustion after removal of sulfur compounds, whereas the tabular values for oil and coal are reported for fuels with no sulfur recovery either before or after combustion. Nevertheless, gas produces far fewer pollutants than its competitors, and demand for gas, the clean fuel, is expected to rise significantly in the near future.

1.1.1 WORLD PICTURE FOR NATURAL GAS

The current status of primary energy sources is summarized in Figure 1.1. Basically, dry natural gas (natural gas with natural gas liquids [NGLs] removed) is on a par with coal in importance.

Six countries possess two thirds of the world's gas reserves (Figure 1.2), with almost half of the reserves located in Iran and Russia. The total reported natural gas reserves (~6,040 Tcf [171 TSm^3] at the beginning of 2005 [Energy Information Administration, 2005c]) do not include discovered reserves that are not economically feasible to bring to market. This "stranded gas" resides in remote regions, where the reserve size does not justify the cost of the infrastructure required to bring it to market. Note that proven reserve estimates are truly

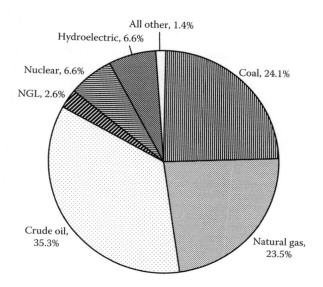

FIGURE 1.1 Primary sources of energy in the world in 2003. Total energy used was 405 quadrillion Btu (Energy Information Administration, 2005b).

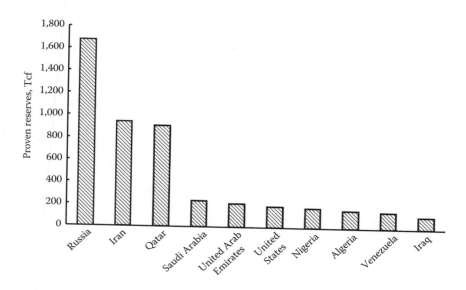

FIGURE 1.2 Major proven natural gas reserves by country. Total proven reserves estimated to be 6,040 Tcf (Energy Information Administration, 2005c).

TABLE 1.2
World Natural Gas Production and Estimated Proven
Reserves at End of 2002

Region	Gross Production[a]	Vented or Flared[a]	Reinjected[a]	Marketed Production[a]	Dry Gas Production[a]	Proven Reserves[b]
North	33,060	176	3.895	28,487	26,893	255,800
America	(936)	(4.98)	(110)	(807)	(762)	(7,243)
	29.5%	6.3%	31.0%	29.5%	29.2%	4.6%
Central and	5,983	350	1,404	4,229	3,722	250,100
South	(169)	(9.91)	(39.76)	(120)	(105)	(7,082)
America	5.3%	12.5%	11.2%	4.4%	4.0%	4.5%
Western Europe	12,333	135	1,236	10,963	10,548	191,600
	(349)	(3.82)	(35.0)	(310)	(299)	(5,426)
	11.0%	4.8%	9.8%	11.4%	11.4%	3.5%
Eastern Europe	27,047	253[c]	1	27,046	27,046	1,964,200
and former	(766)	(7.16)	(0.03)	(766)	(766)	(55,620)
U.S.S.R.	24.1%	9.1%	0.0%	28.0%	29.3%	35.7%
Middle East	12,667	413	2,696	9,558	8,674	1,579,700
	(359)	(11.69)	(76.34)	(271)	(246)	(44,732)
	11.3%	14.8%	21.4%	9.9%	9.4%	28.7%
Africa	9,450	1,241	3,007	5,202	4,741	418,200
	(268)	(35.14)	(85.15)	(147)	(134)	(11,842)
	8.4%	44.5%	23.9%	5.4%	5.1%	7.6%
Asia and	11,637	224	331	11,083	10,528	445,400
Oceania	(330)	(6.34)	(9.37)	(314)	(298)	(12,612)
	10.4%	8.0%	2.6%	11.5%	11.4%	8.1%
World total	112,178	2,792	12,570	96,568	92,152	5,504,900
	(3,177)	(79.06)	(355.94)	(2,735)	(2,609)	(155,881)

[a] Data from Energy Information Administration (2005d).
[b] Data from Energy Information Administration (2004a).
[c] Value given is for 1998 as an estimate because value for 2002 was unreported.
Values are in Bcf (BSm^3) and percentage values are percent of world total.

estimates and vary among sources. Also, proven reserves depend on gas prices; increased gas price causes reserve estimates to rise.

The world production of natural gas is summarized in Table 1.2. Noteworthy are the relationships between production and reserves in North America and Eastern Europe and the high percentage of gas flared or vented in Africa. North America (principally the United States) has the world's second largest production of dry gas and accounts for 29% of world production but possesses only 5% of the reserves. Eastern Europe slightly leads North America in dry gas production but has 36% of

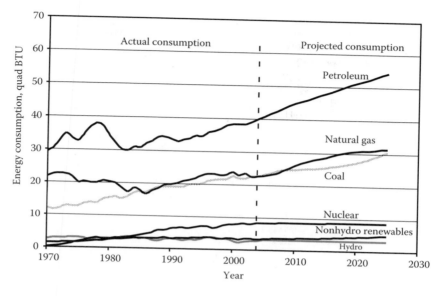

FIGURE 1.3 United States energy consumption by fuel. (Adapted from Energy Information Administration, 2005a.)

the world reserves; three quarters of those reserves are located in Russia. Africa vents or flares 13% of gross production, an exceptionally high number considering that the world average, excluding Africa, is an estimated 2.3%. The disproportionately high loss in Africa is caused by the lack of infrastructure in many of the developing nations. Nigeria alone flares 2 MMscfd (56 MSm³/d*), which is equivalent to the total annual power generation in sub-Saharan Africa. An effort is underway to reduce flaring and to convert much of the gas to LNG for export (Anonymous, 1999).

1.1.2 NATURAL GAS IN UNITED STATES

Natural gas plays an extremely important role in the United States and accounts for approximately 23% of the total energy used. Figure 1.3 shows the relationship among energy sources in the United States, as well as projected growth through 2025. Gas is presently second only to petroleum, and the difference in demand for gas over coal is expected to increase substantially with time. Of interest is the prediction that energy from nuclear and hydroelectric sources will be flat, and nonhydroelectric renewables are not expected to play a significant role through 2025.

The distribution of natural gas from the wellhead through consumption is shown in Figure 1.4. The numbers reveal some significant points. First, substantial amounts of the gross gas produced (14%) are returned to the reservoir for repressurization of the field. Second, the loss of gas because of venting or flaring is quite small,

---

* In this book the symbol M represents 1000 for both engineering and SI units.

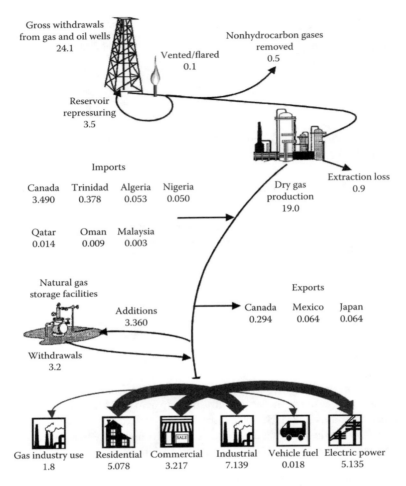

FIGURE 1.4 Natural gas supply and disposition in the United States in 2003. Values shown are in Tcf. (Adapted from Energy Information Administration, 2005d.)

only 0.4% of the gross withdrawal. Third, the nonhydrocarbon gases removed (2.5% of gross) occur in sufficient quantities to render the gas unmarketable, and the extraction losses (4.1% of gross) refer to liquids (NGL) removed from the gas and sold separately. Fourth, the imports that account for approximately 18% of the consumption come predominately from Canada.

In November, 2005, the average wellhead, city gate, and residential prices were $9.84, $11.45 and $15.80 per thousand cubic feet, respectively (Energy Information Agency, 2006 i).

Table 1.3 shows that in the area of production and reserves, imports account for approximately 19% of consumption, but, of that amount, LNG imports are only 2.9% of total consumption. Also worthy of note is that proven reserves in 2004 constituted only an 8-year supply at the current rate of consumption.

TABLE 1.3
Natural Gas in the United States, 2004

U.S. Production	18 Tcf (510 BSm³)
U.S. Consumption	22.4 Tcf (634 BSm³)
U.S. Imports	4.2 Tcf (120 BSm³)
U.S. Exports	0.85 Tcf (24 BSm³)
Wellhead price	$5.49/Mscf
Average city-gate price[a]	$6.65/Mscf
Average price to residential customers	$10.74/Mscf
Average price to commercial customers	$9.26/Mscf
Average price to industrial customers	$6.41/Mscf
Average price to electrical utilities	$5.56/Mscf
LNG imports	0.65 Tcf (18 BSm³)
Number of producing gas and gas condensate wells (2003)	393,327
Pipeline miles (2003)	306,000 (492,000 km)
Pipeline capacity (2003)	178 Bcfd (5 BSm³/day)
Dry natural gas proven reserves	189 Tcf (5.35 BSm³)

[a] City gate is the point where the gas is transferred from the pipeline to the distribution facilities.
Source: Energy Information Administration (2005g).

1.1.3 Nonconventional Gas Reserves in United States

At present, the two major potential nonconventional gas sources are coal bed methane (CBM) and naturally occurring gas hydrates. The United States Geological Survey (USGS) estimates 700 Tcf (20 TSm³) of CBM in the United States, but only 100 Tcf (3 TSm³) are recoverable with existing technology (Nuccio, 2000). The most active region is the Powder River Basin area of Wyoming and Montana. Environmental concerns may limit production (National Petroleum Technology Office, 2004).

Naturally occurring gas hydrates (see Chapter 3) form on the ocean bottom and in sediments of permafrost regions, such as northern Canada and Alaska. The USGS estimates about 320,000 Tcf (9,000 TSm³) of methane in hydrates in the United States; one half of that reserve is in offshore Alaska (Collett, 2001). An estimated 45 Tcf (1.2 TSm³) in gas hydrates is on the North Slope of Alaska, where oil is currently produced. These reserves would be the most economically attractive to produce because the hydrates are concentrated, and much of the infrastructure for gas processing already exists. However, for the gas to reach the market, a pipeline must be built.

1.2 SOURCES OF NATURAL GAS

Conventional natural gas generally occurs in deep reservoirs, either associated with crude oil (associated gas) or in reservoirs that contain little or no crude oil (nonassociated gas). Associated gas is produced with the oil and separated at the casinghead or wellhead. Gas produced in this fashion is also referred to as casinghead gas, oil well gas, or dissolved gas. Nonassociated gas is sometimes referred to as gas-well gas or dry gas. However, this dry gas can still contain significant amounts of NGL components. Roughly 93% of the gas produced in the United States is nonassociated (Energy Information Administration, 2004b). A class of reservoirs, referred to as gas condensate reservoirs, occurs where, because of the high pressures and temperatures, the material is present not as a liquid or a gas but as a very dense, high-pressure fluid.

Figure 1.5 shows a simplified flow of material from reservoir to finished product and provides an overall perspective of the steps involved in taking natural

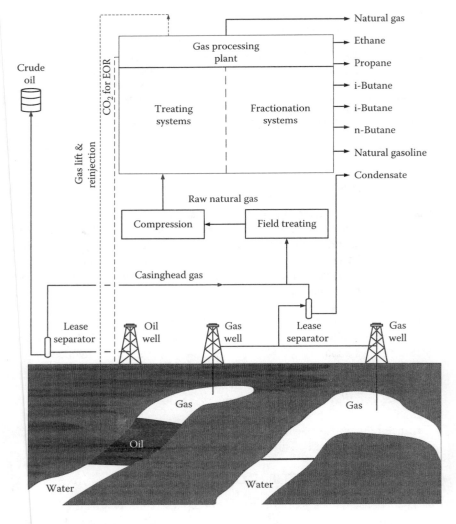

FIGURE 1.5 Schematic overview of natural gas industry. (Adapted from Cannon, 1993.)

gas from the wellhead to the customer. The chapters that follow provide more detail on the various steps. Note that Figure 1.5 oversimplifies the gas gathering systems. These systems typically are complex, and they bring gas from many fields and leases to gas plants.

Some gas plants receive feeds from refineries. These streams differ from natural gases in that they can contain propylene and butylene. They may also contain trace amounts of undesirable nitrogen compounds and fluorides. This book considers only the processing of gas and liquids coming directly from gas and oil leases.

1.3 NATURAL GAS COMPOSITIONS

1.3.1 TRADITIONAL NATURAL GAS

Traditional natural gases, that is, associated and unassociated gas from wells, vary substantially in composition. Table 1.4 shows a few typical gases. Water is almost always present at wellhead conditions but is typically not shown in the analysis. Some gas fields, however, contain no water. Unless the gas has been dehydrated before it reaches the gas processing plant, the common practice is to assume the entering gas is saturated with water at the plant inlet conditions.

TABLE 1.4
Typical Gas Compositions

	Canada (Alberta)	Western Colorado	Southwest Kansas	Bach Ho Field[a] Vietnam	Miskar Field Tunisia	Rio Arriba County, New Mexico	Cliffside Field, Amarillo, Texas
Helium	0.0	0.0	0.45	0.00	0.00	0.0	1.8
Nitrogen	3.2	26.10	14.65	0.21	16.903	0.68	25.6
Carbon dioxide	1.7	42.66	0.0	0.06	13.588	0.82	0.0
Hydrogen sulfide	3.3	0.0	0.0	0.00	0.092	0.0	0.0
Methane	77.1	29.98	72.89	70.85	63.901	96.91	65.8
Ethane	6.6	0.55	6.27	13.41	3.349	1.33	3.8
Propane	3.1	0.28	3.74	7.5	0.960	0.19	1.7
Butanes	2.0	0.21	1.38	4.02	0.544	0.05	0.8
Pentanes and heavier	3.0	0.25	0.62	2.64	0.630	0.02	0.5

[a] Tabular mol% data is on a wet basis (1.3 mol% water)

Source: U.S. Bureau of Mines (1972) and Jones et al. (1999).

1.3.2 IMPORTANT IMPURITIES

A number of impurities can affect how the natural gas is processed:

Water. Most gas produced contains water, which must be removed. Concentrations range from trace amounts to saturation.

Sulfur species. If the hydrogen sulfide (H_2S) concentration is greater than 2 to 3%, carbonyl sulfide (COS), carbon disulfide (CS_2), elemental sulfur, and mercaptans* may be present.

Mercury. Trace quantities of mercury may be present in some gases; levels reported vary from 0.01 to 180 $\mu g/Nm^3$. Because mercury can damage the brazed aluminum heat exchangers used in cryogenic applications, conservative design requires mercury removal to a level of 0.01 $\mu g/Nm^3$ (Traconis et al., 1996)

NORM. Naturally occurring radioactive materials (NORM) may also present problems in gas processing. The radioactive gas radon can occur in wellhead gas at levels from 1 to 1,450 pCi/l (Gray, 1990).

Diluents. Although the gases shown in Table 1.4 are typical, some gases have extreme amounts of undesirable components. For example, according to Hobson and Tiratso (1985), wells that contain as much as 92% carbon dioxide (Colorado), 88% hydrogen sulfide (Alberta, Canada), and 86% nitrogen (Texas) have been observed.

Oxygen. Some gas-gathering systems in the United States operate below atmospheric pressure. As a result of leaking pipelines, open valves, and other system compromises, oxygen is an important impurity to monitor. A significant amount of corrosion in gas processing is related to oxygen ingress.

1.3.3 COAL BED METHANE

Coal beds contain large amounts of natural gas (usually referred to as coal bed methane, or CBM) that is adsorbed on the internal surfaces of the coal or absorbed within the coal's molecular structure. This gas can be produced in significant quantities from wells drilled into the coal seam by lowering the reservoir pressure. As is the case with conventional natural gas, the composition of the CBM produced varies widely. In addition to methane, these gases may contain as much as 20% ethane and heavier hydrocarbons, as well as substantial levels of carbon dioxide. However, a typical CBM analysis would reveal water saturation, up to 10% carbon dioxide, up to 1% nitrogen, no or very small amounts of ethane and heavier hydrocarbons, and a balance of methane. Because water is normally

---

* Mercaptans are highly reactive and odiferous, organic compounds with the formula RSH, in which R represents an alkane group. Natural gases typically contain methyl through amyl mercaptans. The ethyl and propyl mercaptans are added to natural gas and propane as odorants. They received their name from being reactive with mercury. The compounds readily oxidize in the presence of air and metal to form disulfides that are nearly odorless.

TABLE 1.5
Quality of Proven Natural Gas Reserves of the Lower 48
United States in 1998

	Bcf (BSm³)	%
High quality	87,679 (2,464)	59
Subquality	60,698 (1,699)	41
High N_2 only	15,617 (424)	11
High CO_2 only	17,932 (481)	12
High H_2S only	5 691 (161)	4
High N_2 & CO_2	1,577 (29)	1
High N_2 & H_2S	600 (17)	0
High CO_2 & H_2S	12,697 (340)	9
High N_2, CO_2, & H_2S	6,585 (170)	4
Total	148,377 (4,191)	100

Source: Meyer (2000).

present in the reservoir, it is produced in significant amounts along with the CBM, and this produced water can pose a significant problem because it may contain large quantities of dissolved solids that make it unfit for domestic or agricultural uses (National Petroleum Technology Office, 2004).

1.3.4 SUBQUALITY GAS

The Gas Research Institute (Meyer, 2000) classified natural gases from the lower 48 states as high quality and subquality. Subquality is divided into seven categories, depending on the amount of N_2, CO_2, and H_2S present. For their definition of subquality. The gas contains more than 2% CO_2, 4% N_2, and 4 ppmv H_2S. Table 1.5 summarizes the evaluation for proven raw reserves.

1.4 CLASSIFICATION

Natural gases commonly are classified according to their liquids content as either lean or rich and according to the sulfur content as either sweet or sour. This section provides some quantification of these qualitative terms.

1.4.1 LIQUIDS CONTENT

Gas composition plays a critical role in the economics of gas processing. The more liquids, usually defined as C_2+, in the gas, the "richer" the gas. Extraction of these liquids produces a product that may have a higher sales value than does natural gas.

To quantify the liquids content of a natural gas mixture, the industry uses GPM, or gallons of liquids recoverable per 1,000 standard cubic feet (Mscf) of gas. (In metric units, the quantity is commonly stated as m³ of liquid per 100 m³ of gas.) The term usually applies to ethane and heavier components but sometimes applies instead to propane and heavier components. Determination of the GPM requires knowledge of the gas composition on a mole basis and the gallons of liquid per lb-mole. See Appendix B for the gallons per lb-mole for ethane and higher hydrocarbons. Note that ethane is not a liquid at 60°F (15.5°C), so the value is a hypothetical value accepted throughout the industry. Also, the actual volume of liquid obtained from a gas will be less than the GPM value because complete recovery of ethane and propane is impractical for two reasons:

1. Cost. The low temperature and high compression energy required generally makes recovery of more than about 90 to 95% of the ethane, 98% of the propane, and 99% of the butanes uneconomical. Higher ethane recovery plants also have higher recovery of propane and heavier components.
2. Heating value specifications. As discussed below, a specification applies to the heating value of gas. Unless the gas contains no nonflammable diluents (i.e., N_2 and CO_2), additional hydrocarbons must be in the gas to obtain the required heating value.

Example 1.1 Calculate the GPM of the Alberta gas given in Table 1.4.
Computation of the GPM requires summation of the product of the number of moles of each component in 1,000 scf of gas by the gallons of liquid per mole for that component.

Basis: 1,000 scf of gas.

A lb-mole of gas at standard conditions has a volume of 379.49 ft³. This volume translates into 1,000/379.49, or 2.6351 lb-moles for 1,000 scf. This value is multiplied by the mole fraction of each component in the gas and by the gallons of liquid for each component. Table 1.6 summarizes the calculations.

TABLE 1.6
Calculation of GPM of Alberta Gas

	Mole %	Moles	Gal/Mole	GPM
Helium	0	0.0000	0	0
Nitrogen	3.2	0.0843	0	0
Carbon dioxide	1.7	0.0448	0	0
Hydrogen sulfide	3.3	0.0870	0	0
Methane	77.1	2.0317	0	0
Ethane	6.6	0.1739	10.126	1.761
Propane	3.1	0.0817	10.433	0.852
Butanes	2.0	0.0527	12.162	0.641
Pentanes and heavier	3.0	0.0791	13.713	1.084
Totals	100.0	2.6351		4.338

For this example, the Gal/mole for butanes was taken as the average of isobutane and n-butane; the value for C_5+ was taken to be that of pure n-pentane. The resulting GPM for this gas is 4.34.

The rich and lean terms refer to the amount of recoverable hydrocarbons present. The terms are relative, but a lean gas will usually be 1 GPM, whereas a rich gas may contain 3 or more GPM. Thus, the gas described above is considered fairly rich.

1.4.2 SULFUR CONTENT

Sweet and sour refer to the sulfur (generally H_2S) content. A sweet gas contains negligible amounts of H_2S, whereas a sour gas has unacceptable quantities of H_2S, which is both odiferous and corrosive. When present with water, H_2S is corrosive. The corrosion products are iron sulfides, FeS_x, a fine black powder. Again, the terms are relative, but generally, sweet means the gas contains less than 4 ppmv of H_2S. The amount of H_2S allowable in pipeline-quality gas is between 0.25 and 1.0 grains per 100 scf (6 to 24 mg/Sm^3, 4 to 16 ppmv).

1.5 PROCESSING AND PRINCIPAL PRODUCTS

The two primary uses for natural gas are as a fuel and as a petrochemical feedstock, and consequently, the three basic reasons for processing raw natural gas are the following:

- Purification. Removal of materials, valuable or not, that inhibit the use of the gas as an industrial or residential fuel
- Separation. Splitting out of components that have greater value as petrochemical feedstocks, stand alone fuels (e.g., propane), or industrial gases (e.g., ethane, helium)
- Liquefaction. Increase of the energy density of the gas for storage or transportation

Depending on the situation, a process may be classified as either separation or purification. For example, if a small amount of H_2S is removed, incinerated, and vented to the atmosphere, the process is purification, but if large amounts of H_2S are removed and converted to elemental sulfur, often a low-priced commodity, the process is considered separation. Figure 1.6 provides an overview of the materials present in natural gas and the slate of possible products from the gas plant.

Although the principal use of natural gas is the production of pipeline-quality gas for distribution to residential and industrial consumers for fuel, a number of components in natural gas are often separated from the bulk gas and sold separately.

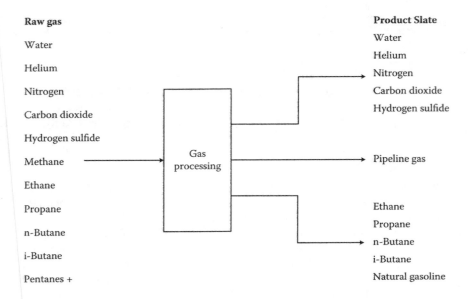

FIGURE 1.6 Generic raw gas and product slate.

1.5.1 METHANE

The principal use of methane is as a fuel; it is the dominant constituent of pipeline quality natural gas. Considerable quantities of methane are used as feedstock in the production of industrial chemicals, principally ammonia and methanol.

1.5.2 ETHANE

The majority of the ethane used in the United States comes from gas plants, and refineries and imports account for the remainder. In addition to being left in the gas for use as a fuel, ethane is used for the production of ethylene, the feedstock for polyethylene.

1.5.3 PROPANE

Gas plants produce about 45% of the propane used in the United States, refineries contribute about 44%, and imports account for the remainder. The principal uses are petrochemical (47%), residential (39%), farm (8%), industrial (4%), and transportation (2%) (Florida Propane Gas Council, 2005). A special grade of propane, called HD-5, is sold as fuel.

1.5.4 ETHANE–PROPANE MIX

When NGL is fractionated into various hydrocarbon streams, the butanes along with part of the propane are sometimes separated for use in local markets because they are transportable by truck. The remaining light ends, an ethane–propane mix (E-P mix), is then pipelined to a customer as a chemical or refining feedstock.

1.5.5 ISOBUTANE

Approximately 42% of the United States supply of isobutene comes from gas plants, refineries supply about 5% (this percentage does not include consumption of isobutane within the refinery), and imports are responsible for about 12%. The remaining isobutane on the market is furnished by isomerization plants that convert n-butane to isobutane. The three primary markets for isobutane are as a feedstock for MTBE (methyl tertiary butyl ether) production (which is being phased out), as a feedstock in the production of reformulated gasoline, and as a feedstock for the production of propylene oxide.

1.5.6 n-BUTANE

Gas plant production of n-butane accounts for about 63% of the total supply, refineries contribute approximately 31%, and imports account for the remainder. Domestic usage of n-butane is predominantly in gasoline, either as a blending component or through isomerization to isobutane. Specially produced mixtures of butanes and propane have replaced halocarbons as the preferred propellant in aerosols.

1.5.7 NATURAL GAS LIQUIDS

Natural gas liquids (NGL) include all hydrocarbons liquefied in the field or in processing plants, including ethane, propane, butanes, and natural gasoline. Such mixtures generated in gas plants are usually referred to as "Y-grade" or "raw product."

1.5.8 NATURAL GASOLINE

Natural gasoline, a mixture of hydrocarbons that consist mostly of pentanes and heavier hydrocarbons and meet GPA product specifications, should not be confused with natural gas liquids (NGL), a term used to designate all hydrocarbon liquids produced in field facilities or in gas plants.

The major uses of natural gasoline are in refineries, for direct blending into gasoline and as a feedstock for C_5/C_6 isomerization. It is used in the petrochemical industry for ethylene production.

1.5.9 SULFUR

Current sulfur production in the United States is approximately 15,000 metric tons per day (15 MMkg/d); about 85% comes from gas processing plants that convert H_2S to elemental sulfur. Some major uses of sulfur include rubber vulcanization, production of sulfuric acid, and manufacture of black gunpowder (Georgia Gulf Sulfur Corporation, 2005).

1.6 PRODUCT SPECIFICATIONS

1.6.1 NATURAL GAS

The composition of natural gas varies considerably from location to location, and as with petroleum products in general, the specifications for salable products from gas processing are generally in terms of both composition and performance criteria. For natural gas these criteria include Wobbe number, heating value, total inerts, water, oxygen, and sulfur content. The first two criteria relate to combustion characteristics. The latter three provide protection from pipeline plugging and corrosion.

Specifications have historically been established in contract negotiations and no firm, accepted standards exist for all products. Consequently, specifications for pipeline quality gas listed in Table 1.7 are typical but not definitive.

TABLE 1.7
Specifications for Pipeline Quality Gas

Major Components	Minimum Mol%	Maximum Mol%
Methane	75	None
Ethane	None	10
Propane	None	5
Butanes	None	2
Pentanes and heavier	None	0.5
Nitrogen and other inerts	None	3
Carbon dioxide	None	2–3
Total diluent gases	None	4–5

	Trace components
Hydrogen sulfide	0.25–0.3 g/100 scf
	(6–7 mg/m^3)
Total sulfur	5–20 g/100 scf
	(115–460 mg/m^3)
Water vapor	4.0–7.0 lb/MM scf
	(60–110 mg/m^3)
Oxygen	1.0%

	Other characteristics
Heating value	950–1,150 Btu/scf
(gross, saturated)	(35,400–42,800 kJ/m^3)
Liquids	Free of liquid water and hydrocarbons
	at delivery temperature and pressure
Solids	Free of particulates in amounts deleterious
	to transmission and utilization equipment

Source: Engineering Data Book (2004).

Hydrocarbon dew point is becoming an issue in some situations. The problem arises from trace condensation in pipelines, which can cause metering problems.

1.6.2 LIQUID PRODUCTS

As with gases, specifications for liquid products are based upon both composition and performance criteria. For liquid products, the performance specifications include Reid vapor pressure, water, oxygen, H_2S, and total sulfur content. Safety considerations make vapor pressure especially important for the liquid products because of regulations for shipping and storage containers. Table 1.8 gives major component and vapor-pressure specifications for common liquid products. Table 1.9 presents upper limits of common contaminants, but actual specifications vary,

TABLE 1.8
Major Components and Vapor Pressures of Common Liquid Products

Liquid Product	Composition[a]		Vapor Pressure[b] at 100°F, psig, max(at 37.8°C, kPa, max)
High-ethane raw	C_1	1–5 wt%	—
streams	C_2–C_5	balance	
Ethane–propane	C_1	0.6–1 wt%	—
mixes	C_2	22.5–40.5	
	C_3	59.5–77.5	
	C_4+	0.2–4.5	
High-purity	C_1	1.5–2.5 wt%	—
ethane	C_2	90–96	
	C_3	6–5	
	C_4+	0.5–3.0	
Commercial	Predominantly		208
propane	C_3 and $C_3=$		(1,434)
Commercial	Predominantly		70
butane	C_4 and $C_4=$		(483)
Commercial	Predominantly		208
butane–propane	C_4 and C_3		(1,434)
mixes			
Propane HD-5	C_3	> 90 liq vol%	208
	$C_3=$	< 5	(1,434)
	C_4+	< 2.5	

[a] Throughout the book C_1, C_2 etc, refer to methane, ethane, etc. The "=" denotes an olefin. The term C_4+ denotes propane and heavier compounds.
[b] Vapor pressure as defined by D1267-02 Standard Test Method for Gage Vapor Pressure of Liquefied Petroleum (LP) Gases (LP-Gas Method).

Source: Engineering Data Book (2004).

TABLE 1.9
**Maximum Levels of Major Contaminants of Common Liquefied Products
Concentrations are in ppmw unless specified otherwise.**

	H_2S	Total Sulfur[a]	CO_2	O_2	H_2O
High-ethane raw streams	50	200	3,500		No free[b]
Ethane–propane mixes	#1[a]	143	3,000	1,000	No free[b]
High–Purity ethane	10	70	5,000	5	No free[b]
Commercial propane	#1	185	—	—	Pass test[c]
Commercial butane	#1	140	—	—	—
Commercial butane–propane mixes	#1	140	—	—	—
Propane HD-5	#1	123	—	—	Pass test[c]

[a] Concentration acceptable provided the copper strip test, which detects all corrosive compounds, is passed. The #1 represents the passing score on the copper-strip test, D1838-05 Standard Test Method for Copper Strip Corrosion by Liquefied Petroleum (LP) Gases. Eckersley and Kane (2004) discuss sample handling problems related to the test.
[b] Limit is no free water present in product.
[c] Moisture level must be sufficiently low to pass the D2713-91(2001) Standard Test Method for Dryness of Propane (valve freeze method), which corresponds to roughly 10 ppmw.
Source: Engineering Data Book (2004).

depending upon contractual agreement. Water content specifications are less stringent for propane and butane because liquid pressures are lower, and hydrate formation is not such a threat. However, as Table 1.9 indicates, the water level in some propane products must pass a dryness test, which ensures that the water content is sufficiently low (< 25 ppmw) to avoid hydrate formation when water is vaporized through an orifice (see Chapter 3). Complete specifications for these products and others are available in GPA standards.

1.7 COMBUSTION CHARACTERISTICS

1.7.1 HEATING VALUE

One of the principal uses of natural gas is as a fuel, and consequently, pipeline gas is normally bought and sold (custody transfer) on the basis of its heating value. Procedures for calculating the heat effect in any chemical reaction are found in standard texts on thermodynamics (e.g., Smith et al., 2001).

Determination of the heating value of a fuel involves two arbitrary but conventional standard states for the water formed in the reaction:

1. All the water formed is a liquid (gross heating value, frequently called higher heating value [HHV])
2. All the water formed is a gas (net heating value, frequently called lower heating value [LHV])

The gas industry always uses the gross heating value in custody transfer. Obviously, the numerical difference between the two heating values is the heat of condensation of water at the specified conditions. Both states are hypothetical because the heating value is normally calculated at 60°F and 1 atm (15.6°C and 1.01 atm), standard conditions for the gas industry, and, thus at equilibrium, the water would be partially liquid and partially vapor. A common practice is also to assume ideal gas behavior, and consequently, the heating values commonly calculated and reported are representative of, but not identical to, the values obtained when the fuel is burned in an industrial or residential furnace.

Heating values for custody transfer are determined either by direct measurement, in which calorimetry is used, or by computation of the value on the basis of gas analysis. The method is set in the sales contract. The formulas for the calculation of ideal gas gross heating values, on a volumetric basis are (Gas Processors Association, 1996)

$$H_v^{id}(dry) = \sum_{i=1}^{n} x_i H_{vi}^{id} \tag{1.1}$$

$$H_v^{id}(sat) = (1 - x_w) \sum_{i=1}^{n} x_i H_{vi}^{id} \tag{1.2}$$

The equations assume that the gas analysis is given on a dry basis, that H_v^{id} is the ideal gross heating value (see Appendix B), and that the mole fraction of water is x_w when the gas is saturated at the specified conditions. The mole fraction can be calculated from

$$x_w = \frac{P_w^{Sat}}{P_b} \tag{1.3}$$

The vapor pressure of water at 60°F (15.6°C), the common base temperature, is 0.25636 psia (1.76754 kPa). The most commonly used base pressures, P_b, and the values of $(1 - x_w)$ are listed below.

P_b(psia)	$1 - x_w$
14.50	0.9823
14.65	0.9825
14.696	0.9826
14.73	0.9826
15.025	0.9829

The situation regarding water is further complicated by the fact that gas analyses are normally given on a dry basis, even though the gas may be partially or fully saturated with water. Consequently, heating value may be calculated on a dry basis, wet (saturated) basis, or, if the humidity is known, a partially saturated basis.

TABLE 1.10
Calculations with Heating Values Obtained from Appendix B

	Mole %	H_{vi}^{id} Btu/scf	$x_i H_{vi}^{id}$
Helium	0	0.0	0
Nitrogen	3.2	0.0	0
Carbon dioxide	1.7	0.0	0
Hydrogen sulfide	3.3	637.1	21.0
Methane	77.1	1010.0	778.7
Ethane	6.6	1769.7	116.8
Propane	3.1	2516.2	78.0
Butanes as isobutane	2.0	3252.0	65.0
Pentanes and heavier as hexane	3.0	4756.0	142.7
Totals	100.0		1202.2

A complete discussion of heating value calculations, including correction of the ideal gas values to the real-gas state by use of calculated compressibility factors (z), is available in GPA Standard 2172-96 (Gas Processors Association, 1996).

Example 1.2 Calculate the heating value of the Alberta gas given in Table 1.4. Assume the heating value for the butanes to be that of isobutene, and for the C_5+ fraction, use pure hexane.

Table 1.10 shows the calculations with heating values obtained from Appendix B.

This mixture has a gross heating value of 1,202.2 Btu/scf (44,886 kJ/Sm³).

Note that credit is not given for the heating value associated with H_2S in contractual situations. It is unlikely that a gas stream with 3.3% H_2S would be burned.

1.7.2 WOBBE NUMBER

In gas appliances, maintenance of the same combustion characteristics are desirable when one gas composition is switched to another. Several factors must be considered, but one of the more important considerations is maintenance of the same heat release at the burner for a given pressure drop through a control valve. This combustion characteristic is measured by the Wobbe number, defined as the gross heating value (Btu/scf) of the gas divided by the square root of the specific gravity (the ratio of the density of the gas divided by the density of air; both densities evaluated at the same pressure and temperature). Two gases with the same Wobbe number are interchangeable as far as heat release at the burner is concerned.

WB = (gross heating value)/(specific gravity)1/2,

with the specific gravity correcting for flow through an orifice. The Wobbe number normally has a value between 1,100 and 1,400. The Wobbe number is calculated from the gross heating value (Btu/scf) and specific gravity of the mixture, not from an average of the Wobbe numbers of the constituents of the mixture. In Europe a value of 1400 is generally required.

Some typical Wobbe numbers are

Methane	1,360 Btu/scf
Ethane	1,740 Btu/scf
Propane	2,044 Btu/scf
80% methane + 20% ethane	1,443 Btu/scf
95% methane + 5% ethane	1,381 Btu/scf

Wobbe numbers are often adjusted by blending the natural gas with air. For example, one distribution company maintains a Wobbe number between 1,130 and 1,280 for gas distribution to its residential customers by blending air with the natural gas in three blending stations. A typical set of operating conditions for one of the air-blending stations is given below.

Inlet Wobbe	1,335
Inlet Btu/scf	1,080 (40,324 kJ/Sm^3)
Inlet specific gravity	0.654
Normal outlet Wobbe	1,210
Peak day outlet Wobbe	1,280

Because air blending also changes the heating value of the gas, a balance must be maintained between the Wobbe number and the gross heating value.

The Wobbe number is more commonly used in Europe. However, it will become more important in the United States in the future as LNG importation increases. Heating values for LNG can be much higher than typical sales gas from a gas plant in the United States.

REFERENCES

Anonymous, Harnessing Abundant Gas Reserves, Africa Recovery, United Nations, 13 (1) 1999, http://www.un.org/ecosocdev/geninfo/afrec/vol13no1/jun99.htm, Retrieved August 2005.

BP Statistical Review of World Energy 2005.

http://www.bp.com/genericsection.do?categoryId=92&contentId=7005893

Cannon, R.E., *The Gas Processing Industry, Origins and Evolution*, Gas Processors Association, Tulsa, OK, 1993.

Collett, T., Natural Gas Hydrates: Vast Resource, Uncertain Future, U.S. Geological Survey Fact Sheet FS-021-01, 2001, http://pubs.usgs.gov/fs/fs021-01/, Retrieved October 2005.

Eckersley, N. and Kane, J.A., Designing customized desulfurization systems for the treatment of NGL streams, in the Proceedings of the Laurance Reid Gas Conditioning Conference, Norman, OK, 2004.

Energy Information Administration, U.S. Department of Energy, Natural Gas 1998, Issues and Trends, 1999, www.eia.doe.gov/oil_gas/natural_ gas/ analysis_publications/ natural_gas_1998_issues_and_trends/it98.html, Retrieved September 2005.

Energy Information Administration, U.S. Department of Energy, Natural Gas Annual 2003, 2004a. www.eia.doe.gov/pub/oil_gas/natural_gas/data_publications/natural_gas_ annual/current/pdf/nga03.pdf, Retrieved June 2005.

Energy Information Administration, U.S. Department of Energy, 2004 United States Total 2002, Distribution of Wells by Production Rate Bracket, 2004b, http://www.eia. doe.gov/pub/oil_gas/petrosystem/us_table.html Retrieved June 2005.

Energy Information Administration, U.S. Department of Energy, Annual Energy Outlook 2005 Market Trends—Energy Demand 2005, 2002, Figure 52, Industrial Delivered Energy Consumption By Fuel, 1970–2025, 2005a, www.eia.doe.gov/oiaf/ aeo/excel/figure52_data.xls, Retrieved June 2005.

Energy Information Administration, U.S. Department of Energy, Annual Energy Review, Table 11.1, World Primary Energy Production By Source, 1970–2002, 2005b, www.eia.doe.gov/aer/txt/ptb1101.html, Retrieved June 2005.

Energy Information Administration, U.S. Department of Energy, International Energy Annual 2003 2005, 2005c, www.eia.doe.gov/emeu/international/reserves.html, Retrieved June 2005.

Energy Information Administration, U.S. Department of Energy, International Energy Annual 2003, Table 4.1, World Natural Gas Production, 2002, 2005d, www. eia. doe.gov/pub/international/iea2003/table41.xls, Retrieved June 2005.

Energy Information Administration, U.S. Department of Energy, International Energy Annual 2003, Figure 2, 2002, 2005e, www.eia.doe.gov/pub/international/iea 2003 /table41.xls, Retrieved June 2005.

Energy Information Administration, U.S. Department of Energy, International Energy Outlook 2004, Figure 40, World Natural Gas Resources by Region As of January 1, 2004, 2005f, www.eia.doe.gov/oiaf/ieo/excel/figure_40data.xls, Retrieved June 2005.

Energy Information Administration, U.S. Department of Energy, Natural Gas Navigator, 2005g, tonto.eia.doe.gov/dnav/ng/ng_sum_lsum_dcu_nus_a.htm, Retrieved June 2005.

Energy Information Administration, U.S. Department of Energy, *Changes in U.S. Natural Gas Transportation Infrastructure in 2004,* 2005h, www.eia.doe.gov/Retrieved March, 2006.

Energy Information Administration, www.eia.doe.gov/ Retrieved March, 2006.

Engineering Data Book, Product Specifications, 12th ed., Sec. 2, Gas Processors Supply Association, Tulsa, OK, 2004.

Florida Propane Gas, Safety, Education and Research Council, www.propanefl.com, Retrieved August 2005.

Gas Processors Association, GPA Standard 2172-96, Calculation of Gross Heating Value, Relative Density, and Compressibility of Natural Gas Mixtures from Compositional Analysis, Tulsa, OK, 1996.

Georgia Gulf Sulfur Corporation, www.georgiagulfsulfur.com, Retrieved August 2005.

Gray, P., Radioactive materials could pose problems for the gas industry, *Oil and Gas J.*, 88 (26) 45, 1990.

Hobson, G.D., and Tiratso, E.N., *Introduction to Petroleum Geology*, Gulf Publishing, Houston, TX, 1985.

Jones, S., Lee, S., Evans, M., and Chen, R., Simultaneous Removal of Water and BTEX from Feed Gas for a Cryogenic Plant, in Proceedings of the Seventy-Eighth Annual Convention of the Gas Processors Association, Tulsa, OK, 1999, 108.

Meyer, H.S., Volume and distribution of subquality natural gas in the United States, *GasTIPS*, 6, 10, 2000.

National Petroleum Technology Office, U.S. Department of Energy, Coal Bed Methane Primer, New Source of Natural Gas—Environmental Implications Background and Development in the Rocky Mountain West, 2004, www.mines.edu/ research/ PTTC/CBM/primer/CBM%20PRIMER%20FINAL.pdf, Retrieved June 2005.

Natural Gas Suppliers Association, History, 2004, http://www.naturalgas.org/overview/ history.asp, Retrieved August 2005.

Nuccio, V., Coal-Bed Methane: Potential And Concerns, U. S. Geological Survey Fact Sheet FS-123-00, 2000, http://pubs.usgs.gov/fs/fs123-00/, Retrieved October 2005.

Smith, J.M., Van Ness, H.C., and Abbott, M.M., *Introduction to Chemical Engineering Thermodynamics*, 6th ed., McGraw-Hill, New York, 2001.

Traconis, B., Mierez, Y.D., and Jimenez, A., Mercury Removal Systems at Santa Barbara Extraction Plant, in Proceedings of the Seventy-Fifth Annual Convention of the Gas Processors Association, Tulsa, OK, 1996, 123.

U.S. Bureau of Mines, Analyses of Natural Gases, 1972, Information Circular # 8607, 1972.

BIBLIOGRAPHY

For additional information on the natural gas industry and gas processing, the following books are recommended.

Campbell, J.M., Gas *Conditioning and Processing, Volumes 1 and 2*. Campbell Petroleum Series, J.M. Campbell Co., Norman, OK, 1976.

Cannon, R.E., *The Gas Processing Industry, Origins and Evolution*, Gas Processors Association, Tulsa, OK, 1993.

Ikoku, C.U., *Natural Gas Engineering*, PennWell Publishing Co., Tulsa, OK, 1980.

Katz, D.L., Cornell, D., Kobayashi, R., Poettmann, F.H., Vary, J.A., Elenbass J.R., and Weinaug, C. F., *Handbook of Natural Gas Engineering*, McGraw-Hill, New York, 1959.

Katz, D.L. and. Lee, R.L, *Natural Gas Engineering, Production and Storage*, McGraw-Hill, New York, 1990.

Kohl, A., and Nielsen, R., *Gas Purification*, 5th ed., Gulf Publishing, Houston, TX, 1997.

WEB SITES

Energy Information Administration: www.eia.doe.gov. This site provides extensive reports, tabulation of data, and statistics regarding reserves, production, transportation, storage, and consumption of energy both in the United States and worldwide.

Gas Processors Association: www.gasprocessors.org. This site is the home page of the major industrial organization dealing with all aspects of gas and gas liquids processing.

Gas Industry Consortium: www.naturalgas.org. This site is the home page of the Natural Gas Supply Association. It provides a good overview of natural gas from exploration to consumption. The site www.ngsa.org gives an overview of the organization and items of importance to the natural gas industry.

International Energy Agency: www.iea.org. This site is the home page of the organization that provides a wide array of data regarding all forms of energy.

National Energy Technology Laboratory, Strategic Center for Natural Gas: www.netl.doe. gov/scngo/. This site is the starting point for learning about energy-related research programs of the U.S. Department of Energy.

2 Overview of Gas Plant Processing

2.1 ROLES OF GAS PLANTS

Gas plants play a variety of roles in the oil and gas industry. The desired end product dictates the processes required. Some primary purposes of gas plants include:

- Dehydration of gas to reduce corrosion and to prevent gas hydrate formation. These plants commonly are found on offshore platforms, where associated gas is separated from oil and dehydrated. Depending upon pipeline infrastructure, the gas may be recombined with the oil before it is put into a pipeline to shore.
- Associated oil stabilization. One of the world's largest gas plants is on the North Slope of Alaska. The sole purpose of the facility is to strip associated gas from the oil so that the oil is safe to put into the Aleyska pipeline. The produced gas is reinjected into the formation to enhance oil recovery. This gas will be sold if a gas pipeline is built between the North Slope and the lower 48 states.
- Carbon dioxide or nitrogen recovery for enhanced oil recovery (EOR). These plants separate natural gas from the CO_2 or N_2; the natural gas is marketed and the CO_2 or N_2 are reinjected into formations. In N_2 projects, an air plant may be constructed on site to provide additional N_2 at the beginning of the project.
- Upgrading subquality gas. To make the gas marketable, the undesired diluents N_2, H_2S, and CO_2 are removed. Of the three components, N_2 is the most difficult to remove because it requires cryogenic processing when large volumes are processed.
- Helium recovery. Few plants are dedicated primarily to helium recovery. Therefore, this facility is typically an addition to a gas plant. Natural gas is the primary source of helium. In the United States, the highest helium concentrations are in eastern Colorado, southwestern Wyoming, western Kansas, and the panhandles of Oklahoma and Texas.
- Liquefaction. Some gas plants are dedicated to the production of hydrocarbon liquids and a natural gas stream to make liquefied natural gas (LNG). These plants are in locations with large gas reserves and no pipelines to market. The only large-scale LNG production facility in the United States is in Alaska. However, the United States has more than 60 peak shaving plants that produce LNG from pipeline gas for storage purposes (Energy Information Administration, 2003).

However, the major role of gas plants in the United States is to process both associated and nonassociated gas to produce high-quality natural gas and hydrocarbon liquids. Sale of liquids provides a significant portion of the income from these plants. Plants optimize profits by adjusting the fraction of liquids recovered while meeting the specifications for the natural gas.

2.2 PLANT PROCESSES

Figure 2.1 is a block schematic of a gas plant that has all of the more common process elements. An overview of each element is given here and details are provided in the subsequent chapters. To show how the process elements are integrated, Chapter 15 discusses several gas plants with different configurations.

We are unaware of any plants that have all of the process steps shown, and the less common processes are shaded. Which processes are in a plant is dictated by the feed gas composition and conditions, along with the desired product streams.

2.2.1 FIELD OPERATIONS AND INLET RECEIVING

All plants have field operations and a network of pipelines that feed the raw natural gas and liquids into the plant. Field operations may include dehydration, CO_2 and H_2S removal, and compression. These processes are discussed more below. Unless the gas is completely free of any liquids, once it enters the plant, the gas and liquids go into inlet receiving, where the initial gas–liquid separation is made. Condensed water, hydrocarbon liquids, and solids are removed. Water and solids are processed for disposal, and the hydrocarbon liquids go on to liquids processing, as discussed below. Chapter 3 discusses both field operations and inlet receiving. The chapter also discusses gas hydrates and how to avoid their formation.

2.2.2 INLET COMPRESSION

Most plants have inlet compression, but compression requirements vary. High pressure is critical, as it drives the cryogenic-liquids recovery process. For inlet pressures of around 1,000 psi (70 bar) or higher, only gas coming from the liquids processing step needs compression. However, most onshore gas plants in the United States have some low-pressure gas streams entering the plant. Chapter 4 discusses compression. The same types of compressors are used for field and outlet compression.

2.2.3 GAS TREATING

Most plants have a gas treating step to remove the acid gases H_2S and CO_2, along with other sulfur impurities. Most plants use water-based absorbents to remove the impurities, but other solvents and processes are used. Chapter 5 discusses the various methods for removal of these components.

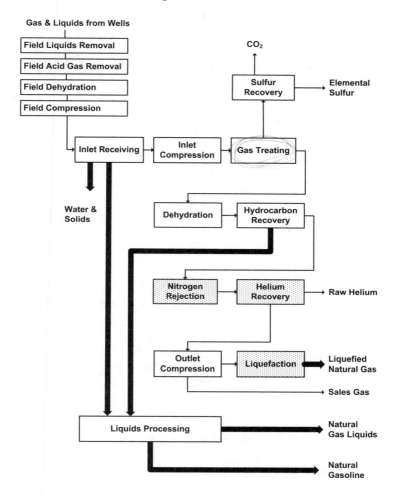

FIGURE 2.1 Block schematic of various processes of a gas plant. Processes required depend upon feed and product slate. Less common process steps for facilities in the United States are shaded. The heavier lines denote liquid streams.

2.2.4 DEHYDRATION

Nearly all plants utilize a dehydration step because the gas that leaves the gas treating step is usually water saturated. Even if no water-based gas treating is required, most gas streams contain too much water to meet pipeline specifications or to enter the cryogenic section of the plant. Field operations sometimes dry the gas to avoid gas hydrate formation as well as to reduce corrosion. Chapter 6 provides details on the commonly used dehydration processes.

2.2.5 HYDROCARBON RECOVERY

Any plant that processes natural gas to produce hydrocarbon liquids (NGL) or LNG utilizes a hydrocarbon recovery step. This step usually involves cryogenic separation to recover the ethane and heavier hydrocarbons. Hydrocarbon recovery often plays an important role in field operations, where it is used for fuel gas conditioning and to alter gas condensation temperatures. Details of these processes are provided in Chapter 7.

2.2.6 NITROGEN REJECTION

Although a less common process in the gas industry, nitrogen rejection will become more important as we shift to lower-quality gas feedstock. This process is typically cryogenic, although membrane and absorbent technology are becoming attractive. Chapter 8 presents details on nitrogen removal.

2.2.7 HELIUM RECOVERY

Helium recovery is uncommon, unless the helium content is above 0.5 vol%. A brief description of this process is given in Chapter 9. The chapter also discusses treatment of other trace components, including BTEX (benzene, toluene, ethylbenzene, and xylenes) emissions and mercury. BTEX is primarily an environmental concern because of possible emissions from glycol dehydration units. Although at extremely low concentrations in the gas, elemental mercury can cause mechanical failure in aluminum heat exchangers.

2.2.8 OUTLET COMPRESSION

Most plants must compress the gas before it goes to the pipeline. The majority of plants that have cryogenic hydrocarbon recovery use turboexpanders to provide refrigeration in the cryogenic section. Work generated in expansion is used to recompress the outlet gas. However, additional compression is usually required. Chapter 7 discusses turboexpanders coupled with compressors, and Chapter 4 covers stand-alone compressors.

2.2.9 LIQUIDS PROCESSING

Liquids processing occurs whenever NGL is a product. The processing required in the step depends upon the liquid content of the inlet gas and desired product slate. Chapter 10 covers the common processing components for sweetening, drying, and fractionating the liquids.

2.2.10 SULFUR RECOVERY

Any plant at which H_2S removal is required, utilizes a sulfur-recovery process if venting the H_2S will exceed environmental limits. The common recovery processes, including tail gas cleanup, are discussed in Chapter 11.

2.2.11 STORAGE AND TRANSPORTATION

Although not shown in Figure 2.1, one other "process" tied in with gas plants is storage and transportation. Chapter 12 discusses the storage and transportation of natural gas and NGL products.

2.2.12 LIQUEFACTION

As noted earlier, liquefaction is not common in the United States, but is becoming more important worldwide as more gas is imported to industrial nations from remote locations. Also, liquefaction plays an increasingly important role as a means for gas storage. Chapter 13 discusses liquefaction processes. Because of the unique properties of liquefied natural gas (LNG), the chapter includes storage, transportation, and vaporization processes as well. These aspects are important to countries, such as the United States, that import LNG.

2.3 IMPORTANT SUPPORT COMPONENTS

Some important components of all gas plants are omitted from detailed discussion in this book. These components include utilities, process control, and safety systems. The Engineering Data Book (2004a) provides more details on these topics.

2.3.1 UTILITIES

Utilities include power, heating fluids (steam and oil), cooling water, instrument air, nitrogen-purge gas, and fuel gas. Most gas plants purchase electrical power, but some generate at least part of their power on site. Cogeneration plants are becoming more attractive options for reduction of operating costs, especially when gas turbines are used for driving compressors. Compared with refineries and most chemical complexes, steam and hot oil are not extensively used in gas plants. Their primary uses are for regeneration of solvents and some reboilers. Cooling water is used primarily in heat exchangers on compressors. The Engineering Data Book (2004c) has an excellent discussion on water treating chemistry.

An uninterrupted source of clean, dry instrument air is critical to plant operations because the air drives most automated valves. Typical pressures are around 100 psig (7 barg). Plants use one or more backup compressors to ensure that air is always available. Many operations have molecular sieve driers (see Chapter 6) to avoid potential line freezes in cold weather. In areas of high instrument flow rates, air receivers permit large flow rates without pressure drops (Engineering Data Book, 2004c).

Nitrogen is used as a purge gas around rotating seals, as well as to purge and blanket vessels. The required purity depends upon the application but usually is not high. In many cases, the enriched nitrogen is obtained from membrane or pressure swing adsorption (PSA) separation (see Chapter 9 for details). If large volumes are required, cryogenic fractionation of air is the most economical process (Engineering Data Book, 2004c).

Gas plants use the gas they process to fuel the facilities. Boilers, hot-oil furnaces, and reciprocating compressors can use low pressure gas. The primary concern is having a particulate-free gas and a roughly constant heating value. The Engineering Data Book (2004c) states that for gas turbines, which are discussed in Chapter 4, the required fuel-gas pressure may be as high as 600 psia (41 bar).

2.3.2 PROCESS CONTROL

Process control has always played a role in gas plants but has become more important over the years as companies try to reduce labor costs. Most plants use good digital control systems (DCS) for individual units to provide both process control and operations history. Since the 1990s, advanced process control (APC) systems, which "sit" on top of existing DCS systems, provide sophisticated plant control. APC uses multivariable algorithms that are trained in the plant to optimize operations. Another aspect of process control commonly used is SCADA (supervisory control and data acquisition). One important use of SCADA is the monitoring of field operations, with the capability of controlling dehydration equipment, flow valves, and compressor stations from the gas plant.

Automation requires accurate input data to make the proper control decisions. Plants usually have a full-time instrument technician to maintain and calibrate the many temperature, pressure, and flow sensors, as well as instruments required for compositional and trace-component analysis.

2.3.3 SAFETY SYSTEMS

Safety systems are critical to all gas plants. These systems include the emergency shutdown of inlet gas, as well as relief valves and vent systems leading to the flare. The Engineering Data Book (2004b) provides criteria for sizing relief systems and flares. Proper sizing of relief valves, rupture disks, and piping is important to ensure that operating systems can be vented quickly. The design often is complicated by the need for two phase flows through valves and lines.

Pipe flares are probably the most common flare in gas plants. In normal operations, the flare flame is barely detectable. If the plant is venting mostly methane, the flare flame is bright but smokeless. If methane plus heavier hydrocarbons are flared, the flame will smoke. To make the flare smokeless, steam or high-flow-rate air or fuel gas is added. If the fuel has a low Btu content (e.g., tail gas from a sulfur recovery unit), fuel gas is added to ensure complete combustion.

2.4 CONTRACTUAL AGREEMENTS AND ECONOMICS

When operational changes are under consideration, the customary analysis for optimization of the balance between capital expenditures and operating costs applies. However, contractual agreements complicate a gas plant economics study whenever the producer and processor are not the same company. Five basic kinds of contracts are commonly used between producers and processors (Kuchinski, 2005):

1. Fee based
2. Percentage of proceeds
3. Wellhead purchase
4. Fixed efficiency
5. Keep whole

2.4.1 FEE-BASED CONTRACTS

In fee based contracts, the producer pays the gas processor a set fee on the basis of gas volumes produced. The processor may obtain additional income by charging fees for additional services, such as gathering, field compression, pipeline transmission, and marketing. In these contracts, the processors income is independent of gas and NGL prices.

2.4.2 PERCENTAGE OF PROCEEDS CONTRACTS

In percentage of proceeds (POP) contracts, the two parties agree to what percentage of the proceeds from the sale of the gas and liquids is to be retained by the producer. Typically, the producer retains more than 70% of the proceeds from the sale of all products. In the case of multiple producers, each has a percentage share of the proceeds, allocated on the basis of each producer's contribution to the proceeds. Allocations are computed on the basis of the Btu content of the gas delivered at the wellhead for a producer divided by the sum of the Btu content of the gas from all producers. Producers and processors share the effect of gas and NGL price fluctuations.

2.4.3 WELLHEAD PURCHASE CONTRACTS

In wellhead purchase contracts, the processor executes a contract to purchase total Btus from the producer at a negotiated price usually based against an index. This purchase is a straight-forward purchase, and the processor's profits depend upon the cost of gathering and production and the selling price of the gas and liquids.

2.4.4 FIXED EFFICIENCY CONTRACTS

In fixed efficiency contracts, the processor agrees to provide a certain percentage recovery (efficiency) of the heavier than methane components from the gas and to pay the producer on the basis of the market value of the theoretical liquid production and resultant residue gas. The processor makes money by processing at a higher efficiency (higher fraction of the liquids removed from the feed stream). Processor profits hinge on actually having higher recoveries and a favorable price margin.

2.4.5 KEEP WHOLE CONTRACTS

In keep whole contracts, the processor agrees to process or condition the producer's gas for sale in the natural gas market and to return to the producer 100% of the Btu content of the raw gas (keep the producer whole on Btus) in exchange for retaining ownership of all liquids extracted from the gas. The processor usually retains all liability for fuel, processing costs, and the purchase of replacement of Btus extracted as a liquid product.

These contracts are more complex, more favorable to the producer, and more risky to the processor. The producer, in essence, sells the whole hydrocarbon stream, at the price of natural gas on a Btu basis, to the processor. The processor makes or loses money, depending on the price difference (price margin) between natural gas and the NGL, which the processor sells.

Most contracts contain penalties for variations from contracted liquid content, impurities, and delivery pressure. Contracts may be set to allow for incremental variations from base composition. Contracts are commonly a combination of two or more of the five basic types.

How costs are shared between producer and processor for capital projects depends upon the nature of the project and the contract. New capital items that benefit both parties may be cost shared. However, maintenance, replacements, and costs of environmentally driven projects are borne by the processor as a cost of staying in business. Situations arise in which costs are too high for the processor to absorb alone, and producers must decide to either share the costs or cease production.

The combination of the varied and complex contractual agreements, the proprietary nature of economic data, and the sometimes biased data in the literature limits the discussion of economics in this book. Chapter 14 provides some capital cost data, but otherwise, only qualitative economic comparisons are provided in the other chapters.

To show how the various processing components tie together, Chapter 15 briefly describes three gas plants. The plants differ in both feed and product slate.

REFERENCES

Energy Information Administration, Office of Oil and Gas, U.S. Department of Energy, U.S. LNG Markets and Uses, Department of Energy, January 2003, www.eia.doe.gov/pub/oil_gas/natural_gas/feature_articles/2003/lng/lng2003.pdf, Retrieved June 2005.

Engineering Data Book, 12th ed., Gas Processors Supply Association, Tulsa, OK, 2004a.

Engineering Data Book, 12th ed., Sec. 5, Relief Systems, Gas Processors Supply Association, Tulsa, OK, 2004b.

Engineering Data Book, 12th ed., Sec. 18, Utilities, Gas Processors Supply Association, Tulsa, OK, 2004c.

Kuchinski, J., private communication, 2005.

3 Field Operations and Inlet Receiving

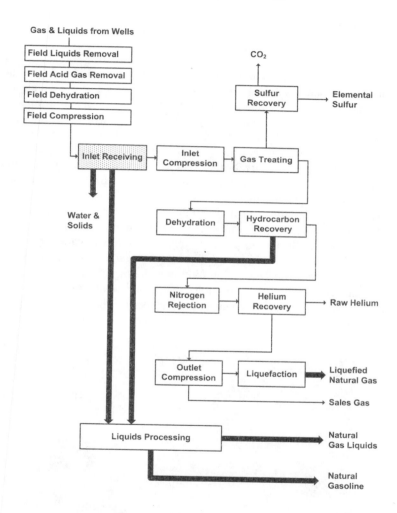

3.1 INTRODUCTION

Gas plant operators usually take responsibility for the gas after it leaves the gas–liquid separator at the wellhead because well operations normally are not under their direct supervision. This chapter touches on wellhead operations and

33

then discusses the field operations and equipment for moving the gas into the gas plant and then through inlet receiving, the first of many gas–liquid separations in the typical gas plant. The chapter includes a discussion of gas hydrates because their control and prevention is critical to field operations.

3.2 FIELD OPERATIONS

3.2.1 WELLHEAD OPERATIONS

Figure 3.1 shows a typical modern gas well, with the wellhead, gas–liquid separator, instrument shed, and condensate tank. Figure 3.2 shows the electronic flow meter, along with the solar collector that powers electronic transmission of the data. The wellhead is in the background. Most companies use electronic metering. However, some wellheads still have circular chart recorders to measure flow rates. The meter is where the gas processor purchases the gas from the producer and the lease royalty owner. Often, the producer and gas processor belong to the same company. If the gas is nonassociated, hydrocarbon liquids knocked out in the separator may be remixed with the gas or stored in a tank and removed by truck. Usually, no compression is applied before the gas is sold.

If the gas is being stripped from oil, separators knock out both oil and water. For high-pressure wells, the oil passes through up to three separators to recover the light ends. Because the last-stage separator is near ambient conditions, the gas may be compressed before it flows into a gathering line to the gas plant.

Frequently, several wells from one lease are tied to one separations unit to reduce the number of separators, compressors, and meters. This practice is especially

FIGURE 3.1 Gas well with condensate tank and separator.

FIGURE 3.2 Electronic flow-measurement transmitting equipment. (Courtesy of The Williams Companies, Inc. of Tulsa, Oklahoma.)

true in offshore operations because producers need to minimize the number of platforms and platform weight for cost purposes. Gas dehydration and sometimes "dew pointing," that is, reduction of the amount of heavier hydrocarbons, is done at the lease. Chapters 6 and 7 discuss dehydration and dew-pointing methods, respectively.

3.2.2 PIPING

For offshore production, many wells are tied back to a platform, and then gas from multiple platforms are tied together into large pipelines that go to the gas plant, which usually is onshore. Pipelines from onshore wells, especially if operated by small independent producers, form an extensive network of small lines from individual wells that tie into increasingly larger lines. The smaller gathering lines may be aboveground or buried, whereas larger lines are always buried. Aboveground lines are much easier to maintain but are exposed to the atmosphere. Surface coating for corrosion prevention and possibly insulation, may be required.

For low-pressure gathering systems with small amounts of liquid present, the processor commonly uses a "drip system." This system involves burying a vessel, frequently a larger-diameter pipe, below low points in the gathering line. Liquids drain from the gathering line into the vessel. The vessel is emptied periodically and the liquid is trucked to the gas plant for processing (McCartney, 2005).

As onshore gas fields are being depleted, wellhead and line pressures are becoming increasingly subambient. Subambient lines are a major concern for gas processors because of potential air intake from leaks. As noted in Chapter 1, oxygen is an unwelcome component because it enhances corrosion, adversely affects several plant processes, and can cause the sales gas to be subquality if the concentration exceeds 1 vol% (Engineering Data Book, 2004a). The oxygen problem is severe enough in some areas that personnel are dedicated to "chasing air" to find the leaks. They analyze pipeline contents for oxygen and work back into the field to find the source. Once found, the line or well is shut down for repairs. Since the late 1990s, gas processors have installed SCADA (supervisory control and data acquisition) systems to monitor both plant and field operations. These systems provide gas plant operators with the capability to remotely shut down wells and close lines, which may be in desolate country, miles from the plant.

3.2.3 COMPRESSION STATIONS

Figure 3.3 shows an onshore compression station, commonly called a booster station. Figure 3.4 is a schematic of common operations at a booster station. (Booster stations are rarely used in offshore production unless the line is roughly

FIGURE 3.3 Booster station on gathering system. (Courtesy of The Williams Companies, Inc. of Tulsa, Oklahoma.)

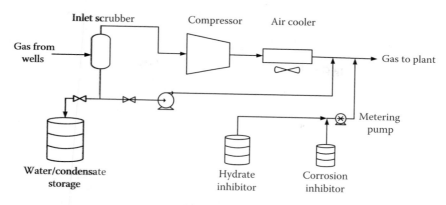

FIGURE 3.4 Schematic of common operations at a compression station on a gathering system. Condensed liquids may be stored or returned to line. Some locations may have dehydration or sweetening facilities.

100 miles or more offshore.) In addition to compressing the gas to move it into the gas plant, the stations usually have inlet suction separators (scrubbers) to remove condensed liquids. The condensates are either reinjected downstream of the compressor discharge or stored before being trucked to the gas plant. Some booster stations include dehydration facilities (see Chapter 6) to reduce corrosion and mitigate hydrate formation. When required, chemical inhibitors for corrosion and gas hydrate prevention are injected into the discharged gas; metering pumps are used to control injection rates.

How to handle condensed liquids at booster stations is an economic decision. Liquids collection requires tankage and trucking to move the liquids to the plant. The benefits are higher line capacity, reduced corrosion rates, and reduced chemical hydrate and corrosion inhibitor injection rates. Also, pigging requirements (see next section) are reduced, as well as the potential for large liquid slugs entering the plant. Putting liquids back into the pipeline eliminates the need for tanks and trucking but increases chemical costs, puts more importance on good pigging, and probably increases pipeline maintenance costs.

Calculation of the pressure drop in these two-phase lines is complex and not covered here. The Engineering Data Book (2004c) discusses two phase flow and gives some calculation methods. However, the best results are obtained from computer models.

Compressor power requirements depend upon gas flow and pressure ratio (see Chapter 4). Booster-station horsepower ranges from the teens to the thousands. Compressors are driven by electric motors, internal-combustion engines, and gas turbines. Horsepower requirements and availability of fuel dictate the best choice. Some compressor drivers are internal-combustion engines that have been in service for 50 or more years and have been rebuilt many times. Compressor engines are usually fueled with natural gas. The best fuel is sales gas from the plant because it is clean and has a constant heating value. However, this type of fueling requires running pipe between the plant and booster stations. Raw gas can be used, but may be too rich for the engines. Fuel-gas conditioners employ low-temperature separators (LTS) to condense heavier hydrocarbons from the gas and lower the heating value (see Chapter 7). If rich gas is used with internal combustion engines, the timing of the engine must be retarded to prevent detonation problems (Kuchinski, 2005). Electric motors are often used in remote sites when the pipeline gas cannot be used or where emissions from a gas engine burning rich fuel are an environmental concern.

3.2.4 Pigging

Pigging is the process of forcing a solid object through a pipeline. Two papers by Webb (1978a, 1978b) provide an extensive discussion of pigging for both liquid and gas pipeline. The process involves inserting the pig, via a pig launcher, into the pipelines and removing it by use of a pig receiver. Pigging is used to perform any of the following functions:

1. Provide a barrier between liquid products that use the same pipeline
2. Check wall thickness and find damaged sections of lines
3. Remove debris such as dirt and wax from lines
4. Provide a known volume for calibrating flow meters
5. Coat inner pipe walls with inhibitors
6. Remove condensed hydrocarbon liquids and water in two phase pipelines

For field operations, the last function is the most important. Gathering systems typically are in the two phase stratified flow region, where the liquid flow rate is much slower than the gas flow rate. Thus, liquid accumulates in low spots in the line. Field operations must follow a rigorous pigging schedule to prevent the plant from being hit by large slugs of liquid that would flood inlet receiving and carry liquids into the gas-processing units. Fortunately, plant operators usually know when a "killer pig" is coming, and they draw down liquid levels in inlet receiving. To protect the plant from large liquid surges, operators respond by shutting in gas, which shuts down field compressors and upsets the plant with the potential for producing off-spec products. Varnell and Godby (2003) discuss procedures to minimize the potential for these problems.

Figure 3.5 shows examples of pigs and spheres used in two phase service. Pigs are usually made of polyurethane foam. The smaller, projectile-shaped pigs in the front of the picture and the larger ones with similar shape are used for cleaning out soft deposits, gauging lines, and removing water after a line is hydrotested. The pig in the center of the back row has metal brushes for removing hard solids. The back two pigs on the left are designed for bidirectional flow. Larger pigs have a metal mandrel holding the cups and disks, which can be replaced after wear (Girard, undated).

FIGURE 3.5 Different shapes of foam spheres and pigs used to clean gathering lines. (Courtesy of Girard Industries.)

Foam spheres, like those shown in the left-front of Figure 3.5, are more commonly used in gathering lines because multiple spheres can be loaded into a pig launcher for remote launching. Spheres for nominal 2 to 4 inch (50 to 100 mm) diameter lines are solid, whereas larger ones are inflated, usually with water or ethylene glycol–water mixtures.

Spheres are launched from a pipe branch that can be isolated and purged for loading. Figure 3.6 shows a sphere launcher. The launchers typically hold three to five spheres and are inclined so that spheres can be inserted remotely by

FIGURE 3.6 Remote automated sphere launcher. (Courtesy of Pearl Development Company.)

gravity feed. At the end of the run, the sphere is caught in a sphere catcher (usually called a pig catcher even if spheres are used). Both launchers and receivers are on the straight run of a pipe tee. Gas released during venting, before emptying of the catcher, goes to low-pressure fuel lines or is recompressed for processing (Webb, 1978b).

Pressure differential drives the sphere through the line. However, it usually stops intermittently when it comes to low points where pools of liquid accumulate. Field and plant operators learn the typical run times for spheres in a line, and in-line detectors signal a sphere's location.

3.3 GAS HYDRATES

A field of technology, called flow assurance, exists for ensuring that hydrocarbons flow unimpeded by line blockage from wells to the point of processing. The three areas of concern are:

- Wax and asphaltene solids deposition
- Scale (inorganic salt) deposition
- Gas hydrate solids formation

For gas gathering systems, wax and asphaltene deposition is normally not a serious problem and can be remediated by pigging. Scale is a common issue at the wellhead but should not be one in the pipeline. However, gas hydrates strike fear into the hearts of flow assurance people for two reasons. First, hydrate plugs can occur within minutes without prior warning. The other solids take weeks, months, or years to cause plugging and are usually detected by increased line-pressure drop. Second, although hydrate formation can be inhibited in a number of ways, injection pump failure, separator failure, and process upsets can suddenly make pipeline contents vulnerable to hydrate formation. Sloan (2000) provides a good overview of hydrate problems and how to address them, especially for offshore situations, where hydrate plugging can be a major problem.

3.3.1 PROPERTIES

Gas hydrates are a class of solid, nonstoichiometric compounds called clathrates. They form when a host material, water for hydrates through hydrogen bonding, forms a caged structure that contains guest molecules, such as methane. Both host and guest must be present for the solid to form, but not all of the cages will be occupied. Gas hydrates should not be confused with salt hydrates, which form stoichio-metric compounds. For details of hydrate structure, composition, and phase behavior see, for example, Sloan (1998) and Carroll (2003).

Gas hydrates were a laboratory curiosity until Hammerschmidt (1939) noted that they caused pipeline plugging. As Figure 3.7 shows, hydrates form at temperatures well above the freezing point of water. Figure 3.7 shows a break in all of the curves at 32°F (0°C) because the hydrates are in equilibrium with

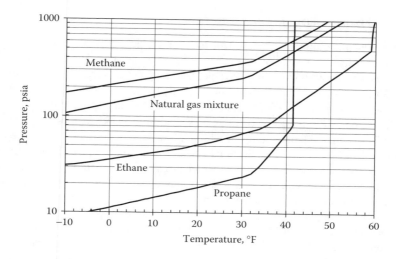

FIGURE 3.7 Hydrate formation conditions for pure methane, ethane, and propane (data from Sloan, 1998) and natural gas mixture with composition equal to that of Rio Arriba County, New Mexico. Mixture composition given in Table 1.4; line computed using CSMHYD98. See text for details.

gas and ice at the lower temperatures, and with gas and liquid water at the higher temperatures. For the ethane hydrate line, an abrupt change occurs at 57°F (14°C). At this point, the hydrate is in equilibrium with gaseous ethane and both liquid ethane and liquid water (i.e., the hydrate formation line intersects the vapor pressure line, which is not shown). At higher temperatures, the ethane hydrate is in equilibrium with liquid ethane and liquid water. The same explanation applies to the propane. For gas-processing applications, the intersection with the vapor-pressure line puts an upper temperature limit on hydrate formation because the line is essentially vertical at pressures commonly seen in gas processing. No similar point exists for methane, because methane has a critical temperature of about −180°F (−82°C) (see Appendix B).

Figure 3.7 also illustrates the compositional dependence on hydrate formation with the predicted hydrate formation conditions for the Rio Arriba County, New Mexico, gas mixture (see Table 1.4). This mixture contains only 1.33 mol% ethane and 0.19 mol% propane, but the hydrate formation pressure at a given temperature drops by 30% compared with pure methane. At a given pressure, pure methane forms hydrates at the lowest temperature. Adding ethane through butane, hydrogen sulfide, or carbon dioxide raises the formation temperature significantly. However, nitrogen will lower the hydrate formation temperature of the mixture. Heavier compounds, C_5+, are too large to fit into the cages formed by natural gas and have an insignificant effect as a diluent because their concentration is so low. Helium also acts as a diluent.

3.3.2 HYDRATE FORMATION PREDICTION

Thermodynamics provides a powerful tool for prediction of the temperature and pressure for hydrate formation on the basis of gas composition. However, even when hydrates are thermodynamically possible, they may never form. Hydrate formation kinetics is complex and poorly understood, partly because the crystal growth process is random. To be safe, operating conditions should be outside the hydrate region.

Since the 1970s, a method based upon statistical thermodynamics has been used to predict formation temperatures to within a few degrees Fahrenheit. It is the most successful and widely used application of statistical thermodynamics in industry. The model is complex and requires a computer program to use. However, commercial simulators and stand-alone programs are available that use improved models, incorporate the effect of thermodynamic inhibitors, and accurately predict phase behavior for all phases present. A book by Sloan (1998) includes a program, CSMHYD98, that uses the statistical thermodynamic model, along with a guide to its use.

Before the statistical mechanical model, two empirical methods were widely used. One employed K-values similar to those used for vapor–liquid equilibrium calculations, and the other is based upon gas gravity. The K-value method gives slightly better results for sweet natural gas with specific gravities between 0.6 and 0.8, whereas it is much better for sour gases. We discuss only the gas gravity correlation, which is adequate for quick calculations. (The Engineering Data Book [2004d] provides details of both empirical methods.) If more accurate values are needed, computer programs that employ the statistical mechanical model should be used.

Figure 3.8 shows hydrate formation prediction curves for natural gases as a function of gas specific gravity. For below 1,000 psi (70 bar), the figure can be approximated by

$$t(°F) = -16.5 - 6.83/(SpGr)^2 + 13.8 \ ln[P(psia)] \tag{3.1a}$$

$$t(°C) = -6.44 - 3.79/(SpGr)^2 + 7.68 \ ln[P(bara)] \tag{3.1b}$$

The specific gravity is defined as the ratio of the mass of a given volume of a gas to that of an equal volume of air; both volumes are measured at 14.7 psia and 60°F (1.01 bar and 15.6°C). For an ideal gas, the specific gravity is molar mass of the gas divided by the molar mass of air (28.96).

Example 3.1 Estimate the hydrate-formation temperature at 325 psia (22.4 bar) for the gas with the composition in Table 3.1. Compare the results from Figure 3.8 and Equation 3.1.

Use of either Figure 3.8 or Equation 3.1 requires knowledge of the specific gravity of the gas, so this value is calculated first, as shown in Table 3.1. The specific gravity = molar mass$_{gas}$ /molar mass$_{air}$ = 20.08/28.96 = 0.693.

By use of Equation 3.1, we obtain

$t = -16.5 - 6.83/(0.693)^2 + 13.8 \times ln(325) = 49°F$. Figure 3.8 gives 50°F, and the experimentally reported value for a 0.7-gravity gas is 50°F (Engineering Data Book,

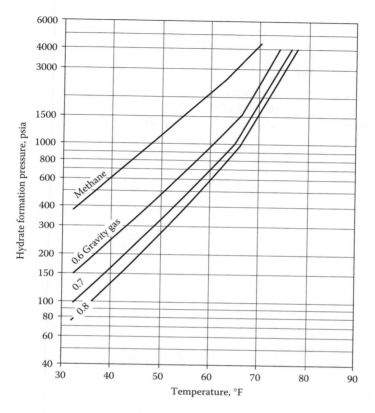

FIGURE 3.8 Pressure–temperature curves for estimation of hydrate-formation conditions as a function of gas gravity. These curves should be used for approximation purposes only. (Adapted from Engineering Data Book, 2004d.)

TABLE 3.1
Gas Composition for Example 3.1

Component	Volume Fraction[a]	Molar Mass	lb/lb-Mol Mixture (kg/kg Mol Mixture)
CO_2	0.002	44.01	0.09
N_2	0.094	28.01	2.63
C_1	0.784	16.04	12.58
C_2	0.060	30.07	1.80
C_3	0.036	44.10	1.59
iC_4	0.005	58.124	0.29
nC_4	0.019	58.124	1.10
Totals	1.000		20.08

[a.] For ideal gases volume, fracion is the same as mole fraction

2005d). The computer program CSMHYD98 predicts hydrates form at 49°F. The agreement is exceptionally good between the methods and experimental data for this example.

Gas hydrates form anywhere in a pipeline or process stream, but they are particularly likely to form downstream of orifices or valves because of Joule-Thomson expansion effects. The Engineering Data Book (2004d) provides charts for estimation of how much pressure drop is allowed to avoid hydrate formation during expansion. However, the charts are only estimates, and calculations by a computer program should be used to determine the possibility of hydrate formation.

3.3.3 HYDRATE INHIBITION

Three ways exist to avoid hydrate formation in natural gas streams:

- Operate outside the hydrate formation region.
- Dehydrate the gas.
- Add hydrate inhibitors.

The first option is ideal but usually impractical. If practical and economically feasible, dehydration is the next best option. Because of the high gas volumes and high pressures, offshore operations frequently use dehydration (see Chapter 6), but it is less common in onshore field operations. The third best, but the most commonly used option, is addition of inhibitors. Whenever operating conditions are prone to hydrate formation and liquid water may form in the lines, field operations control hydrate formation by use of chemical inhibitors. They are added at the wellhead and at booster stations through positive displacement pumps so that the addition rate can be accurately controlled.

Regarding dehydration, how dry a gas should be to prevent hydrate formation is uncertain. The gas obviously should have a water dew point below the lowest operating temperature to avoid water condensation. Thermodynamics predicts that hydrates can form even when the gas phase is unsaturated with water. This problem has occurred in several instances in pipelines in which the gas residence time (Sloan, 2005) was unusually long, but it is extremely rare. This occurrence is explainable through mechanistic arguments. Nominally, six water molecules are required for each guest molecule. However, 20 or more water molecules are needed to form a cage around a gas molecule, and many cages must combine to form the hydrate lattice. Therefore, the likelihood of sufficient number of water molecules at concentrations of parts per thousand coming together to form the lattice is low.

Use of chemical inhibitors is the least attractive hydrate inhibition method for several reasons:

- The proper inhibitor dosage must be known to avoid plugging or needless chemical costs, but oftentimes it is determined empirically.
- The chemical cost, although it is usually a small fraction of overall operating costs.

- The reliability of inhibitor injection can be a problem because of malfunctioning injection pumps and depleted inhibitor reservoirs, especially at remote sites.
- The possible interaction between hydrate inhibitors and other additives reduces the effectiveness of some of additives, an effect that is usually determined empirically.

Despite these distractions, multimillions of dollars are spent each year on hydrate inhibitors because they require minimal capital cost to implement and because injection rates are readily altered according to flow and operating conditions.

Chemical hydrate inhibitors come in three types:

1. Antiagglomerates (AA)
2. Kinetic (KHI)
3. Thermodynamic

The first two types are technologies developed in the late 1990s. Antiagglomerates prevent small hydrate particles from agglomerating into larger sizes to produce a plug. The inhibitors reside in the liquid hydrocarbon phase and are most often used in pipelines where gas is dissolved in oil. They require testing to ensure proper concentrations.

Kinetic inhibitors slow crystal formation by interfering with the construction of the cages. Their advantage is that they can be used at concentrations in the 1 wt% range in the aqueous phase, and they are nonvolatile. Their disadvantage is that the proper dosage must be determined empirically, as too much inhibitor may enhance hydrate formation rates. These inhibitors are limited to a subcooling (difference between desired operating temperature and hydrate formation temperature at constant pressure) of 28°F (15.5°C). However, Mehta and Klomp (2005) recommend a maximum subcooling of 20°F (11°C) for these kinds of inhibitors. Kinetic inhibitors are being used in many offshore operations and will become more widely applied as experience with their use increases.

Thermodynamic inhibitors, mainly methanol and ethylene glycol, are widely used. They are essentially antifreeze. Inorganic salts are effective but rarely used, and further discussion relates only to methanol and ethylene glycol.

The required dosage of thermodynamic inhibitors is predictable, but the concentrations can be high, over 50 wt% of the water phase. A number of empirical correlations, on the basis of thermodynamic properties of solutions, predict the amount of any hydrate inhibitor required to depress hydrate formation temperatures. The two most commonly used are discussed here (more rigorous methods are used with the computer models discussed above). Hammerschmidt (1939) proposed the following equation:

$$\Delta t(°F) = \frac{2335X_i}{MW_i(1 - X_i)} \tag{3.2a}$$

or

$$X_i = \frac{\Delta t(°F)MW_i}{\Delta t(°F)MW_i + 2335},\qquad (3.2b)$$

where t is the hydrate-depression temperature,°F, X_i is the mass fraction of inhibitor in the free-water phase, and MW_i is the molecular weight of the inhibitor. The Engineering Data Book (2004d) recommends use of this equation for methanol concentrations of 20 to 25 wt% of the water phase; it recommends the equation for ethylene glycol concentrations up to 60 to 70 wt%. However, it notes that the constant 2,335 may be too low in field situations, and a constant as high as 4,000 may be appropriate. The smaller figure is best used first, and then lower concentration levels used on the basis of operating experience.

Nielsen and Bucklin (1983) proposed

$$\Delta t(°F) = -129.6 \ln x_W,\qquad (3.3)$$

where x_W is the mole fraction of water in the aqueous phase. The paper provides details on the equation's derivation and gives example calculations. It also discusses methanol injection systems. The Engineering Data Book (2004d) recommends use of this equation for methanol concentrations up to 50 wt%.

3.3.3.1 Methanol vs. Ethylene Glycol

Methanol is more widely used than ethylene glycol, and both have advantages and disadvantages, primarily on the basis of their physical properties. Both inhibitors are hydrophilic and remain predominantly with a condensed water phase, even if a condensed hydrocarbon phase is present. However, methanol is volatile. Because the gas volume greatly exceeds the water volume in most gathering systems, methanol vaporization losses must be considered. Figures 3.9 and 3.10 provide a means for estimation of vaporization losses and losses in the hydrocarbon phase. For ethylene glycol, vaporization losses are negligible, and a solubility loss of 0.3 lb/1,000 gal (0.07 kg/m³) of NGL is a common estimate for design purposes (Engineering Data Book, 2004d). These inhibitor losses to the condensate assume the condensate is low in aromatics. If aromatic concentrations are atypically high, inhibitor losses can be significantly higher.

Ethylene glycol is relatively easy to recover, as it remains with the aqueous phase and can be concentrated by evaporation of the water. Separation of methanol from water requires distillation. Some gas plants have recovery facilities, but most ship the water–methanol mixtures elsewhere for recovery or disposal. For onshore operations, the methanol content in condensate typically presents little or no problems. However, refineries impose a penalty fee for high concentrations of methanol in hydrocarbon liquids that arrive from offshore Gulf of Mexico because they must remove the methanol before processing the liquids.

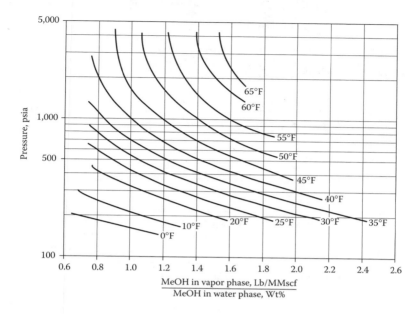

FIGURE 3.9 Ratio of methanol-vapor composition to methanol-liquid composition. (Adapted from Engineering Data Book, 2004d), gpsa fig 20-65.

The high vapor pressure of methanol also presents a safety hazard. This problem is a concern for offshore operations where large quantities of methanol must be stored. Both inhibitors are toxic and restrictions apply to their disposal, both on land and at sea. One quirk of ethylene glycol is that it interferes with the oil and grease test performed on water to be discharged from offshore platforms, giving false high values.

Ethylene glycol is viscous and must be either diluted, usually with water, or kept in a warm storage vessel in cold weather (Appendix B contains a chart that shows the viscosity as a function of temperature and concentration). Viscosity is not an issue with methanol.

Example 3.2 A sweet gas with a specific gravity of 0.73 leaves a gas–liquid separator at 100°F and 600 psia saturated with water. The gas drops to 35°F before reaching the next booster station at 500 psia. Assume no hydrocarbon condensate has formed in the line. Calculate how much methanol must be added to prevent hydrate formation between the separator and the booster station per MMscf. Repeat the calculation for ethylene glycol, which is added in an 80 wt% mixture with water.

Use of Equation 3.1a or Figure 3.8 shows that the hydrate-formation temperature at 600 psia is 59°F. This value means a 59°F − 35°F = 24°F subcooling into the hydrate region, and hydrate formation is probable without inhibition.

Determination of the inhibitor rate requires:

1. Determination of the amount of liquid water formed
2. Calculation of the required amount of inhibitor in the water phase
3. Calculation of the required amount of inhibitor in the gas phase

1. The water content of the gas leaving the separator is 95 lb/MMscf and 13 lb/MMscf at 35°F and 600 psia (see Figure 6.1 to obtain these data). Therefore, 95 lb – 13 lb = 82 lb of water per MMscf drops out in the line, assuming worse-case conditions.

Methanol Requirement

2. To estimate the concentration of methanol, rearrange Equation 3.3 to give
$ln\ x_W = -t(°F)/129.6 = -24/129.6 = -0.185$
$x_W = 0.831$ or $x_{MeOH} = 0.169$ mole fraction. This value is $(0.169 \times 32)/(0.831 \times 18 + 0.169 \times 32) = 26.6$ wt% or 0.362 lb methanol/lb water. Thus, the methanol needed in the water phase is $0.362 \times 82 = 29.7$ lb/M Mscf.
3. To estimate the methanol in the vapor phase use Figure 3.9, which, at 35°F and 600 psia, gives 1.22 lb methanol vaporized per wt% methanol in the aqueous phase, or $1.22 \times 26.6 = 32$ lb methanol per MMscf. Thus, the vapor phase consumes more methanol than does the aqueous phase. The total amount of methanol required is $(26 + 32) = 58$ lb/MMscf. If a condensate phase was present as well, the losses estimated by use of Figure 3.10 would have to be added into that phase.

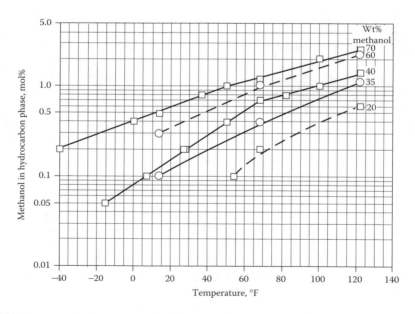

FIGURE 3.10 Solubility of methanol in paraffinic hydrocarbons as a function of temperature at various methanol concentrations. (Adapted from Engineering Data Book, 2004d.)

This amount of methanol will be much less than what goes into the aqueous and vapor phase in gas lines.

Ethylene Glycol Requirement

2. Use Equation 3.2b with the constant 3,225 and a molar mass of ethylene glycol of 62 to obtain the mass fraction of pure glycol required:

$$X_i = \frac{24 \times 62}{24 \times 62 + 2235} = 0.40$$

However, the glycol is diluted to 80 wt%. To obtain the mass of inhibitor solution added per unit mass of free water present to obtain the desired concentration, we use $X_0/(X_i - X_0)$, where X_0 is the weight fraction of inhibitor in the solution to be added. The amount of inhibitor solution added per pound of free water initially present is then $0.8/(0.8 - 0.40) = 2.00$, and the total amount of ethylene glycol solution added is $2.00 \times 82 = 164$ lb/MMscf. At these conditions, glycol loss into the vapor phase is negligible, so the total amount of solution required is 164 lb/MMscf.

3.4 INLET RECEIVING

Gas and liquids that enter the gas plant pass emergency shutdown valves, which isolate the plant from incoming streams and pig receivers, and then go to inlet receiving, where condensed phases drop out. Gas from inlet receiving goes to inlet compression (if necessary), and the liquids go to storage for processing (see Chapter 10). The initial gas–liquid separation occurs in a slug catcher. Slug catchers are critical because downstream gas processing units rely on a continuous gas stream free of liquids, even when surges of liquid enter the plant. A slug catcher is a gas–liquid separator sized to hold the biggest slug a plant will experience. Depending upon slug catcher design, inlet receiving handles just slugs or combines slug catching with liquid storage. Kimmitt et al.(2001) detail the process to properly size and design a slug catcher, and the Engineering Data Book (2004b) provides details on separator design. This section provides an overview of separator principles and then discusses two slug catcher configurations.

3.4.1 SEPARATOR PRINCIPLES

Called scrubbers, knockout pots, inlet receivers, and just separators, effective phase separators protect downstream equipment designed to process a single phase. It is the critical first step in most processes in gas plants and typically is a simple vessel with internal components to enhance separation. Bacon (2001) provides an extensive discussion on designing separators and provides four design examples. (Bacon (2001) provides a spread sheet which includes the design calculations on the conference proceedings CD.) In this section, the emphasis is on gas–liquid separation. Because of its importance in gas processing, a short section on liquid–liquid separation is included. Finally, typical residence times are given for both gas–liquid and liquid–liquid separators in various types of service to check vessel sizing.

3.4.1.1 Gas-Liquid Separation

Separator vessel orientation can be vertical or horizontal. Vertical separators are most commonly used when the liquid-to-gas ratio is low or gas flow rates are low. They are preferred offshore because they occupy less platform area. However, gas flow is upwards and opposes the flow of liquid droplets. Therefore, vertical separators can be bigger and, thus, more costly than horizontal separators. Inlet suction scrubbers at compressor stations are usually vertical. Horizontal separators are favored for large liquid volumes or if the liquid-to-gas ratio is high. Lower gas flow rates and increased residence times offer better liquid dropout. The larger surface area provides better degassing and more stable liquid level as well. Figure 3.11 shows a schematic of gas–liquid separators and indicates the four areas or types of separation:

> Primary separation
> Gravity settling
> Coalescing
> Liquid collecting

Primary Separation—Primary separation is accomplished by utilizing the difference in momentum between gas and liquid. Larger liquid droplets fail to make the sharp turn and impinge on the inlet wall. This action coalesces finer droplets so that they drop out quickly. Although inlet geometries vary, most separators use this approach to knock out a major portion of the incoming liquid.

Gravity Settling—Gravity settling requires low gas velocities with minimal turbulence to permit droplet fallout. The Engineering Data Book (2004b) provides detailed information for settling calculations. This section summarizes useful equations for quick estimation of separator performance. The calculations assume the droplets to be rigid spheres. The terminal-settling velocity, V_T, for a sphere falling through a stagnant fluid is governed by particle diameter, density differences, gas viscosity, and a drag coefficient that is a function of both droplet shape and Reynolds number. Here the Reynolds number is defined as

$$N_{Re} = D_P V_T \rho_g / \mu_g, \tag{3.4}$$

where D_P is particle diameter, ρ_g is the density, and μ_g is the viscosity. Thus, calculations for V_T are an iterative process.

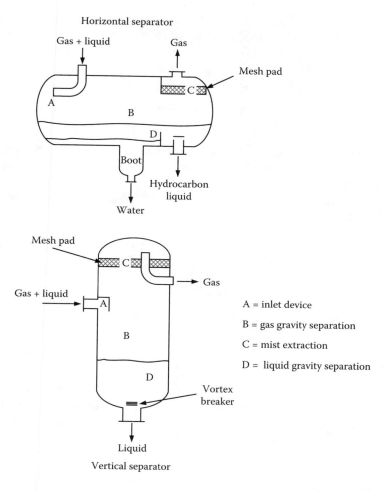

FIGURE 3.11 Gas–liquid separators. (Adapted from Engineering Data Book, 2004b.)

For large particles (1,000 to ~70,000 micron), the terminal velocity is computed by the equation

$$V_T = 1.74 \sqrt{\frac{gD_p(\rho_l - \rho_g)}{\rho_g}}, \tag{3.5}$$

where g is the gravitational constant. This equation, known as Newton's law, applies when the Reynolds number is greater than 500.

If the particle size is too large, excessive turbulence occurs and Equation 3.5 fails. The upper limit is found by use of the equation

$$D_P = K_{CR}\left[\frac{\mu_g^2}{g\rho_g(\rho_l - \rho_g)}\right]^{\frac{1}{3}} \tag{3.6}$$

with $K_{CR} = 18.13$ and 23.64 for engineering and metric units, respectively, and is based upon a Reynolds number of 200,000, which is the upper limit for Newton's law to hold.

At the other extreme, where the flow is laminar ($N_{Re} < 2$), Stokes' law applies. The terminal velocity is

$$V_T = \frac{1,488gD_P^2(\rho_g - \rho_l)}{18\mu_g} \tag{3.7a}$$

$$V_T = \frac{1,000gD_P^2(\rho_g - \rho_l)}{18\mu_g} \tag{3.7b}$$

where Equation 3.7a is in English units and Equation 3.7b is in SI. Stokes' law applies to particles in the 3 to 100 micron range. To find the maximum size particle in this flow regime, use $K_{CR} = 0.0080$ in Equation 3.6, which corresponds to an N_{Re} of 2. Particles smaller than 3 microns will not settle because of Brownian motion. Unfortunately, droplets that condense from a vapor tend to be in the 0.1 to 10 micron range; the majority are around 1 micron. Entrained droplets are 100 times larger.

To reduce turbulence, the settling section may contain vanes. They also act as droplet collectors to reduce the distance droplets must fall.

Coalescing—The coalescing section contains an insert that forces the gas through a torturous path to bring small mist particles together as they collect on the insert. These inserts can be mesh pads, vane packs, or cyclonic devices. Table 3.2 lists some of the features of the wire mesh and vane pack mist extractors. Cyclonic devices are proprietary devices, and features are application dependent.

Mesh pads are either wire or knitted mesh, usually about 6 inches (15 cm) thick, and, preferably, are mounted horizontally with upward gas flow, but they can be vertical. They loose effectiveness if tilted. Mesh pads tend to be more effective at mist removal than vane packs but are subject to plugging by solids and heavy oils.

Figure 3.12 shows several elements of a vane pack, which are corrugated plates, usually spaced 1 to 1.5 inches (2.4 to 3.8 cm) apart, that force the gas and mist to follow a zigzag pattern to coalesce the mist into larger particles as they hit the plates. Coalesced drops collect and flow out the drainage traps in the plates. Although not as effective at removing small drops, they are ideal for "dirty" service because they

TABLE 3.2
Features of Mist Extracting Devices

	Wire Mesh	Vane Pack
Gas capacity factor, C in Equation 3.8, ft/sec (m/sec)	0.22 – 0.39 (0.067 – 0.12)	Horizontal flow 0.9 – 1.0 (0.27 – 0.30) Vertical flow 0.4 – 0.5 (0.12 – 0.15)
Droplet efficiency	99 – 99.5% removal of 3- to 10-micron droplets	99% removal of 10- to 40-micron droplets
Turndown range, % of design gas rate	30 – 110	Rapid decrease in efficiency with decreased gas flow
Pressure drop, inches of water (kPa)	< 1 (0.25)	0.5 to 3.5 (0.12 to 0.87)

Source: Engineering Data Book (2004b).

will not plug. However, solids can collect on the back edge of the vane and plug the drainage ports. Liquid then collects and is re-entrained. This problem is resolved by putting coarse filters upstream of the vane pack (McCartney, 2005).

Figure 3.13 shows qualitatively the range for mist pads and vane packs. The data are based upon an air–water system and differs from natural gas data because of density and surface tension. However, Figure 3.13 shows the regions where

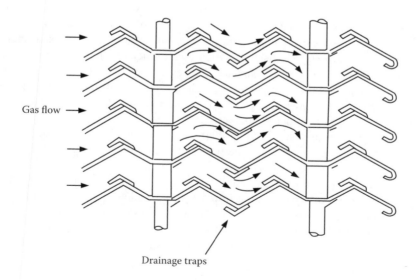

Gas flow →

Drainage traps

FIGURE 3.12 Sketch of vane pack mist extractor element with liquid drainage traps. (Adapted from Engineering Data Book, 2004b.)

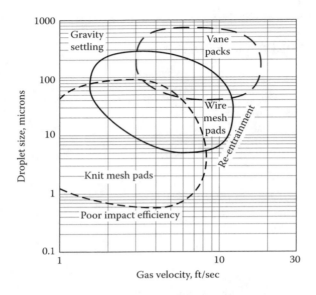

FIGURE 3.13 Approximate operating ranges for different kinds of demisters. Data are based upon water and air. (Courtesy of ACS Industries, 2005.)

each demister type is effective. Note that these devices fail to coalesce droplets below around 0.5 micron, and each has both upper and lower velocity limits. The lower limit is caused by too low a velocity to force sufficient impinging of the droplets on the solid surface to provide coalescing. At high velocities, the coalesced droplets are stripped from the solid by the high velocity gas.

Cyclones use centrifugal force to enhance separation of condensed phases from gas. Their advantage is that they are extremely efficient at high-gas throughput, which means smaller vessel diameter. However, they tend to have a narrow operating range, relatively high pressure drop, and difficulties handling liquid slugs. Sterner (2001) presents details on various cyclone separators as well as a brief discussion on mist pads and vane packs.

The Engineering Data Book (2004b) and Bacon (2001) provide design calculations for wire mesh and vane pack coalescing units, but cyclonic device design is proprietary. Below is a short discussion on the determination of separation performance for mesh pads and vane packs.

The two parts of separation performance are droplet-removal efficiency and gas-handling capacity. Manufacturers provide charts on removal efficiency as a function of droplet size at design conditions. Gas handling capacity can be estimated by use of the following equations:

$$V_t = C \sqrt{\frac{(\rho_l - \rho_g)}{\rho_g}},$$ (3.8)

where C is the gas capacity factor given in Table 3.2. The required mist extractor, A, area is given by

$$A = Q_A/V_t, \tag{3.9}$$

where Q_A is the actual volumetric flow rate. The Engineering Data Book (2004b) tabulates the percentage reduction in C as a function of pressure, but it can be estimated by

$$C(\text{ft/sec}) = 100 - 1.35 \, P(\text{psig})^{0.42} \tag{3.10a}$$

$$C(\text{m/sec}) = 100 - 4.12 \, P(\text{barg})^{0.42}, \tag{3.10b}$$

where P is in psig and barg for Equation 3.10a and Equation 3.10b, respectively. This equation should not be used above 1,150 psig (80 barg).

To remove droplet sizes below 3 microns, filter separators and coalescing filters are required. They are separate units from the gas–liquid separators discussed here. These filtration units frequently are used upstream of gas treating units to prevent hydrocarbon liquids and compressor oil from entering amine absorbers and causing foaming problems (see Chapter 5). They are upstream of dehydration units to protect solid adsorbants from amine solution (see Chapter 6).

Filter separators, like other gas–liquid separators, can be horizontal or vertical; the horizontal unit is more common (Engineering Data Book, 2004b). The units contain a filter coalescing section, followed by a final mist extraction section. The first section either knocks out the fine mist or coalesces it into larger droplets. The filter bundle resembles a heat exchanger bundle, with the gas entering on the shell side. The gas, with coalesced droplets, leaves the tubes and passes through a final mist extractor that contains either mist pads or vane packs. Pressure drop through a clean separator is usually 1 to 2 psi (0.07 to 0.14 bar), and filters are changed when the pressure drop rises to about 10 psi (0.7 bar).

Coalescing filters are similar to filter separators but normally are vertical and designed for very high gas-to-liquid ratios. However, they contain only filter cartridges. Details of these units are proprietary. Manufacturers must be contacted for details of applicability and operating ranges.

Liquid Collection—The liquid collection section acts as a holder for the liquids removed from the gas in the above three separation sections. This section also provides for degassing of the liquid and for water and solids separation from the hydrocarbon phase. The most common solid is iron sulfide from corrosion, which can interfere with the liquid–liquid separation. If a large amount of water is present, separators often have a "boot," as shown in the horizontal separator, at the bottom of the separator for the water to collect. The Engineering Data Book (2004b) estimates that retention times of 3 to 5 minutes are required for hydrocarbon–water separation by settling.

3.4.1.2 Liquid-Liquid Separators

Liquid–liquid separators are similar to gas–liquid separators, and the configuration depends upon whether the separator has two or three phases present. If three phases are present, the configuration shown in Figure 3.11 is typical. A coalescing screen may be in the liquid section to coalesce the droplets. If a vapor phase is not present, the mixture enters at one end, goes through a coalescing medium, and then enters the section where gravity separation occurs.

The same equations described for gas–liquid gravity settling apply. However, the particles typically follow Stokes law. Liquids present two difficulties usually not encountered with gas–liquid separations: relatively small differences in density (use of change in direction provides little or no benefit) and possibly emulsions. Both of these factors force longer residence times. The Engineering Data Book (2004b) provides equations for both horizontal and vertical vessel sizing. However, for quick sizing, a calculation that bases the volume on suggested residence times is frequently adequate.

3.4.1.3 Residence Time for Various Separator Applications

This section provides typical residence times for a variety of gas–liquid and liquid–liquid separations seen in both field and plant operations. The residence time is simply the volume of the phase present in the vessel divided by the volumetric flow rate of the phase. Table 3.3 provides typical retention times for common gas–liquid separations. Table 3.4 gives residence times for liquid–liquid separations.

3.4.2 SLUG CATCHER CONFIGURATIONS

This section briefly describes two kinds of slug catchers, manifolded piping and inlet vessels. Kimmitt et al. (2001) provides detailed design information. The

---

TABLE 3.3
Typical Retention Times for Gas-Liquid Separations

Type of Separation	Retention Time (Minutes)
Natural gas condensate separation	2 – 4
Fractionator feed tank	10 – 15
Reflux accumulator	5 – 10
Fractionation column sump	2[a]
Amine flash tank	5 – 10
Refrigeration surge tank	5
Refrigeration economizer	3
Heat medium oil surge tank	5 – 10[b]

[a] If the fractionator column sump is feeding a downstream fractionator column, it should be sized as a feed tank (McCartney, 2005).

[b] This vessel must have adequate space to allow for expansion of the heat medium from ambient to operating temperature (McCartney, 2005).

Source: Engineering Data Book (2004b).

TABLE 3.4
Typical Retention Times for Liquid-Liquid Separators

Type of Separation	Retention Time (Minutes)
Hydrocarbon–water	
Above 35° API hydrocarbon	3 – 5
Below 35° API hydrocarbon	
> 100°F (38°C)	5 – 10
80°F (27°C)	10 – 20
60°F (16°C)	20 – 30
Ethylene glycol-hydrocarbon separators (cold separators)	20 – 60
Amine-hydrocarbon	20 – 30
Caustic-propane	30 – 45
Caustic-heavy gasoline	30 – 90

Source: Engineering Data Book (2004b).

most difficult part of a slug catcher design is the proper sizing. Sizing requires knowledge of the largest expected liquid slug, as liquid pump discharge capacity on the slug catcher will be trivial compared with the sudden liquid influx.

3.4.2.1 Manifolded Piping

One reason piping is used instead of separators is to minimize vessel wall thickness. This feature makes piping attractive at pressures above 500 psi (35 bar). The simplest slug-catcher design is a single-pipe design that is an increased diameter on the inlet piping. However, this design requires special pigs to accommodate the change in line size. Figure 3.14 shows a schematic of typical multipipe harp design for a slug catcher. Figure 3.15 shows a view from the end of a multipipe slug catcher. The number of pipes varies, depending upon the required volume and operating pressure. Also, some designs include a loop line, where some of the incoming gas bypasses the slug catcher. Primary separation occurs when the gas makes the turn at the inlet and goes down the pipes. Liquid distribution between pipes can be a problem, and additional lines between the tubes are often used to balance the liquid levels. In harp designs, the pipes are sloped so that the liquid drains toward the outlet.

Gravity settling occurs as the gas flows to the vapor outlet on the top while the liquid flows out the bottom outlet. Pipe diameters are usually relatively small (usually less than 48 inches [120 cm]), so settling distances are short.

Because manifolded piping is strictly for catching liquid slugs, demisters are usually installed downstream in scrubbers. Likewise, liquid goes to other vessels, where degassing and hydrocarbon–water separation occurs.

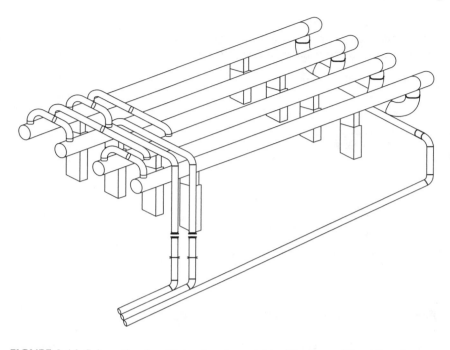

FIGURE 3.14 Schematic of multiple-pipe slug catcher. (Courtesy of Pearl Development Company.)

FIGURE 3.15 Multipipe slug catcher with liquid volume of 500 Bbls (80 m³). Nominal inlet gas rate is 206 MMscfd (5.8 MMSm³/d). (Courtesy of Duke Energy Field Services.)

Kimmitt et al. (2001) note several advantages to the pipe design, including the fact that design specifications are based upon pipe codes instead of vessel codes. Also, the slug catcher can be underground, which reduces maintenance costs and insulation costs if the slug catcher would otherwise need to be heated.

3.4.2.2 Inlet Vessels

These slug catchers, commonly called inlet receivers, are simply gas–liquid separators that combine slug catching with liquid storage. They are usually employed where operating pressures are relatively low or where space is a problem. Horizontal vessels are preferred, unless area is limited (as on offshore platforms), because they provide the highest liquid surface area. Usually two or three vessels are manifolded together to permit larger volumes and to allow servicing of one vessel without plant disruption. Length-to-diameter ratios are typically 3:1 to 5:1 to maintain a low gas velocity through the gravity-settling section.

With their large volume, inlet vessels usually act as primary storage for inlet liquids as well. The vessels can accommodate demisters, so additional separation may be unnecessary before the gas goes to compression.

3.4.2.3 Comparison of Slug Catcher Configurations

Below is a comparison of piping, vertical vessels, and horizontal vessels for slug catching according to various considerations:

- Land or surface requirements. If no land constraints apply, piping is attractive. If constraints are severe, as on offshore platforms, vertical vessels are preferred. Otherwise horizontal vessels are the best choice.
- Operating pressure. If inlet pressures are greater than about 500 psi (35 bar), significant savings in material costs can be achieved by use of the smaller diameter piping slug catcher.
- Gas–liquid separation capability. Horizontal vessels provide the best separation, whereas piping provides the least because its main function is to catch liquid slugs. The large liquid surface area of horizontal vessels provides the best degassing. Piping has the shortest gas residence time when liquid levels are properly maintained in the vessels. However, with piping, small diameter gas scrubbers can be used for demisting.
- Liquid storage. Horizontal vessels can act as primary liquid storage, whereas liquids from vertical vessels and piping must be sent to another vessel. Regardless of slug catcher used, liquids will go to low-pressure flash drums to recover light ends.

Gas plants routinely receive gas from more than one inlet line. In many cases, the inlet line pressures differ. To minimize required compression on the gas that leaves inlet receiving, slug catchers operate at each of the inlet line pressures.

3.5 SAFETY AND ENVIRONMENTAL CONSIDERATIONS

Some common safety issues involved with field operations include the following:

- Leaking pipelines. Leaks present a danger of fire if an ignition source is available and of poisoning if the gas contains high hydrogen sulfide concentrations. Respirators may be required for work in a sour gas field.
- Plugged pipelines. Plugs can be caused by solids blockage, such as hydrates or, occasionally, a stuck pig. Clearing a hydrate plug should be done by depressurization of both sides of the plug to prevent it from dislodging and becoming a projectile that potentially damages the line. Dislodging a hydrate plug in subsea lines can be complex. See Sloan (2000) for more details.

Spheres rarely stick in well-maintained pipelines. The two most common reasons for sticking are gas bypass and line obstruction. Technical help from vendors should be sought. One should contact a sphere manufacturer if common company practice fails.

Other than the liquid slugging problem already mentioned, few safety considerations apply to inlet receiving. One hazardous operation in this area is retrieval of pigs from the pig receiver.

Another consideration is the impact of liquid slugs on the piping. Often, inlet slug catching equipment is designed to field codes, which allow thinner piping than would be used in a gas processing plant. If liquid flows exceed design rates, the required pipe anchors and wall thickness should be checked to ensure the unit can handle the increased pipe stresses induced by potentially larger slugs of liquid (McCartney, 2005).

From an environmental perspective, the most obvious problem is leaking pipelines. However, if hydrate inhibitors are used, berms are required around the storage tanks to prevent soil contamination; in the case of methanol, the potential for fire exists as well. A major issue is control of exhausts from compressors. Carbon monoxide and NO_X are the major concerns for the internal-combustion engines that drive reciprocating compressors, and NO_X is the main pollutant from turbine-driven centrifugal compressors (see Chapter 4). If raw gas is burned, sulfur removal or sulfur oxide emissions may be required.

In situations in which the plant is required to shut-in, gas may have to be vented or flared at remote booster stations. Rich gas in gathering systems can to be heavier than air at lower temperatures, which makes venting hazardous. If the gas is sour, configurations that permit the gas to go to flares may be required.

REFERENCES

Bacon, T.R, Fundamentals of Separation of Gases, Liquids, and Solids, Proceedings of the Laurance Reid Gas Conditioning Conference, Norman, OK, 2001.

Carroll, J.J., *Natural Gas Hydrates: A Guide for Engineers*, Gulf Professional Publishing, Houston, TX, 2003.

Engineering Data Book, 12th ed., Sec. 2, Product Specifications, Gas Processors Supply Association, Tulsa, OK, 2004a.

Engineering Data Book, 12th ed., Sec. 7, Separation Equipment, Gas Processors Supply Association, Tulsa, OK, 2004b.

Engineering Data Book, 12th ed., Sec. 17, Fluid Flow and Piping, Gas Processors Supply Association, Tulsa, OK, 2004c.

Engineering Data Book, 12th ed., Sec. 20, Dehydration, Gas Processors Supply Association, Tulsa OK, 2004d.

Girard Industries, undated. http://www.girardind.com/index.htm, Retrieved June 2005.

Hammerschmidt, E.G., Preventing and removing gas hydrates formations in natural gas pipelines, *Oil Gas J*. 37 (52) 66, 1939.

Kimmitt, R.P., Root, C.R., and Rhinesmith, R.B., Proven Methods for Design and Operation of Gas Plant Liquid Slug Catching Equipment, Proceedings of the Eightieth Annual Convention of the Gas Processors Association, Tulsa, OK, 2001.

Kuchinski, J., private communication, 2005.

McCartney, D., private communication, 2005.

Mehta, A.P. and Klomp, U.C., An Industry Perspective on the State of the Art of Hydrates Management, Proceedings of the Fifth International Conference on Gas Hydrates, Trondheim, 2005, 1089.

Nielsen, R.B and Bucklin, R.W, Why not use methanol for hydrate control? *Hydroc. Proc.*, 62 (4) 71, 1983.

Sloan, E.D., *Clathrate Hydrates of Natural Gases*, 2nd ed. Marcel Dekker, New York, 1998.

Sloan, E.D., *Hydrate Engineering, Monograph*, Vol. 21, Society of Petroleum Engineers, Richardson, TX, 2000.

Sloan, E.D., private communication, 2005.

Sterner, A.J., Developments in Gas-Liquids Separation Technology, Proceedings of the Eightieth Annual Convention of the Gas Processors Association, Tulsa, OK, 2001.

Varnell, B. and Godby, S., Managing Plant Inlet Liquid Slugs, Proceedings of the Eighty-Second Annual Convention of the Gas Processors Association, Tulsa, OK, 2003, 238.

Webb, B.C., Guidelines set out for pipeline pigging, *Oil Gas J.*, 76 (46) 196, 1978a.

Webb, B.C., More guidelines given for gas liquid pipeline pigging, *Oil Gas J.*, 76 (48) 74, 1978b.

WEB SITES

Society of Petroleum Engineers (SPE): www. spe. org/ spe/ jsp/ basic/ 0,,1104 _ 1714 _ 1003866,00.html. Site provides history and overview of well-site operations that relate to gas production.

Occupational Safety and Health Administration (OSHA): www. osha. gov/ SLTC/ etools/ oilandgas/illustrated_glossary.html: Site provides excellent glossary of terms used at drilling sites. Not directly related to gas processing.

Colorado School of Mines Hydrate Research Program: www.mines.edu/research/chs/. Site gives details of current research activities and provides the free downloading of the hydrate prediction program CSMHYD98 to compute hydrate-formation conditions.

4 Compression

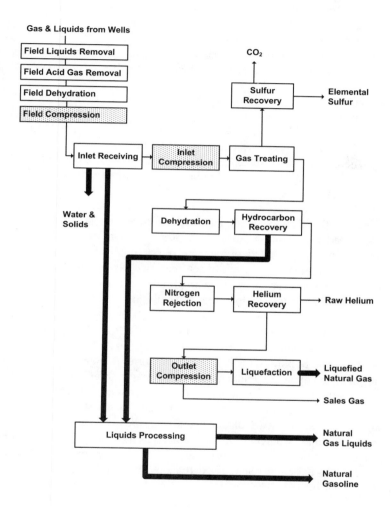

Gas & Liquids from Wells

- Field Liquids Removal
- Field Acid Gas Removal
- Field Dehydration
- Field Compression

Inlet Receiving

Water & Solids

Inlet Compression

Gas Treating

CO₂

Sulfur Recovery → Elemental Sulfur

Dehydration

Hydrocarbon Recovery

Nitrogen Rejection

Helium Recovery → Raw Helium

Outlet Compression

Liquefaction → Liquefied Natural Gas

→ Sales Gas

Liquids Processing → Natural Gas Liquids

→ Natural Gasoline

4.1 INTRODUCTION

Pressure plays a major role in gas processing, as it moves gas from the field, through the gas plant, and into the sales gas line. Pressure provides the source for cooling the gas to low temperatures. Process streams that undergo compression include the following:

- Gas leaving inlet receiving (Chapter 3). To maximize liquids recovery, the gas must be at 850 to 1,000 psi (60 to 70 bar) when it enters the hydrocarbon recovery section but at 600 to 650 psi if only primarily propane and heavier components are recovered (see Chapter 7).
- Gas coming from stabilizing the liquids in liquids processing (see Chapter 10).
- Gas from other vapor recovery facilities that collect low-pressure gas. Low-pressure gas not used as fuel gas is compressed for hydrocarbons recovery.
- Gas exiting hydrocarbon recovery and entering the pipeline (see Chapter 7). For gas plants that process more than 5 MMscfd (140 MSm^3/d) (Crum, 1981), use of turboexpanders to cool the gas for liquid hydrocarbons recovery is economically attractive. Mechanical energy generated by the turboexpander provides part of the power for outlet compression. However, additional outlet compression is required to bring the sales gas to pipeline pressure.

Typically, the only other demands for compression are in recompressing adsorption bed regeneration gas and in propane refrigeration in hydrocarbon recovery. Some plants avoid propane refrigeration by having high inlet gas pressures. This option is attractive when the inlet gas contains relatively small amounts of C_2+ hydrocarbons or when high ethane recoveries are not required.

Gas and refrigeration compressors typically are the largest capital expense in the construction of a new gas processing plant. They account for up to 50 to 60% of the total installed cost of the facility (McCartney, 2005). They also tend to have the largest maintenance expense in the facility.

As in the case of pumps, the two broad categories of compressors are positive displacement and dynamic. Fundamentally, the major difference between the two categories is the former is a volume displacement device, whereas the latter is a pressure (or pump head) device, with performance dependent upon flow and outlet pressure demands. Positive displacement compressors increase pressure by decreasing volume, whereas dynamic compressors turn gas velocity (kinetic energy) into pressure. Whereas positive displacement compressor performance is insensitive to the gas being compressed, dynamic compressors are limited to gases with molecular weights of about 10 or higher (Jandjel, 2000). This limitation is not a concern in gas plant service, unless helium-rich gas from a helium-recovery process needs compression.

After reviewing some fundamentals regarding compression, this chapter presents an overview of the various types of compressors typically found in gas plants and lists their strengths and weaknesses. More discussion on compressors other than reciprocating compressors is presented because they are less familiar to most engineers. This discussion is followed by ways to estimate power requirements for a given set of flow conditions or to estimate flow conditions for a given compressor design. Jandjel (2000) provides a good overview of compressors for gas and air service, and the Engineering Data Book (2004a) gives details on power requirements. Unfortunately, Jandjel (2000) provides no additional references.

4.2 FUNDAMENTALS

This section reviews some basic thermodynamic relations in compression and then defines some of the common efficiencies used to rate compressors.

4.2.1 THERMODYNAMICS OF COMPRESSION

To compute the reversible shaft work, w_S, of a continuous open process that increases the absolute pressure from P_1 to P_2 requires evaluation of the integral (see a standard thermodynamics textbook [e.g., Smith et al., 2001]).

$$w_S = -\Delta h = -\int_{P_1}^{P_2} v dP, \qquad (4.1)$$

where all variables are based upon energy per unit mass. Because the gas is compressible, the integral depends upon the PV path. Minimizing the work requires minimizing the volume during compression, which implies that the process must be isothermal. For an ideal gas, where $V = RT/P$, the reversible isothermal shaft work is given by

$$w_S = -RT \; ln(P_2/P_1)/MW, \qquad (4.2)$$

where R is the gas constant, MW is the molecular mass, and T is the absolute temperature. Reversible isothermal work is the minimum work requirement for a compression process. Note the negative sign indicates that work is being done on the system and that all pressures are in absolute units.

Achievement of isothermal compression requires infinitely good heat transfer to remove all heat of compression. This case represents an unrealistic case, and isothermal work calculations are rarely used in practice.

The other extreme is no heat transfer (adiabatic). All heat generated in compression goes into the gas phase, which maximizes the volume and, thus, requires the maximum work for compression. Compressors, although operating irreversibly, more closely follow this path because compression takes place rapidly, with little time allowed for heat transfer. For reversible adiabatic (isentropic) compression of an ideal gas, the PV path follows

$$Pv^\gamma = \text{constant} \qquad (4.3)$$

and the work is calculated by

$$w_S = \frac{\gamma RT_1}{MW(\gamma - 1)}\left[1 - \left(\frac{P_2}{P_1}\right)^{(\gamma-1)/\gamma}\right], \tag{4.4}$$

where T_1 is the inlet temperature and is the ratio of the molar heat capacities:

$$\gamma = (C_P/C_V) = [C_P/(C_P - R)] \tag{4.5}$$

The value of γ is both temperature and composition dependent. Molar gas mixture heat capacities are computed by

$$C_{PM} = \sum_i x_i C_{P_i} \tag{4.6}$$

and C_{VM} is computed the same way. Appendix B gives heat capacity values for natural gas components as a function of temperature. Common practice is to determine the heat capacity ratio at the average of suction and discharge temperatures.

Equations 4.2 and 4.4 show that the work required is a function of the ratio of absolute pressures, sometimes referred to as the "compression ratio." The actual pressures have a significant effect on compressor design and cost. They have only a secondary effect on the work in that they change the heat capacities and γ. This effect is commonly ignored because the calculations provide only estimates. Equations of state are needed to obtain more accurate work and power requirements.

Example 4.1 Assume an ideal gas, and compute the ratio of heat capacities for the southwest Kansas gas in Table 1.4 (Chapter 1) at 100°F. Assume butanes to be all n-butane and C_5+ fraction to be hexane (Table 4.1).

TABLE 4.1
Data for Example 4.1

	Mol %	C_P at 100°F Btu/lb-mol-°R	γC_P
Helium	0.45	4.97	0.022
Nitrogen	14.65	6.96	1.020
Methane	72.89	8.65	6.305
Ethane	6.27	12.92	0.810
Propane	3.74	18.20	0.680
Butanes	1.38	24.32	0.336
Pentanes and heavier	0.62	40.81	0.253
C_{PM}			9.426

The C_{VM} for the mixture is $C_{PM} - R = 9.43 - 1.99 = 7.44$ Btu/lb-mol-°R and $\gamma = 9.43/7.44 = 1.33$.

The outlet temperature, T_2, for a reversible adiabatic compression of an ideal gas can be calculated by

$$T_2 = T_1(P_2/P_1)^{(\gamma-1)/\gamma}. \tag{4.7}$$

However, the Engineering Data Book (2004a) recommends use of the following empirical equation to obtain a more representative discharge temperature

$$T_2 = T_1\left(1 + \left[\left(\frac{P_2}{P_1}\right)^{(\gamma-1)/\gamma} - 1\right]\right)/\eta_{IS}, \tag{4.8}$$

where η_{IS} is the entropic efficiency discussed below. In actuality, compressors have some heat loss and are irreversible. The PV path followed in this case is

$$Pv^\kappa = \text{constant}, \tag{4.9}$$

where κ is the empirically determined polytropic constant. This constant replaces γ in the above equations, an, thus, the polytropic work is

$$w_S = \frac{\kappa RT_1}{MW(\kappa - 1)}\left[1 - \left(\frac{P_2}{P_1}\right)^{(\kappa-1)/\kappa}\right]. \tag{4.10}$$

Unless the value of κ is known, γ is used for the typical quick calculations on compressors.

All the components in natural gas behave like ideal gases at near-ambient pressure and ambient and above temperatures. However, they become increasing nonideal with increasing pressure. To estimate the work of compression, ideal gas behavior can be assumed. However, to accurately determine the reversible w_S requires use of either an equation of state or thermodynamic charts, such as a temperature–entropy or pressure–enthalpy diagram to accurately calculate h.

Example 4.2 Assume ideal gas behavior, and compute the reversible work required to compress a natural gas mixture from 10 to 60 psig, both isothermally and adiabatically, with an initial temperature of 80°F. Also, calculate the exit temperature for adiabatic compression. The molar mass of the gas is 18, and the ratio of heat capacities is 1.15. Assume the compressor is 100% efficient for this calculation.

Isothermal Compression

From Equation 4.2, the work is

$w_S = -RT \, ln(P_2/P_1)/MW = -1.986 \, (80 + 460) \, ln[(60 + 14.7)/(10 + 14.7)]/18$
$ = -65.9$ Btu/lb

Adiabatic Compression

Use of Equation 4.4 to compute the work, gives

$$w_S = \frac{1.15 \times 1.986 \times (80 + 460)}{18(1.15-1)}\left[1-\left(\frac{60+14.7}{10+14.7}\right)^{(1.15-1)/1.15}\right] = -70.9\,Btu/lb$$

The outlet temperature is given by Equation 4.7:

$$T_2 = (80+460)[(60+14.7)/(10+14.7)]^{(1.15-1)/1.15} = 624°R.$$

$$t_2 = 624 - 460 = 164°F$$

Thus, the adiabatic work of compression is about 8% higher than the isothermal work.

4.2.2 MULTISTAGING

Multistaging of compressors is done for any of several reasons. An obvious reason is that by cooling the gas between stages, the process reduces the gas volume, and, thus, the work required. Although this reason is important, the major reason for multistaging is materials limitations. Assuming γ remained constant in Example 4.1 and the outlet gas pressure was 625 psig, the outlet temperature would be 365°F. Considering materials of construction, seals, and lubricants, the Engineering Data Book (2004a) recommends about 300°F (150°C) as a "good average" for outlet temperatures. If high pressure is involved, 250 to 275°F (120 to 140°C) is recommended. Assuming the initial temperature is near ambient, these maximum temperatures suggest pressure ratios of 3:1 to 5:1.

The minimum work is obtained when each stage of a multistage unit does the same amount of work, and, thus, most compressors will have approximately the same pressure ratio for each stage. In this case, the compression, or pressure ratio, PR, for m stages is computed by

$$PR = (P_2/P_1)^{1/m} \qquad (4.11)$$

(Note that these compression ratios are always the ratio of absolute pressures.) Then the total work of compression will be the sum of the work in each stage, as computed by Equation 4.4.

Example 4.3 Determine the number of stages required to compress the gas in Example 4.1 from 10 to 625 psig by use of a compression ratio of 3:1. Also, calculate the exit temperature for each stage if the gas enters each stage at 80°F.

To determine the value of m, Equation 4.11 is rearranged to give

$m = \ln(P_2/P_1)/\ln (PR) = \ln[(625 + 14.7)/(10 + 14.7)]/\ln (3) = 2.96$ or 3 stages.

The actual pressure ratio will be $[(625 + 14.7)/(10 + 14.7)]^{1/3} = 2.96$. Equation 4.7 gives a discharge temperature of 161°F.

4.2.3 COMPRESSOR EFFICIENCIES

Several efficiencies are used to define compressor performance. The most common is the reversible adiabatic efficiency, commonly referred to as the isentropic efficiency and defined as

$$\eta_{IS} = w_S/w_{S,ACTUAL} \tag{4.12}$$

with w_S computed by Equation 4.4.

The other commonly used efficiency, especially with centrifugal compressors, is the polytropic efficiency, η_P, defined as

$$\eta_P = [(\gamma-1)/\gamma]/[(\kappa-1)/\kappa]. \tag{4.13}$$

This value approximates the ratio of isentropic work to the polytropic work (i.e., Equation 4.4 divided by Equation 4.10), if the pressure-ratio term is ignored. The polytropic efficiency will always be higher than the isentropic efficiency. It can be used to calculate work when applied to Equation 4.10 and is useful for estimating discharge conditions, as in Equation 4.14.

$$T_2 = T_1(P_2/P_1)^{(1/\eta_P)(\gamma-1)/\gamma} \tag{4.14}$$

Example 4.4 Determine the isentropic efficiency of a compressor if the polytropic efficiency is 77% and the gas has $\gamma = 1.15$. Assume the molar mass of the gas is 18, the inlet temperature is 80°F, and the compression ratio is 3.0.

Conversion from polytropic to adiabatic efficiency requires the following steps:

1. Calculate ideal polytropic work.
2. Calculate the actual work by dividing polytropic work by polytropic efficiency.
3. Calculate isentropic work.
4. Calculate isentropic efficiency from ratio of isentropic work to polytropic work.

Step 1. Both γ and η_P are given so Equation 4.13 can be used to determine κ for use in Equation 4.10. However, η_P can be used in combination with γ in Equation 4.10 to give

$$w_S = \frac{\gamma R T_1 \eta_P}{MW(\gamma-1)}\left[1-\left(\frac{P_2}{P_1}\right)^{(1/\eta_P)(\gamma-1)/\gamma}\right]$$

$$= \frac{1.15 \times 1.986(80+460)\,0.77}{18(1.15-1)}\left[1-(3.0)^{(1/0.77)(1.15-1)/1.15}\right] = -72.1\,\text{Btu/lb}$$

Step 2. The actual work required is

$$w_S = -72.1/0.77 = -93.6 \text{ Btu/lb}$$

Step 3. The isentropic work required is

$$w_S = \frac{1.15 \times 1.986 \times (80 + 460)}{18(1.15 - 1)} \left[1 - (3)^{(1.15-1)/1.15}\right] = -70.3 \text{ Btu/lb}$$

Step 4. The isentropic efficiency is then $-70.3/-93.6 = 0.75$ or 75%, two percentage points below the polytropic efficiency.

A volumetric efficiency also applies to reciprocating compressors. This efficiency is the actual volume of gas in the piston cylinder divided by the piston-displacement volume and includes the "dead" volumes associated with the valves and other mechanical clearances, such as those between the piston and heads. Considering only piston displacement, the volumetric efficiency, η_V, is computed by

$$\eta_V = 100 - (P_2/P_1) - C \left[(z_2/z_1) (P_2/P_1)^{1/\gamma} - 1\right], \qquad (4.15)$$

where z is the compressibility factor ($z = PV/RT$), and C is the percent clearance given by

$$C = 100 \text{ (clearance volume)/(piston displacement volume)} \qquad (4.16)$$

Ideally, vendor information on a compressor's volumetric efficiency is available. However, once a compressor is overhauled, the data probably are inaccurate (McCartney, 2005). Even if the compressor is several years old and has not been overhauled, the data for that particular unit is important to have as manufacturers modify efficiency data for newer models. The preferred way to obtain the current volumetric efficiency on older units is to obtain a *PV* curve from a compressor analyzer and back calculate the efficiency.

If valid volumetric efficiency information on the compressor is unavailable, several corrections to the volumetric efficiency must be made (Engineering Data Book, 2004a):

1. Subtract 4% for volumes around suction and discharge valving.
2. Subtract 5% for nonlubricated compressors to account for slippage losses.
3. Subtract 4% for compressing propane or butane.

The first correction always applies, and if the last two apply, they are cumulative. Therefore, the volumetric efficiency could be as much as 13% lower than calculated by Equation 4.15.

The volumetric efficiency is required to estimate compressor capacity and equivalent capacity. Section 4.4.1 discusses these calculations.

4.3 COMPRESSOR TYPES

Only reciprocating compressors were available when the natural gas industry started. Although still used, they are being replaced by other compressor types. In the 1980s, centrifugal compressors came into use. Rotary screw compressors came into use in the 1990s. However, reciprocating compressors will remain dominant in gas processing for many years to come. Table 4.2 (compiled from numerous sources) lists the cost-effective ranges for various compressor types used in gas plants, along with isentropic efficiency ranges.

Figure 4.1 shows the range of discharge pressures as a function of actual inlet-gas flow rates at suction temperature and pressure (acfm) for the compressor types discussed here. The entries serve as guidelines, as the values vary among different sources. Other factors besides those listed in Table 4.2 must be considered when the various compressor types are evaluated. These considerations are discussed below for the two compressor categories.

Compressors are driven by steam or gas-combustion turbines, electrical motors, or internal-combustion engines running on fuel gas. Lower horsepower

TABLE 4.2
Typical Cost Effective Ranges of Compressors Used in Gas Processing

	Inlet Flow Rate[a] acfm (m³/h)	Maximum Pressure psig (barg)		Isentropic Efficiency, %
		Inlet	Discharge	
Reciprocating				
Single stage	1 – 300 (2 – 500)	No limit	< 3,000 (200)	75 – 85
Multistage	1 – 7,000 (2 – 12,000)	No limit	< 60,000 (4,000)	
Centrifugal				
Single stage	50 – 3,000 (80 – 5,000)	No limit	1,500 (100)	70 – 75
Multistage	500 – 200,000 (800 – 350,000)	No limit	10,000 (700)	
Oil-free rotary screw	<40,000 (70,000)	<150 (10)	<350 (25)	70 – 85
Oil-injected rotary screw	< 10,000 (20,000)	< 400 (30)	< 800 (60)	70 – 85

[a] Compressor-gas volumes are based upon actual gas volumes at suction temperature and pressure.

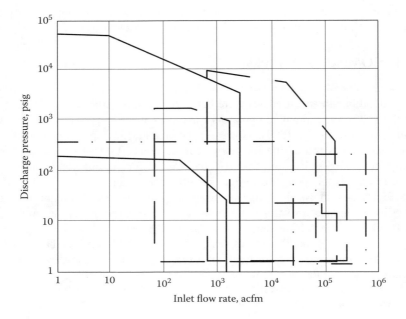

FIGURE 4.1 Ranges of discharge pressure as a function of actual flow rates for various compressors. Two ranges exist for the reciprocating and centrifugal compressors. Lower range denotes single-stage units and higher range denotes multiple-stage units. Legend: — reciprocating, - - centrifugal, - · — · rotary screw, - · · · — axial. (Adapted from Engineering Data Book, 2004a.)

and remote units tend to use electric motors. For high capacity applications, gas turbines are often used.

4.3.1 Positive Displacement Compressors

A variety of positive displacement compressors are used. This section discusses those that have a flow capacity of at least 10 acfm (actual cubic feet per minute) (20 m^3/h) and discharge pressures greater than 100 psig (7 barg). This excludes diaphragm compressors, which have lower flow rates, and various rotary-vane compressors, which have lower discharge pressures.

4.3.1.1 Reciprocating Compressors

Almost every onshore gas plant and field operation uses reciprocating compressors. Figure 4.2 shows a reciprocating compressor on a gas line. In addition to being an established and known technology, they offer the highest efficiency, and their performance is unaffected by gas composition. With well maintained ring seals, they produce an oil-free output. Also, they can handle smaller gas volumes and higher discharge pressures than other types of compressors. Only reciprocating

FIGURE 4.2 Reciprocating compressor at a gas wellhead. (Courtesy of Nuovo Pignone Spa, Italy.)

and centrifugal compressors have sufficiently high discharge pressures to meet the gas processing and gas-transmission requirements. Many reciprocating units have been replaced by centrifugal compressors for high-volume service, but reciprocating units remain as backups.

Reciprocating compressors are frequently multistaged and will have either single-action or double-action pistons (Engineering Data Book, 2004a). Both single-action and double-action pistons may be on a single frame.

Unfortunately, reciprocating compressors have a number of drawbacks; the major one is reliability. Their runtime is 90 to 95% compared with about 99% for other compressor types (Jandjel, 2000). However, units can have runtimes as high as 99%. This reliability tends to be found in plants where operators can watch the units and on electrically driven compressors, as gas-fired engines are more prone to failure than the compressor (McCartney, 2005).

Their efficiency advantage decreases when the throughput is reduced. Remedies for reduced volumes include

- Use of variable speed drives
- Adjustment of suction pressure
- Use of inlet-valve unloaders (which prevent the valves from completely seating)
- Recirculation of gas to compensate for lower input rates

Also, the compression ratio typically is limited to about 3:1 to 4:1 for each stage. A combination of high-temperature limitations and rod strength are the constraining factors for the low ratio.

Reciprocating compressors tend to be noisier than other types and are the only ones requiring pulsed flow dampeners (snubbers). These compressors are susceptible to condensed phases which erode cylinder walls, destroy lubricating fluids and damage the rings. Finally, for the same volume of throughput, they occupy much more space than the other types. In spite of the drawbacks, the versatility of reciprocating compressors will keep them in the gas industry for a long time.

Because of their widespread use, the Engineering Data Book (2004a) gives extensive details on reciprocating compressors, including details on controls and pulsation dampeners. The book also includes a troubleshooting guide.

4.3.1.2 Oil-Free Rotary Screw Compressors

Rotary screw compressors use two screws, or lobes (see Figure 4.3), to compress the gas. Gas enters as the threads at the suction side are separating, and it moves down the threads as the screws rotate. Clearances between the threads decrease and compress the gas. The gas exits in an axial port at the end of the screws. A timing drive keeps the two lobes synchronized. The screws run at 3,000 to 8,000 rpm, and the speed is easily varied to provide an efficient means to handle lower flow rates. Unlike reciprocating compressors, essentially all gas is displaced (i.e., the volumetric efficiency is near 100%).

These compressors were first used in the steel industry to compress air that contained up to 200-micron-sized particulates. Thus, they can easily process streams containing condensate droplets. They are extremely reliable.

FIGURE 4.3 Cutaway view of an oil-injected rotary screw compressor. (Courtesy of Ariel Corporation.)

Jandjel (2000) states that condensed vapor with small γ (e.g., isobutane) is sometimes injected as a coolant that permits compression ratios to increase to as high as 8:1.

The primary limitation of oil-free screw compressors is the low discharge pressure. For gas processing, this limitation prevents these compressors from being used for inlet and outlet compression. However, they are used for vapor recovery and compression of low-pressure gases up to the suction pressure of high-pressure compressors. The ability to inject liquid to increase the pressure ratio makes them attractive for propane refrigeration units, although the oil-injected compressors discussed next are widely used as well.

4.3.1.3 Oil-Injected Rotary Screw Compressors

Figure 4.3 shows a cutaway view of an oil-injected rotary screw compressor. Gas enters in the upper right side, flows through the compressor, and exits at the lower left. For oil-injected compressors, the male screw (upper one in Figure 4.3) is driven and the female screw follows. Unlike the oil-free rotary screw compressor, the oil-injected type needs no timing drive. The injected oil provides the seal between each screw and the casing and between the screws. Oil floods the unit at various points along the screws. The oil-injected rotary screw compressor is similar to the oil-free version but has two ports (Bruce, 2000). Like the oil-free compressor, the oil-injected rotary screw compressor has an axial port, which in this case, removes gas and remnants of oil. This compressor type also has a slide valve located on the bottom of the compressor that removes the fluids through a radial port.

The slide valve provides a major advantage for these compressors because its position determines the volumetric throughput. The slide valve may be "fixed" and thus require manual repositioning to adjust volumetric flows or may be adjusted automatically. If the pressure ratio remains constant, some loss in efficiency occurs when the volumetric throughput is adjusted because the optimum volume ratio is a function of pressure ratio (Bruce, 2000). Bruce (2000) provides details on turndown strategies for oil-injected screw compressors.

The oil system is a closed-loop system in these compressors and is a critical component, as the oil provides both sealing and cooling. Discharged gas and oil go to a scrubber that contains a demister to reduce oil concentration to less than 10 ppm (Bruce, 2000). The discharged oil is cooled from $170 - 220°F$ ($75 - 105°C$) to 140 to 160°F (60 to 70°C) before being reinjected. Figure 4.3 shows tubing, which is used to circulate the oil and two "bottles," which are filters to remove particulates from the oil. The tubing that goes to and from the extended piece on the lower right drives the positioning of the slide valve on this particular unit.

Oil injection permits pressure ratios up to 23:1, but discharge pressures are still below the highest pressures required for gas processing. These compressors are well suited for and commonly used in propane refrigeration systems. Jandjel (2000) estimates that these compressors will become the preferred compressors in the future for vapor recovery and fuel gas compression. He cites the following reasons:

- Improved lubricants that are both stable and inert
- Improved filtration systems that remove oil droplets down to less than 1 micron in size and to concentrations less than 1 ppm to meet oil-free specifications
- Improved pressure and flow limits

These improvements address past concerns regarding these compressors. However, oil compatibility with the process components can still be a problem on new installations and selection of the proper oil is an empirical process.

Some newer units have extremely tight clearances between the lobes to increase discharge pressure. A drawback to close tolerances is that if shut down, the compressor must cool before being restarted.

For offshore applications, the oil-injected screw compressor has several advantages over reciprocating compressors in low-pressure service, including:

- Small size, about one quarter of that required by reciprocating compressor for comparable capacity
- Low maintenance
- Ease of adjusting to variable inlet flows
- High reliability

In addition to relatively low discharge pressure, the other disadvantages involve the oil system. The lubricating oil must be compatible with the hydrocarbons present at the operating conditions. Hydrocarbon solubility in the oil must be minimal to prevent foaming problems on the suction side and to avoid loss of the lubricating properties of the oil. Whereas oil-free screw compressors handle "dirty gas" extremely well, oil-injected systems are more susceptible to solids. If solids are a problem, the oil filtration systems must be maintained properly. Energy is spent to compress the oil, but this is work comparable to pumping liquid and is trivial compared with the work of gas compression.

4.3.2 Dynamic Compressors

The two dynamic compressor types are centrifugal and axial. Axial compressors handle large gas volumes (50 to 300 Macfm [80 to 500 Mam³/h]) and have higher efficiencies than do centrifugals. However, they generally have discharge pressures below 200 psig (14 barg) and are restricted to clean gases. Although not used for processing the natural gas, axial compressors will be discussed briefly in the context of their use in the gas turbines that drive centrifugal compressors.

To illustrate how axial and centrifugal compressors work in tandem, Figure 4.4 shows a schematic of a typical gas turbine–driven centrifugal compressor. Figure 4.5 shows a large gas turbine compressor system. The centrifugal compressor is the large cylindrical unit in the center of the picture, with the two large attached pipes. The large ductwork to the left of the compressor is for the turbine exhaust gas, and the one further to the left is intake air to the axial turbine.

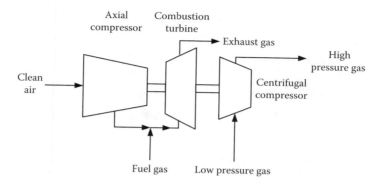

FIGURE 4.4 Schematic of a gas turbine-driven centrifugal compressor.

4.3.2.1 Centrifugal Compressors

Since the 1980s, centrifugal compressors have been replacing reciprocating compressors and are widely used throughout the industry. With their high flow rates and relatively high discharge pressures, they commonly provide both inlet and outlet compression. They also are used extensively for gas pipeline transmission.

FIGURE 4.5 Large gas turbine-driven centrifugal compressor. (Courtesy of Solar Turbines.)

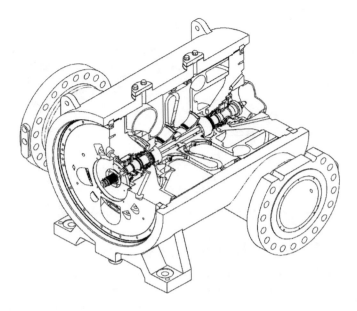

FIGURE 4.6 Cutaway of a centrifugal compressor. (Courtesy of Solar Turbines.)

In gas processing, they are usually powered by either a gas turbine or a turboexpander but can be steam or electrically driven.

Figure 4.6 shows a cutaway view of a multistage centrifugal compressor. Basically, the compressor consists of a rapidly rotating (1,000 to 20,000 rpm for natural gas applications) impeller (rotor), surrounded by a stationary diaphragm. (For more details on these compressors, see the Engineering Data Book [2004a].) Centrifugal compressors work by having gas enter the rotor of each stage near the rotor shaft. The gas is thrust outwards into the channels of the stationary diaphragm, which are called the diffuser or diffuser passages, where the velocity converts to pressure. The diffuser contains vanes to maximize conversion of velocity to pressure and to direct gas into the next stage. Vendors estimate that about one third of the pressure increase occurs in the diffuser, and the rest comes from the rotors.

The rotor can be open or shielded on one or both sides. A survey of literature from compressor manufacturers indicates that most compressors for natural gas service are shielded on both sides (i.e., closed), as shown in Figure 4.6. The blades in the rotor are swept back to improve aerodynamics and mechanical stability. A nice feature of these compressors is that all stages are on the same drive shaft.

Like other compressors, exit temperature limits the pressure ratio in centrifugal compressors. The Engineering Data Book (2004a) recommends a maximum discharge temperature of 400°F (200°C).

A major difference between positive displacement and centrifugal compressors is how discharge pressure affects gas flow. Figure 4.7 shows the interaction

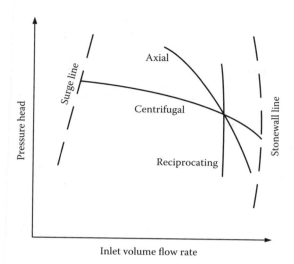

FIGURE 4.7 Relationship between compressor and inlet-volume flow rate for centrifugal, axial, and reciprocating compressors. (Adapted from Engineering Data Book, 2004a.) 13-27

between head (i.e., discharge pressure) and actual flow rate for reciprocating, axial, and variable speed centrifugal compressors. For reciprocating compressors, the gas flow rate is constant and independent of the downstream pressure (constant volume – variable head), which is why relief valves or other devices are necessary to protect the compressor from excessive discharge pressures.

For centrifugal compressors, the downstream pressure greatly affects the throughput (constant head – variable volume). The range over which the compressor functions properly is limited. If downstream pressure becomes too high (or inlet pressure becomes too low), the compressor cannot push the gas out and backflow occurs, which results in a drop in the discharge pressure. Then the compressor raises the discharge pressure only to go into backflow again. This cyclic process is called "surge" and must be avoided because the pressure oscillations cause overheating and vibration that can damage internal bearings. All centrifugal compressors come with surge control systems that protect the compressor by gas recirculation. The Engineering Data Book (2004a) discusses surge-control methods.

The other limit occurs when the gas flow through the compressor reaches sonic velocity and flow cannot be increased. This limit is choked flow, or "stonewall." To increase the flow requires internal modifications.

Centrifugal compressors have other disadvantages. Perhaps the most important is the effect of gas density on discharge pressure. This effect is not a factor where the gas composition remains nearly constant, such as in pipeline transmission boosters. However, if used in applications where gas composition changes over time, the compressors must be restaged periodically to obtain maximum efficiency while maintaining a constant flow rate and discharge pressure. Restaging is expensive in terms of both capital outlay and downtime.

Gases that flow into centrifugal and axial compressors must be free of condensed phases because the particles damage the vanes. Their off-design operation is poor compared with the other compressors, and they are not good for intermittent operations. However, where they can be used, their reliability readily offsets their drawbacks.

4.3.2.2 Axial Flow Compressors

Axial compressors routinely compress large volumes of gas (50 to 500 Macfm [80 to 800 Mam³/h]) at pressures up to about 200 psig (14 barg), although in special applications, they may go to higher pressures. They are smaller and more efficient than centrifugals. As the name implies, axial compressors use stationary and rotating vanes to push the gas down the axis instead of in a radial direction like centrifugal compressors.

Figure 4.7 shows that axial compressors have a similar behavior to centrifugal compressors, but in combination with a gas turbine and centrifugal compressor they operate in a range where neither surge nor stonewalling occurs. They have a higher efficiency than centrifugals and are ideal for moving large volumes of air, such as that needed to fire the gas turbines used to drive compressors and generators. The remainder of this section focuses on axial compressors combined with gas turbines.

A gas-fired turbine works on the same principle as a jet engine. Figure 4.8 shows a cutaway view of an axial compressor in combination with a gas turbine. Clean air enters the axial compressor at the lower left side of the drawing. It is compressed in the 11 stages of the axial compressor to an exit temperature of 650 to 950°F (350 to 500°C). The hot air burns the fuel in the combustion zone

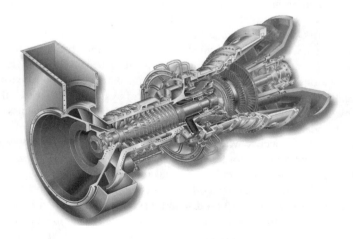

FIGURE 4.8 Cutaway view of a gas-fired turbine. (Courtesy of Nuovo Pignone Spa, Italy.)

and increases the gas temperature to 1,800 to 2,300°F (1,000 to 1,250°C). The high-temperature, high-pressure gas then passes over the three turbine blades to generate the mechanical energy and exits at approximately 950°F (500°C).

Like all heat engines, the turbine thermal efficiency is driven by temperature differences. Gas turbines follow the Brayton cycle (Smith et al., 2001), and ideal efficiencies (defined as the ratio of net work generated divided by heat input) can be around 40%. However, company literature indicates actual overall efficiencies in the 28 to 35% range. Turbine ratings are often expressed in terms of heat rate, which is the ratio of the heat generated to the power produced (Btu/hp-h or kJ/kW-h). Note that the lower heating value is used in this calculation. The Engineering Data Book (2004b) provides additional details on the calculation of heat rates and power on the basis of operating conditions. The book also describes combined-cycle units that use the waste heat to generate steam for additional power generation.

Example 4.5 Compute the overall thermal efficiency of a gas turbine that has a heat rate of 8,500 Btu/hp-h (12,000 kJ/kWh).

The thermal efficiency is the reciprocal of the heat rate. Including the conversion between energy and power (2,544 Btu/hp-h or 3,600 kJ/kW-h), the thermal efficiency is

Thermal Efficiency = 2,544/heat rate (3,600/heat rate)

= 2,544/8,500 (3,600/12,000) = 0.30 or 30%

(If the heat rate is stated in Btu/kWh, the conversion is 3,414 Btu/kW-h.)

For applications in which the speed is nearly constant, all rotating components are on a single shaft. However, if a variable-speed centrifugal compressor is needed, a second shaft will be coupled to the turbine shaft to permit the compressor and turbine to run at different speeds.

The axial compressor–gas turbine unit is known for high reliability. However, the greater number of moving parts and seals, along with high temperature operation, makes the turbine the least reliable of the turbine–centrifugal compressor combination. Air filtration systems are mandatory to protect the axial compressor from particulate matter. In offshore applications, demisters replace the filters. Self-cleaning filters that use pulses of clean air from one filter bank to blowout another filter bank are used in many environments. They have the advantage of being able to de-ice a filter bank at the same time when in cold-weather service. The Engineering Data Book (2004b) provides additional recommendations on filtration systems.

Plant operators watch the combustion zone exhaust temperature closely because it runs close to materials limits. High ambient temperatures sometimes raise the exhaust temperature to a point where gas throughput must be curtailed.

4.4 CAPACITY AND POWER CALCULATIONS

This section provides some methods for estimating the capacity and power requirements of compressors when ideal gases or nearly ideal gases are used. The Engineering Data Book (2004a) provides more detailed calculations. Most of the calculations are independent of compressor type. Screw compressors with liquids present require either polytropic or more rigorous thermodynamic methods to be used. Reciprocating compressors require knowledge of displacement volumes to compute flow rates, as will be discussed below. If accurate calculations are needed, equation of state methods should be used, along with information about the compressor performance provided by the vendor.

4.4.1 CAPACITY

4.4.1.1 General Calculations

In this section, we assume that the gas can be described by

$$PV = nzRT, \tag{4.17}$$

where z is the compressibility factor. (Appendix B provides charts to estimate z.)

Example 4.6 Estimate the compressibility factor, z, for the gas composition given in Table 4.3 at 120°F and 600 psig.

One commonly used method for estimating the compressibility factor for a gas or gas mixture is to use a general compressibility factor chart (see Appendix B), which assumes that the compressibility factor for any gas with a molecular weight less than 40 is a function of the reduced pressure, P_R, and temperature, T_R, of the gas. (For more details on this "corresponding states" method, see a book on physical properties [e.g., Poling et al., 2000]). For a pure gas, the reduced pressure and temperature is obtained by division of the absolute temperature by the critical temperature and the absolute pressure by the critical pressure. For mixtures, the critical properties are obtained by use of pseudocritical pressure and temperature for the mixture defined by (Engineering Data Book, 2004c)

$$P_{CM} = \sum x_i P_{C_i} \text{ and } T_{CM} = \sum x_i T_{C_i}.$$

TABLE 4.3
Data for Example 4.6

Component	Mol fraction, y	P_c, psia	yP_c	T_c, °R	yT_c
Methane	0.88	666	586.1	343	301.8
Ethane	0.06	707	42.4	550	33.0
Propane	0.03	616	18.5	666	20.0
i-butane	0.01	528	5.3	734	7.3
Butane	0.02	551	11.0	765	15.3
Mixture properties			663.3		377.5

The reduced temperature for the mixture is $(120 + 460)/377.5 = 1.54$, and the reduced pressure is $(600 + 14.7)/663.3 = 0.93$. From the compressibility chart in Appendix B, the compressibility factor is 0.91.

For capacity calculations, the actual volumetric flow rate, Q, is based upon inlet conditions to each stage. To compute the actual flow rate from gas flows given as standard flow rates use

$$Q = scfm \left(\frac{14.7}{P_1(psia)} \right) \left(\frac{T_1(°R)}{520} \right) \left(\frac{z_1}{z_R} \right) \tag{4.18a}$$

$$Q = sm^3h \left(\frac{101}{P_1(bar)} \right) \left(\frac{T_1(K)}{288} \right) \left(\frac{z_1}{z_R} \right), \tag{4.18b}$$

where the subscript "1" denotes inlet conditions. The subscript "R" refers to the conditions used to define the reference state, in this case, 14.7 psia (1.01 bara) and 60°F (15.6°C). In these equations, the units must be consistent and the temperatures are absolute (°R or K). If mass flow rates are given, the molar mass (MW), of the gas as well as the mass flow rate, $\dot{w}$, must be known:

$$Q = \frac{10.73\dot{w}(lb/min)T_1(°R)}{MWP_1(psia)} \left(\frac{z_1}{z_R} \right) \tag{4.19a}$$

$$Q = \frac{8.314\dot{w}(kg/hr)T_1(K)}{MWP(Bar)_1} \left(\frac{z_1}{z_R} \right) \tag{4.19b}$$

If the molar flow rate $\dot{n}$, is given, the actual gas flow rate is computed by

$$Q = 379.5\dot{n}(lb-mol/hr) \left(\frac{T_1(°R)}{520} \right) \left(\frac{14.7}{P_1(psia)} \right) \left(\frac{z_1}{z_R} \right) \tag{4.20a}$$

$$Q = 8.314\dot{n}(mole/hr) \left(\frac{T_1(K)}{P_1(Bar)} \right) \left(\frac{z_1}{z_R} \right) \tag{4.20b}$$

4.4.1.2 Reciprocating Compressors

Calculating the throughput of a reciprocating compressor requires knowledge of the volume displaced during each stroke. Normally, the piston displacement, PD, is in ft³/min (m³/h), which is the product of cross-sectional area, stroke travel, and RPM. The equations for computing PD depend upon compressor type and are given below, in dimensioned form (Engineering Data Book, 2004a).

For a single-acting piston compressing only on the outer end (familiar configuration):

$$PD = \frac{\pi D^2 (in^2)}{4} \times RPM \left(\frac{rev}{min}\right) \times Stroke(in) \times \frac{ft^3}{1728 in^3} \qquad (4.21a)$$

$$= 4.55(10^{-4}) \times D^2 \times RPM \times Stroke$$

$$PD = \frac{\pi D^2 (cm^2)}{4} \times RPM \left(\frac{rev}{min}\right) \times Stroke(cm) \times \frac{m^3}{10^6 cm^3} \times \frac{60\,min}{h} \qquad (4.21b)$$

$$= 4.7(10^{-5}) \times D^2 vRPM \times Stroke$$

For a single-acting piston compressing only on the crank end:

$$PD = \frac{\pi (D^2 - d^2)(in^2)}{4} \times RPM \left(\frac{rev}{min}\right) \times Stroke(in) \times \frac{ft^3}{1728 in^3} \qquad (4.22a)$$

$$= 4.55(10^{-4}) \times (D^2 - d^2) \times RPM \times Stroke$$

$$PD = \frac{\pi (D^2 - d^2)(cm^2)}{4} \times RPM \left(\frac{rev}{min}\right) \times Stroke(cm) \times \frac{m^3}{10^6 cm^3} \times \frac{60\,min}{h} \qquad (4.22b)$$

$$= 4.7(10^{-5}) \times (D^2 - d^2) \times RPM \times Stroke$$

For double-acting piston compressing on both the crank and piston end (excluding the tail rod type, which has rods on both side of the piston):

$$PD = \frac{\pi (2D^2 - d^2)(in^2)}{4} \times RPM \left(\frac{rev}{min}\right) \times Stroke(in) \times \frac{ft^3}{1728 in^3} \qquad (4.23a)$$

$$= 4.55(10^{-4}) \times (2D^2 - d^2) \times RPM \times Stroke$$

$$PD = \frac{\pi (2D^2 - d^2)(cm^2)}{4} \times RPM \left(\frac{rev}{min}\right) \times Stroke(cm) \times \frac{m^3}{10^6 cm^3} \times \frac{60\,min}{h} \qquad (4.23b)$$

$$= 4.7(10^{-5}) \times (2D^2 - d^2) \times RPM \times Stroke$$

With the displacement volume, suction pressure and volumetric efficiency (Equation 4.15), the net capacity can be computed. Note that compressor vendors usually use a standard pressure of 14.4 psia* (0.99 bar), instead of 14.7 psia. In dimensioned form, the equivalent volume is

$$MMscfd = \frac{z_{14.4}}{z_1} \frac{\eta_V}{100} \times PD\left(\frac{ft^3}{min}\right) \times \frac{P_1}{14.4} \times \frac{1440\,min}{d} \times \frac{10^{-6}\,MMft^3}{ft^3}$$

$$(4.24a)$$

$$= \frac{\eta_V}{z_1} \times PD \times P_1 \times 10^{-6}, \text{ assuming } Z_{14.4}$$

$$Sm^3d = \frac{z_{1bar}}{z_1} \frac{\eta_V}{100} \times PD\left(\frac{m^3}{h}\right) \times \frac{P_1}{1} \times \frac{24h}{d}$$

$$(4.24b)$$

$$= 0.24 \times \frac{\eta_V}{z_1} \times PD \times P_1, \text{ assuming } z \text{ at } 1 \text{ bar} = 1.$$

4.4.2 Power Requirements

The ideal power requirements for isentropic compression makes use of the shaft work calculated in Equation 4.4 and the isentropic efficiency to give

$$Power = \dot{m}\ z_{avg}\ w_S, \qquad (4.25)$$

where $\dot{m}$ is the mass flow rate, and z_{avg} is the average of the inlet and outlet compressibility factors.

Because they are the more common compressor types, the Engineering Data Book (2004a) gives some useful methods for estimating power requirements for reciprocating and centrifugal compressors. The following paragraphs give these relationships.

4.4.2.1 Reciprocating Compressors

An empirical equation (Engineering Data Book, 2004a) for estimating the power required for reciprocating compressors is

$$Brake\ HP = 22\ F\ PR\ m\ MMacfd \qquad (4.26a)$$

$$Brake\ HP = 0.014\ F\ PR\ am^3/h, \qquad (4.26b)$$

---

* The 14.4 psia value was selected long ago as a matter of convenience to make calculations simpler because a day contains 1,440 minutes; see, for example, Equation 4.24a (McCartney, 2005).

where Brake HP is the actual work delivered to the compressor and the correction factor F is given by

$F = 1.0$ for single-stage compression ($m = 1$)
$F = 1.08$ for two-stage compression ($m = 2$)
$F = 1.10$ for three-stage compression ($m = 3$)

The volume should be based upon inlet temperature and 14.4 psia (1.0 bara).

One should be aware of several items when doing these calculations. First, vendors rate compressors on the basis of 14.4 instead of 14.7 psia. The above equation was developed for large low-speed (300 to 400 rpm) compressors that process gases with a specific gravity of 0.65 and PR values greater than 2.5. For gases with specific gravities of 0.8 to 1.0, use 20 in Equation 4.26a or 0.013 in Equation 4.26b; the value should be 16 to 18 (10 to 12) for PR in the 1.2 to 2 range.

Example 4.7 Use Equation 4.26 to estimate the brake horsepower needed to compress 30 MMscfd of the gas in Example 4.1 from 10 to 625 psig. Assume the intake temperature to be 80°F.

Example 4.3 showed that three stages are necessary to go from 10 to 625 psig; therefore, the factor F is 1.10. Also, the example found the compression ratio, PR, to be 2.96. The molar mass of the gas is 18, so the specific gravity is $18/28.96 = 0.62$. Therefore, the factor of 22 is appropriate. Finally, use Equation 4.18, and assume the compressibility factors to be unity, to correct the volumetric flow rate to 14.4 psia and 80°F:

$$MMcfd = MMscfd \left(\frac{14.7}{P_1(psia)} \right) \left(\frac{T_1(°R)}{520} \right) = 30 \left(\frac{14.7}{14.4} \right) \left(\frac{460 + 80}{520} \right) = 31.8 \, MMcfd$$

Then the estimated horsepower requirement is

Brake HP $= 22$ F PR m MMcfd $= 22 \times 1.10 \times 2.96 \times 3 \times 31.8 = 6,830$ hp (5.1 MW).

4.4.2.2 Centrifugal Compressors

As in the case of centrifugal pumps, head, h_d (the height of a column of the fluid being pumped, [ft-lb$_f$/lb$_m$, N·m/kg]) is commonly used in work with centrifugal compressors, instead of pressure. The isentropic head, h_{dIS}, is given by

$$h_{dIS} = -z_{Avg} w_S = \frac{z_{Avg}}{MW} \frac{\gamma RT_1}{\gamma - 1} \left[\left(\frac{P_2}{P_1} \right)^{(\gamma-1)/\gamma} - 1 \right], \qquad (4.27)$$

where z_{avg} is the average compressibility factor between inlet and discharge temperature and pressure. If polytropic head, h_P, is needed, replace γ with κ. The relationship between isentropic and polytropic head is

$$h_{dP}/h_{dIS} = \eta_p/\eta_{IS}. \tag{4.28}$$

The actual power in compressing the gas is

$$\text{Power} = \dot{m}h_{dIS}/\eta_{IS} \tag{4.29}$$

for isentropic compression. The isentropic power can be calculated by use of the isentropic head and efficiency. For an estimation of the mechanical losses, the Engineering Data Book (2004a) suggests

$$\text{Mechanical losses} = (\text{Power})^{0.4} \tag{4.30}$$

so that the brake HP requirement will be the sum of Equation 4.29 and Equation 4.30.

To provide variable flow rates, centrifugal compressors vary the speed of the compressor. The effect of RPM on compressor performance follows the affinity, or fan laws, which are valid for single-stage compressor or multistage compressor with low compression ratios or very low Mach numbers. The fan laws give

$$\frac{Q_1}{Q_2} = \frac{RPM_1}{RPM_2} \tag{4.31}$$

$$\frac{h_1}{h_2} = \left(\frac{RPM_1}{RPM_2}\right)^2 \tag{4.32}$$

$$\frac{BrakeHP_1}{BrakeHP_2} = \left(\frac{RPM_1}{RPM_2}\right)^3. \tag{4.33}$$

Although these relations are useful, they become less accurate with deviations of more than about ±10% from design conditions.

4.5 COMPARISON OF RECIPROCATING AND CENTRIFUGAL COMPRESSORS

Only reciprocating and centrifugal compressors can meet the demand for discharge pressures in the 1,000 psi (70 bar) range. When gas volumes are large enough for centrifugal compressors to be considered, they have many advantages over reciprocating compressors, including (Engineering Data Book, 2004a):

- Lower initial and maintenance costs
- Higher reliability and longer intervals between downtimes
- Less operating attention
- Greater throughput per unit of platform area
- Adaptability to high speed, low-maintenance drivers

Both efficiency and maintenance must be considered when reciprocating and centrifugal compressors are compared. Reciprocating compressors are far more efficient (see Table 4.2) on the compressor side. (Heat rates of gas turbines and internal-combustion engines widely vary with power output and vendor, which makes a general comparison meaningless. (The Engineering Data Book [2004b] lists these data for major vendors.) If the gas processor shares the fuel costs with the producer (e.g., with POP contracts [see Chapter 2]), the difference in efficiencies is not a major factor for the gas processor, because most of the fuel cost is borne by the producer.

However, the processor typically pays all maintenance costs. These costs for reciprocating compressors run \$40 to \$60/hp/year (\$30 to \$45/kW/yr). For centrifugal compressors, the cost is \$10 to \$15/hp/yr (\$8 to \$12/kW/yr) (McCartney, 2005). If a 100 MMscfd (2.8 Mm³/d) plant has 70,000 hp (50 MW) devoted to inlet and outlet compression, the savings by use of centrifugal compressors can be several million dollars per year.

4.6 SAFETY AND ENVIRONMENTAL CONSIDERATIONS

Apart from the dangers inherent in high-pressure equipment, noise is the main safety concern related to compression, especially for compressors on the scale of centrifugal units. Normally ear protection is worn in the plant, but it is a must around compressors, which produce noise levels in the 85 db to 95 db range.

Exhaust emissions from compressor power drivers are the major environmental concern. Newer internal-combustion engines that drive reciprocating compressors were designed to meet an emission level of 2 g NO_X and CO per horsepower by use of "clean burn or lean burn" units (Kuchinski, 2005). However, tighter controls are being imposed and catalytic converters, which reduce NO_X to N_2 and oxidize CO to CO_2, are being installed on these units. This requirement applies to smaller booster station compressors as well as large-plant units. This requirement is one of the reasons that smaller booster stations use electric motor–driven compressors where reliable power is available.

High air-to-fuel ratios combined with good combustor designs minimize CO emission from gas turbines. Nitrogen oxides are the most significant exhaust pollutant and high-efficiency systems enhance NO_X production. Water is often added to reduce NO_X production. This addition helps reduce the NO_X formed from N_2 but not that formed from organic nitrogen, which might be in the fuel. Water may actually enhance NO_X production from the organic nitrogen, as well as increase CO production (Engineering Data Book, 2004b). Modification of the air-fuel ratio and combustor design helps keep the NO_X exit concentration level in the 10 to 15 ppm range.

REFERENCES

Bruce, J. T., Screw compressors: A Comparison of Applications and Features to Conventional Types of Machines, Proceedings of the Seventy-Ninth Annual Convention of the Gas Processors Association, Tulsa, OK, 2000, 229.

Crum, F.S., Application of J-T Plants for LP-Gas Recovery, Proceedings of the Sixtieth Annual Convention of the Gas Processors Association, Tulsa, OK, 1981, 68.

Engineering Data Book, 12th ed., Sec. 13, Compressors and Expanders, Gas Processors Supply Association, Tulsa, OK, 2004a.

Engineering Data Book, Prime Movers for Mechanical Drives, Gas Processors Supply Association, Tulsa OK, 2004b.

Engineering Data Book, 12th ed., Sec. 23, Physical Properties, Gas Processors Supply Association, Tulsa OK, 2004c.

Jandjel, D.G., Select the right compressor, *Chem. Eng. Progress*, 96 (7) 15, 2000.

Kuchinski, J., private communication, 2005.

McCartney, D.G., private communication, 2005.

Poling, B.E., Prausnitz, J.M., and O'Connell, J.P., *The Properties of Gases and Liquids*, 5th ed., McGraw Hill, New York, 2000.

Smith, J.M., Van Ness, H.C., and Abbott, M.M., *Introduction to Chemical Engineering Thermodynamics*, 6th ed., McGraw-Hill, New York, 2001.

5 Gas Treating

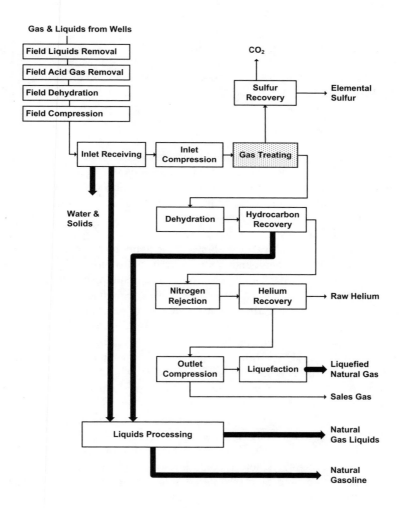

5.1 INTRODUCTION

Gas treating involves reduction of the "acid gases" carbon dioxide (CO_2) and hydrogen sulfide (H_2S), along with other sulfur species, to sufficiently low levels to meet contractual specifications or permit additional processing in the plant without corrosion and plugging problems. This chapter focuses on acid gases

91

because they are the most prevalent. When applicable, discussion is given to other sulfur species.

In this introduction, we briefly address the following questions to set the stage for the remainder of the chapter:

1. Why are the acid gases a problem?
2. What are the acid gas concentrations in natural gas?
3. How much purification is needed?
4. What is done with the acid gases after separation from the natural gas?
5. What processes are available for acid gas removal?

5.1.1 THE PROBLEM

Hydrogen sulfide is highly toxic, and in the presence of water it forms a weak, corrosive acid. The threshold limit value (TLV) for prolonged exposure is 10 ppmv and at concentrations greater than 1,000 ppmv, death occurs in minutes (Engineering Data Book, 2004b). It is readily detectable at low concentrations by its "rotten egg" odor. Unfortunately, at toxic levels, it is odorless because it deadens nerve endings in the nose in a matter of seconds.

When H_2S concentrations are well above the ppmv level, other sulfur species can be present. These compounds include carbon disulfide (CS_2), mercaptans (RSH), and sulfides (RSR), in addition to elemental sulfur. If CO_2 is present as well, the gas may contain trace amounts of carbonyl sulfide (COS). The major source of COS typically is formation during regeneration of molecular-sieve beds used in dehydration (see Chapter 6).

Carbon dioxide is nonflammable and, consequently, large quantities are undesirable in a fuel. Like H_2S, it forms a weak, corrosive acid in the presence of water.

The presence of H_2S in liquids is usually detected by use of the copper strip test (ASTM D1838 Standard test method for copper strip corrosion by liquefied petroleum (LP) gases). This test detects the presence of materials that could corrode copper fittings. One common method of determining ppm levels of H_2S in gases is to use stain tubes, which involves gas sampling into a glass tube that changes color on the basis of H_2S concentration. This method is good for spot and field testing. A newer, continuous method measures reflectance off of a lead acetate–coated tape that darkens in the presence of H_2S. The test (ASTM D4084 Standard test method for analysis of hydrogen sulfide in gaseous fuels (Lead Acetate reaction rate method) 2005) detects H_2S down to 0.1 ppmv. It is specific to H_2S but used to detect total sulfur by hydrogenating the sulphur species in the gas before going to the detector.

5.1.2 ACID GAS CONCENTRATIONS IN NATURAL GAS

Although many natural gases are free of objectionable amounts of H_2S and CO_2, substantial quantities of these impurities are found in both gas reserves and production in the United States. In a survey of U.S. gas resources, Meyer (2000) defined subquality gas as that containing $CO_2 \geq 2\%$, $N_2 \geq 4\%$, or $H_2S \geq 4$ ppmv. These criteria were selected because gases that contain these amounts of impurities generally require upgrading or blending. Using these criteria, Meyer (2000) estimated that 41% of proven gas reserves are subquality and 34% of 1996 gas production in the lower 48 states was subquality. For associated gas reserves, they estimated that 13% was subquality. It is obvious that removal of H_2S and CO_2 is a major concern in gas processing.

5.1.3 PURIFICATION LEVELS

The inlet conditions at a gas processing plant are generally temperatures near ambient and pressures in the range of 300 to 1,000 psi (20 to 70 bar), so the partial pressures of the entering acid gases can be quite high. If the gas is to be purified to a level suitable for transportation in a pipeline and used as a residential or industrial fuel, then the H_2S concentration must be reduced to 0.25 gr/100 scf (6 mg/m^3) (Engineering Data Book, 2004a), and the CO_2 concentration must be reduced to a maximum of 3 to 4 mol%. However, if the gas is to be processed for NGL recovery or nitrogen rejection in a cryogenic turboexpander process, CO_2 may have to be removed to prevent formation of solids. If the gas is being fed to an LNG liquefaction facility, then the maximum CO_2 level is about 50 ppmv (Klinkenbijl et al., 1999) because of potential solids formation.

5.1.4 ACID GAS DISPOSAL

What becomes of the CO_2 and H_2S after their separation from the natural gas? The answer depends to a large extent on the quantity of the acid gases.

For CO_2, if the quantities are large, it is sometimes used as an injection fluid in EOR (enhanced oil recovery) projects. Several gas plants exist to support CO_2 flooding projects; the natural gas and NGL are valuable by-products. If this option is unavailable, then the gas can be vented, provided it satisfies environmental regulations for impurities. Moritis (2001) gives a general discussion of the state of EOR in the United States. Although the United States has not ratified the Kyoto Treaty, which limits CO_2 emissions to the atmosphere, caps may be placed on these emissions because of the preponderance of scientific data that shows CO_2 is a significant contributor to global warming.

In the case of H_2S, four disposal options are available:

1. Incineration and venting, if environmental regulations regarding sulfur dioxide emissions can be satisfied
2. Reaction with H_2S scavengers, such as iron sponge

3. Conversion to elemental sulfur by use of the Claus or similar process (see Chapter 11)
4. Disposal by injection into a suitable underground formation, as discussed by Wichert and Royan (1997) and Kopperson et al. (1998 a,b)

The first two options are applicable to trace levels of H_2S in the gas, and the last two are required if concentrations are too high to make the first two options feasible.

5.1.5 PURIFICATION PROCESSES

Four scenarios are possible for acid gas removal from natural gas:

1. CO_2 removal from a gas that contains no H_2S
2. H_2S removal from a gas that contains no CO_2
3. Simultaneous removal of both CO_2 and H_2S
4. Selective removal of H_2S from a gas that contains both CO_2 and H_2S

Because the concentrations of CO_2 and H_2S in the raw gas to be processed and the allowable acid gas levels in the final product vary substantially, no single process is markedly superior in all circumstances, and, consequently, many processes are presently in use. Table 5.1 summarizes the more important

TABLE 5.1
Acid Gas Removal Processes

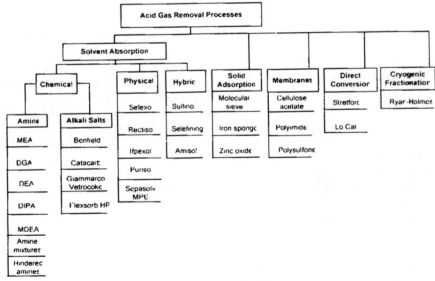

Details of processes are available in Kohl and Nielson (1997).

processes and groups them into the generally accepted categories. Details of processes not discussed in this chapter may be found in the handbook by Kohl and Nielsen (1997).

Some of the more important items that must be considered before a process is selected are summarized from the Engineering Data Book (2004b):

- The type and concentration of impurities and hydrocarbon composition of the sour gas. For example, COS, CS_2, and mercaptans can affect the design of both gas and liquid treating facilities. Physical solvents tend to dissolve heavier hydrocarbons, and the presence of these heavier compounds in significant quantities tends to favor the selection of a chemical solvent.
- The temperature and pressure at which the sour gas is available. High partial pressures (50 psi [3.4 bar] or higher) of the acid gases in the feed favor physical solvents, whereas low partial pressures favor the amines.
- The specifications of the outlet gas (low outlet specifications favor the amines).
- The volume of gas to be processed.
- The specifications for the residue gas, the acid gas, and liquid products.
- The selectivity required for the acid gas removal.
- The capital, operating, and royalty costs for the process.
- The environmental constraints, including air pollution regulations and disposal of byproducts considered hazardous chemicals.

If gas sweetening is required offshore, both size and weight are additional factors that must be considered. Whereas CO_2 removal is performed offshore, H_2S removal is rarely done unless absolutely necessary because of the problems of handling the rich acid gas stream or elemental sulfur.

Selection criteria for the solvent-based processes are discussed by Tennyson and Schaaf (1977) and Figure 5.1 to Figure 5.4 are based on their recommendations. The guidelines in these figures are naturally approximate and should be treated as such. These figures are for solvent-based processes only. Thus, they exclude some commonly used processes such as adsorption and membranes. We slightly modified their recommendations to make it more current, but these guidelines, established in 1977, still are useful. Note that "hybrid" in the figures denotes mixed-solvent systems that contain both amine and a physical solvent.

Table 5.2, adapted from Echterhoff (1991), presents a detailed summary of the more widely used acid gas processes, including degree of purification attainable, selectivity for H_2S removal, and removal of the sulfur compounds COS, CS_2, and mercaptans. As a rule, these sulfur compounds exist at much lower concentrations than do H_2S, but their removal is important if product specifications have an upper limit on total sulfur. Note that in cases of deep liquids recovery, most of the COS, CS_2, and mercaptans present in the feed will stay with the liquid product (NGL). The

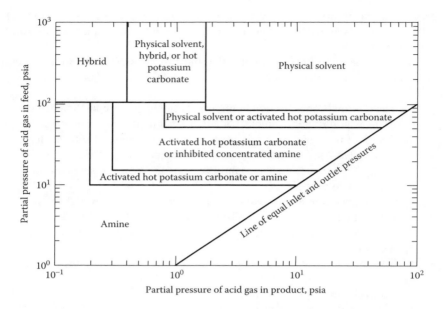

FIGURE 5.1 Process selection chart for CO_2 removal with no H_2S present. (Adapted from Tennyson and Schaaf, 1977.)

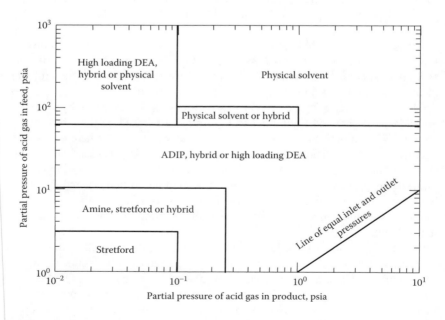

FIGURE 5.2 Process selection chart for H_2S removal with no CO_2 present. (Adapted from Tennyson and Schaaf, 1977.)

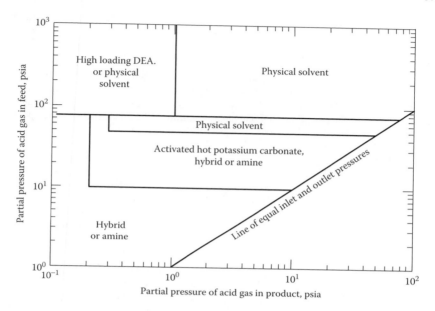

FIGURE 5.3 Process selection chart for simultaneous H_2S and CO_2 removal. (Adapted from Tennyson and Schaaf, 1977.)

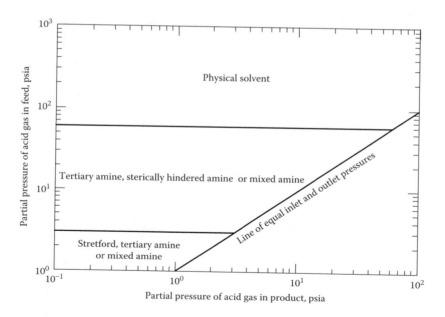

FIGURE 5.4 Process selection chart for selective H_2S removal with CO_2 present. (Adapted from Tennyson and Schaaf, 1977.)

TABLE 5.2
CO$_2$ and H$_2$S Removal Processes for Gas Streams

Process	Normally Capable of Meeting H$_2$S Specification[a]	Removes COS, CS$_2$, and Mercaptans	Selective H$_2$S Removal	Minimum CO$_2$ Level Obtainable	Solution Subject to Degradation? (Degrading Species)
Monoethanol-amine (MEA)	Yes	Partial	No	100 ppmv at low to moderate pressures	Yes (COS, CO$_2$, CS$_2$, SO$_2$, SO$_3$ and mercaptans)
Diethanol amine (DEA)	Yes	Partial	No	50 ppmv in SNEA-DEA process	Some (COS, CO$_2$ CS$_2$, HCN and mercaptans)
Triethanol amine (TEA)	No	Slight	No	Minimum partial pressure of 0.5 psia (3 kPa)	Slight (COS, CS$_2$ and mercaptans)
Methyldiethanol-amine (MDEA)	Yes	Slight	Some	Bulk removal only	No
Diglycol amine (DGA)	Yes	Partial	No	100 ppmv at moderate to high pressures	Yes (COS, CO$_2$, and CS$_2$)
Diisopropanol-amine (DIPA)	Yes	COS only	Yes	Not applicable	Resistant to degradation by COS
Sulfinol	Yes	Partial	Yes (Sulfinol-M)	50 ppmv, 50% slippage while meeting H$_2$S product spec	Some (CO$_2$ and CS$_2$)
Hot potassium carbonate	Yes, with special design features	Partial	No	Not reported	Not reported
Stretford	Yes	No	Yes	No significant amounts of CO$_2$ are removed	Yes (CO$_2$ at high concentrations)

TABLE 5.2 (Continued)
CO$_2$ and H$_2$S Removal Processes for Gas Streams

Process	Normally Capable of Meeting H$_2$S Specification[a]	Removes COS, CS$_2$, and Mercaptans	Selective H$_2$S Removal	Minimum CO$_2$ Level Obtainable	Solution Subject to Degradation? (Degrading Species)
Selexol®	Yes	Slight	Some	Can be slipped or absorbed	No
Rectisol	Yes	Yes	No	1 ppmv	Not reported
Molecular sieves	Yes	Yes (excluding CS$_2$)	Some	Can meet cryogenic spec when CO$_2$ feed content is less than ~2%	Not applicable
Membranes	No	Slight	No	Feed concentration dependent	Not applicable

[a] H$_2$S specification is 25% grain H$_2$S per 100 scf (6 mg/m^3)

Source: Adapted from Echterhoff, 1991.

distribution of these sulfur compounds is discussed in Chapter 10. The amines are susceptible to degradation by O$_2$, which forms undesirable and corrosive compounds.

In this discussion, we consider only the more commonly used amine processes, an alkali salt process (hot potassium carbonate), the Selexol® physical absorption process, molecular sieve adsorption, the Ryan/Holmes cryogenic fractionation process, membranes, and H$_2$S scavengers. For a more comprehensive discussion of acid gas purification, the reader should refer to Kohl and Nielsen (1997) and Engineering Data Book (2004b).

5.2 SOLVENT ABSORPTION PROCESSES

In solvent absorption, the two major cost factors are the solvent circulation rate, which affects both equipment size and operating costs, and the energy requirement for regenerating the solvent. Table 5.3 summarizes some of the advantages and disadvantages of chemical and physical solvents.

TABLE 5.3
Comparison of Chemical and Physical Solvents

Chemical Solvents	
Advantages	**Disadvantages**
Relatively insensitive to H_2S and CO_2 partial pressure	High energy requirements for regeneration of solvent
Can reduce H_2S and CO_2 to ppm levels	Generally not selective between CO_2 and H_2S
	Amines are in a water solution, and thus the treated gas leaves saturated with water

Physical Solvents	
Advantages	**Disadvantages**
Low energy requirements for regeneration	May be difficult to meet H_2S specifications
Can be selective between H_2S and CO_2	Very sensitive to acid gas partial pressure

5.2.1 AMINES

Amines are compounds formed from ammonia (NH_3) by replacing one or more of the hydrogen atoms with another hydrocarbon group. Replacement of a single hydrogen produces a primary amine, replacement of two hydrogen atoms produces a secondary amine, and replacement of all three of the hydrogen atoms produces a tertiary amine. Primary amines are the most reactive, followed by the secondary and tertiary amines. Sterically hindered amines are compounds in which the reactive center (the nitrogen) is partially shielded by neighboring groups so that larger molecules cannot easily approach and react with the nitrogen. The amines are used in water solutions in concentrations ranging from approximately 10 to 65 wt% amines, and a table of their physical properties is included in Appendix B. All commonly used amines are alkanolamines, which are amines with OH groups attached to the hydrocarbon groups to reduce their volatility. Figure 5.5 shows the formulas for the common amines used in gas processing.

Amines remove H_2S and CO_2 in a two step process:

1. The gas dissolves in the liquid (physical absorption).
2. The dissolved gas, which is a weak acid, reacts with the weakly basic amines.

Absorption from the gas phase is governed by the partial pressure of the H_2S and CO_2 in the gas, whereas the reactions in the liquid phase are controlled by

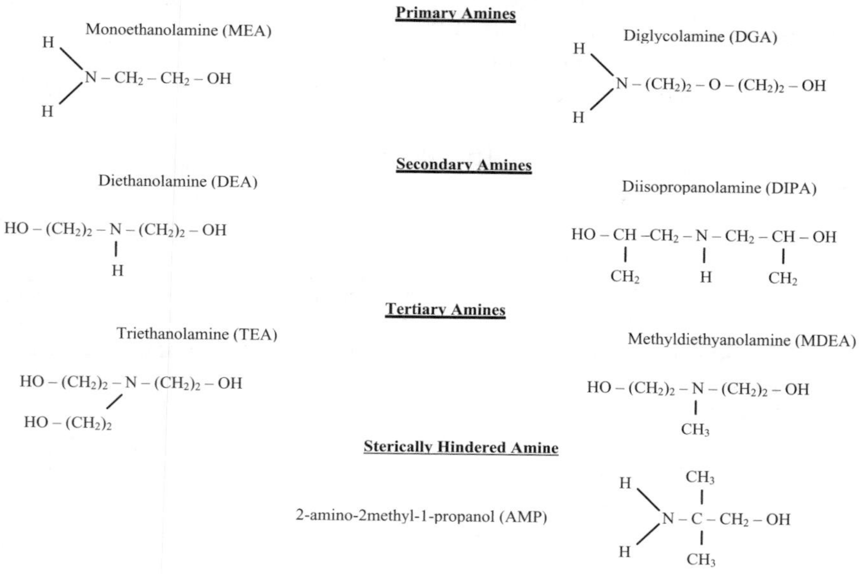

FIGURE 5.5 Molecular structures of commonly used amines.

the reactivity of the dissolved species. The principal reactions (see, e.g., Sigmund et al., 1981) are summarized in the next section.

5.2.1.1 Basic Amine Chemistry

Figure 5.5 shows the structure of the commonly used amines.

Amines are bases, and the important reaction in gas processing is the ability of the amine to form salts with the weak acids formed by H_2S and CO_2 in an aqueous solution. When a gas stream that contains the H_2S, CO_2, or both, is contacted by a primary or secondary amine solution, the acid gases react to form a soluble acid–base complex, a salt, in the treating solution. The reaction between the amine and both H_2S and CO_2 is highly exothermic. Regardless of the structure of the amine, H_2S reacts rapidly with the primary, secondary, or tertiary amine via a direct proton-transfer reaction, as shown in Equation 5.1, to form the amine hydrosulfide:

$$R_1R_2R_3N + H_2S \leftrightarrow R_1R_2R_3NH^+HS^- \tag{5.1}$$

The reaction is shown for a tertiary amine but applies to primary and secondary amines as well. The reaction between the amine and the CO_2 is more complex because CO_2 reacts via two different mechanisms. When dissolved in water, CO_2 hydrolyzes to form carbonic acid, which, in turn, slowly dissociates to bicarbonate.

The bicarbonate then undertakes an acid–base reaction with the amine to yield the overall reaction shown by Equation 5.2:

$$CO_2 + H_2O \leftrightarrow H_2CO_3 \text{ (carbonic acid)} \qquad (5.2a)$$

$$H_2CO_3 \leftrightarrow H^+ + HCO_3^- \text{ (bicarbonate)} \qquad (5.2b)$$

$$H^+ + R_1R_2R_3N \leftrightarrow R_1R_2R_3NH^+ \qquad (5.2c)$$

$$CO_2 + H_2O + R_1R_2R_3N \leftrightarrow R_1R_2R_3NH^+ HCO_3^- \qquad (5.2)$$

This acid–base reaction occurs with any of the alkanolamines, regardless of the amine structure, but the reaction is not as rapid as that of H_2S, because the carbonic acid dissociation step to the bicarbonate is relatively slow.

A second CO_2 reaction mechanism, shown by Equation 5.3, requires the presence of a labile (reactive) hydrogen in the molecular structure of the amine.

$$CO_2 + R_1R_2NH \leftrightarrow R_1R_2N^+HCOO^- \qquad (5.3a)$$

$$R_1R_2N^+HCOO^- + R_1R_2NH \leftrightarrow R_1R_2NCOO^- + R_1R_2NH_2^+ \qquad (5.3b)$$

$$CO_2 + 2R_1R_2NH \leftrightarrow R_1R_2NH_2^+R_1R_2NCOO^- \qquad (5.3)$$

This second reaction mechanism for CO_2, which forms the amine salt of a substituted carbamic acid, is called the carbamate formation reaction and occurs only with the primary and secondary amines. The CO_2 reacts with one primary or secondary amine molecule to form the carbamate intermediate, which in turn reacts with a second amine molecule to form the amine salt. The rate of CO_2 reaction via carbamate formation is much faster than the CO_2 hydrolysis reaction, but slower than the H_2S acid–base reaction. The stoichiometry of the carbamate reaction indicates that the capacity of the amine solution for CO_2 is limited to 0.5 moles of CO_2 per mole of amine if the only reaction product is the amine carbamate. However, the carbamate undergoes partial hydrolysis to form bicarbonate, which regenerates free amine. Hence, CO_2 loadings greater than 0.5, as seen in some plants that employ diethanolamine (DEA), are possible through the hydrolysis of the carbamate intermediate to bicarbonate. We found no systematic study to identify how to obtain the higher loadings.

The fact that CO_2 absorption occurs by two reaction mechanisms with different kinetic characteristics, significantly affects the relative absorption rates of H_2S and CO_2 among the different alkanolamines. For primary and secondary amines, little difference exists between the H_2S and CO_2 reaction rates because of the availability of the rapid carbamate formation for CO_2 absorption. Therefore, the primary and secondary amines achieve essentially complete removal of H_2S and CO_2. However, because the tertiary amines have no labile hydrogen, they cannot form the carbamate. Tertiary amines must react with CO_2 via the slow hydrolysis mechanism in Equation 5.2. With only the slow acid–base reaction available for CO_2 absorption, methyldiethanolamine (MDEA) and several of the formulated MDEA products yield significant selectivity toward H_2S relative to CO_2, and, consequently, all of the H_2S is removed while some of the CO_2 "slips" through with the gas. Because the CO_2 reaction with water to form bicarbonate is slow and the H_2S reaction is fast for MDEA, the H_2S reaction is considered gas-phase limited and the CO_2 reaction is considered liquid-phase limited.

For the reactions discussed above, high pressures and low temperatures drive the reactions to the right, whereas high temperatures and low pressures favor the reverse reaction, which thus provides a mechanism for regeneration of the amine solution

Accurate modeling of the vapor–liquid phase behavior of CO_2 and H_2S in amine solutions for process design requires incorporation of the ionic chemistry into the above equations. As the demand for lower sulfur contents increased, modeling became more important. The Gas Processors Association (GPA) is conducting a multiyear research program to provide the data needed to improve amine modeling, including the widely used model proposed by Kent and Eisenberg (1976). For details of the work, the reader should review the list of GPA research reports available through the GPA or in the Engineering Data Book (2004b).

Brief descriptions of the more commonly used amines follow; more detailed discussions are available in the handbook of Kohl and Nielson (1997) and the Engineering Data Book (2004b). Table 5.4 provides an abbreviated list of the more important operating parameters for the commonly used amines.

5.2.1.2 Monoethanolamine

Monoethanolamine (MEA) is the most basic of the amines used in acid treating and thus the most reactive for acid gas removal. It has the advantage of a high solution capacity at moderate concentrations, and it is generally used for gas

---

TABLE 5.4
Some Representative Operating Parameters for Amine Systems

	MEA	DEA	DGA	MDEA
Wt% amine	15 to 25	25 to 35	50 to 70	40 to 50
Rich amine acid gas loading Mole acid gas/mole amine	0.45 to 0.52	0.43 to 0.73	0.35 to 0.40	0.4 to 0.55
Acid gas pickup Mole acid gas/mole amine	0.33 to 0.40	0.35 to 0.65	0.25 to 0.3	0.2 to 0.55
Lean solution residual acid gas Mole acid gas/mole amine	0.12 ±	0.08 ±	0.10 ±	0.005 to 0.01

Source: Engineering Data Book (2004b).

streams with moderate levels of CO_2 and H_2S when complete removal of both impurities is required.

Monoethanolamine has a number of disadvantages, including:

- A relatively high vapor pressure that results in high vaporization losses
- The formation of irreversible reaction products with COS and CS_2
- A high heat of reaction with the acid gases that results in high energy requirements for regeneration (see Table 5.5)
- The inability to selectively remove H_2S in the presence of CO_2
- Higher corrosion rates than most other amines if the MEA concentration exceeds 20% at high levels of acid gas loading (Kohl and Nielsen, 1997)
- The formation of corrosive thiosulfates when reacted with oxygen (McCartney, 2005)

A slow production of "heat stable salts" form in all alkanol amine solutions, primarily from reaction with CO_2. Oxygen enhances the formation of the salts. In addition to fouling regenerator reboilers, high concentrations of salts can carry over to the contactor and cause foaming, which degrades contactor efficiency. One "advantage" of MEA (and DGA) over other amines is that "reclaimers" are installed in-line for intermittently removing these salts, along with the irreversible reaction products formed from COS and CS_2 (see Section 5.2.1.10 Amine Reclaiming).

5.2.1.3 Diglycolamine

Compared with MEA, low vapor pressure allows Diglycolamine [2-(2-aminoethoxy) ethanol] (DGA) to be used in relatively high concentrations (50 to 70%), which results in lower circulation rates. It is reclaimed onsite to remove heat stable salts and reaction products with COS and CS_2.

5.2.1.4 Diethanolamine

Diethanolamine (DEA), a secondary amine, is less basic and reactive than MEA. Compared with MEA, it has a lower vapor pressure and thus, lower evaporation losses; it can operate at higher acid gas loadings, typically 0.35 to 0.8 mole acid gas/mole of amine versus 0.3 to 0.4 mole acid-gas/mole; and it also has a lower energy requirement for reactivation. Concentration ranges for DEA are 30 to 50 wt% and are primarily limited by corrosion. DEA forms regenerable compounds with COS and CS_2 and, thus, can be used for their partial removal without significant solution loss. DEA has the disadvantage of undergoing irreversible side reactions with CO_2 and forming corrosive degradation products; thus, it may not be the best choice for high CO_2 gases. Removal of these degradation products along with the heat stable salts must be done by use of either vacuum distillation or ion exchange. The reclaiming may be done offsite or in portable equipment brought onsite.

5.2.1.5 Methyldiethanolamine

Methyldiethanolamine (MDEA), a tertiary amine, selectively removes H_2S to pipeline specifications while "slipping" some of the CO_2. As noted previously, the CO_2 slippage occurs because H_2S hydrolysis is much faster than that for CO_2, and the carbamate formation reaction does not occur with a tertiary amine. Consequently, short contact times in the absorber are used to obtain the selectivity. MDEA has a low vapor pressure and thus, can be used at concentrations up to 60 wt% without appreciable vaporization losses. Even with its relatively slow kinetics with CO_2, MDEA is used for bulk removal of CO_2 from high-concentration gases because energy requirements for regeneration are lower than those for the other amines. It is not reclaimable by conventional methods (Veroba and Stewart, 2003).

5.2.1.6 Sterically Hindered Amines

In acid gas removal, steric hindrance involves alteration of the reactivity of a primary or secondary amine by a change in the alkanol structure of the amine. A large hydrocarbon group attached to the nitrogen shields the nitrogen atom and hinders the carbamate reaction. The H_2S reaction is not significantly affected by amine structure, because the proton is small and can reach the nitrogen. However, CO_2 removal can be significantly affected if the amine structure hinders the fast carbamate formation reaction and allows only the much slower bicarbonate formation (reactions 5.2a and 5.2b).

5.2.1.7 Mixed Amines

The selectivity of MDEA can be reduced by addition of MEA, DEA, or proprietary additives. Thus, it can be tailored to meet the desired amount of CO_2 slippage and still have lower energy requirements than do primary and secondary amines. The following section looks at the difference in heats of reaction.

5.2.1.8 Heats of Reaction

The magnitude of the exothermic heats of reaction, which includes the heat of solution, of the amines with the acid gases is important because the heat liberated in the reaction must be added back in the regeneration step. Thus, a low heat of reaction translates into smaller energy regeneration requirements. Table 5.5 summarizes the important data for the common amines. The values in Table 5.5 are approximate because heats of reaction vary with acid gas loading and solution concentration. Jou et al. (1994) and Carson et al. (2000) report measured heats of reaction for CO_2 in MEA, DEA, and MDEA.

5.2.1.9 Process Flow Diagram

A typical diagram for the removal of acid gases from natural gas by use of MEA is shown in Figure 5.6. The diagram is simplified, and all operating conditions are representative, not definitive. The sour gas feed enters the bottom of the contactor

TABLE 5.5

Average Heats of Reaction[a] of the Acid Gases in Amine Solutions

Amine	H₂S, Btu/lb (kJ/kg)	CO₂, Btu/lb (kJ/kg)
MEA	610 (1420)	825 (1920)
DEA	555 (1290)	730 (1700)
DGA®	674 (1570)	850 (1980)
MDEA	530 (1230)	610 (1420)

[a] The heats of reaction include both heat of solution and heat of reaction.

Source: Engineering Data Book (2004b).

at pressures to 1,000 psi (70 bar) and temperatures in the range of 90°F (32°C). The sour gas flows upward, countercurrent to the lean amine solution which flows down from the top. The lean amine that returns to the contactor is maintained at a temperature above the vapor that exits the contactor to prevent any condensation

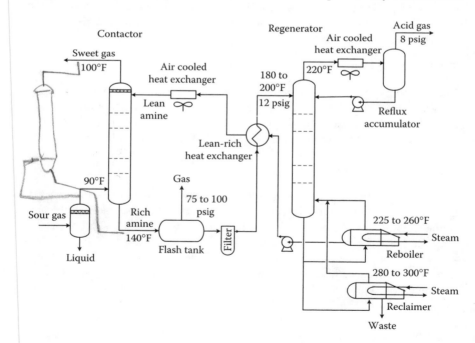

FIGURE 5.6 Process flow diagram for amine treating by use of MEA. Contactor commonly operates at pressures up to 1,000 psi (70 bar). Flow rates to reclaimer are 1 to 3% of amine circulation rate. (Adapted from Smith, 1996.)

of heavier liquid hydrocarbons. Intimate contact between the gas and amine solution is achieved by use of either trays or packing in the contactor.

The contactor operates above ambient temperature because of the combined exothermic heat of absorption and reaction. The maximum temperature is in the lower portion of the tower because the majority of the absorption and reaction occurs near the bottom of the unit. The temperature "bulge" in the tower can be up to about 180°F (80°C). The treated gas leaves the top of the tower water saturated and at a temperature controlled by the temperature of the lean amine that enters, usually around 100°F (38°C).

The rich amine leaves the bottom of the unit at temperatures near 140°F (60°C) and enters a flash tank, where its pressure is reduced to 75 to 100 psig (5 to 7 barg) to remove (flash) any dissolved hydrocarbons. The dissolved hydrocarbons are generally used as plant fuel. If necessary, a small stream of lean amine is contacted with the fuel gas to reduce the H_2S concentration. The rich amine then passes through a heat exchanger and enters the solvent regenerator (stripper) at temperatures in the range of 180 to 220°F (80 to 105°C). The reboiler on the stripper generally uses low-pressure steam. The vapor generated at the bottom flows upward through either trays or packing, where it contacts the rich amine and strips the acid gases from the liquid that flows down. A stream of lean amine is removed from the stripper, cooled to about 110°F (45°C), and reenters the contactor at the top to cool and condense the upward-flowing vapor stream. The vapor, which consists mostly of acid gases and water vapor, exits the top of the stripper and is generally processed for sulfur recovery.

The lean amine exits the bottom of the stripper at about 260°F (130°C) and is pumped to the contactor pressure, exchanges heat with the rich amine stream, and is further cooled before it enters the top of the contactor. Both the treated gas that leaves the gas contactor and the acid gas to the Claus unit are water saturated.

Figure 5.6 shows an MEA unit. Smith (1996) states the major pieces of equipment and the operating conditions are also representative of DEA and DGA units, with the exception of the reclaimer. DEA units do not have reclaimers, and DGA reclaimers operate at slightly higher temperatures of 375 to 385°F (190 to 195°C).

Lyddon and Nguyen (1999) evaluate several alterations to the basic flow scheme for amine units that can increase gas throughput. When additional capacity exists in the regenerator but not in the absorber, and additional pressure drop is acceptable, static mixers may help. Blending sour gas and lean amine provides essentially one additional absorber stage. Two configurations were considered. In one case, a side stream of sour gas passes through a static mixer. Following separation from the amine, the gas is blended with the sweet gas from the absorber. In the other case, the whole stream goes through the static mixer upstream of the inlet separator before it enters the absorber. In both cases, the rich amine from the static mixer goes to the regenerator. Lyddon and Nguyen (1999) report one case in which use of the bypass static mixers allowed a 17% increase in plant capacity.

5.2.1.10 Amine Reclaiming

Amines react with CO_2 and contaminants, including oxygen, to form organic acids. These acids then react with the basic amine to form heat stable salts (HSS). As their name implies, these salts are heat stable, accumulate in the amine solution, and must be removed. For MEA and DGA solutions, the salts are removed through the use of a reclaimer (refer to Figure 5.6), which utilizes a semicontinuous distillation process, as described by Smith (1996). The reclaimer is filled with lean amine, and a strong base, such as sodium carbonate or sodium hydroxide, is added to the solution to neutralize the heat stable salts. A slipstream of 1 to 3% of the circulating amine is then continuously added to the reclaimer while the mixture is heated. Water and amine vapor are taken off the top, which leaves the contaminants in the liquid bottoms. Heating is continued until the temperature is approximately 300°F (150°C) for MEA or 360 to 380°F (180 to 195°C) for DGA. The cycle is then stopped and the bottoms that contain the contaminants (dissolved salts, suspended solids) are removed. DEA does not form a significant amount of nonregenerable degradation products, and it requires more difficult reclaiming through vacuum distillation or ion exchange. Further details on reclaimers can be found in the Engineering Data Book (2004b), Smith (1996), Simmons (1991), and Holub and Sheilan (2000).

5.2.1.11 Operating Issues

A comprehensive discussion of operating issues is given by Holub and Sheilan (2000) and is beyond the scope of this book. The brief discussion here of their work highlights a few of the most important considerations.

Corrosion—Some of the major factors that affect corrosion are:

- Amine concentration (higher concentrations favor corrosion)
- Rich amine acid gas loading (higher gas loadings in the amine favor corrosion)
- Oxygen concentration
- Heat stable salts (higher concentrations promote corrosion and foaming)

In addition to destroying vessels and piping, the corrosion products can cause foaming.

Solution Foaming—Foaming of the liquid amine solution is a major problem because it results in poor vapor–liquid contact, poor solution distribution, and solution holdup with resulting carryover and off spec gas. Among the causes of foaming are suspended solids, liquid hydrocarbons, surface active agents, such as those contained in inhibitors and compressor oils, and amine degradation products, including heat stable salts. One obvious cure is to remove the offending materials; the other is to add antifoaming agents.

Heat Stable Salts—As mentioned above, these amine degradation products can cause both corrosion and foaming. They are normally dealt with through the use of amine reclaimers (see Section 5.2.1.10, Amine Reclaiming).

5.2.2 ALKALI SALTS

The hot potassium carbonate process for removing CO_2 and H_2S was developed by the United States Bureau of Mines and is described by Benson and coworkers in two papers (Benson et al., 1954, 1956). Although the process was developed for the removal of CO_2, it can also remove H_2S if H_2S is present with CO_2. Special designs are required for removing H_2S to pipeline specifications or to reduce CO_2 to low levels.

The process is very similar in concept to the amine process, in that after physical absorption into the liquid, the CO_2 and H_2S react chemically with the solution. The chemistry is relatively complex, but the overall reactions are represented by

$$K_2CO_3 + CO_2 + H_2O \leftrightarrow 2KHCO_3 \tag{5.4a}$$

$$K_2CO_3 + H_2S \leftrightarrow KHS + KHCO_3 \tag{5.4b}$$

The process has several variations, and a simplified diagram of the single-stage process is shown in Figure 5.7.

In a typical application, the contactor will operate at approximately 300 psig (20 barg), with the lean carbonate solution entering near 225°F (110°C) and leaving

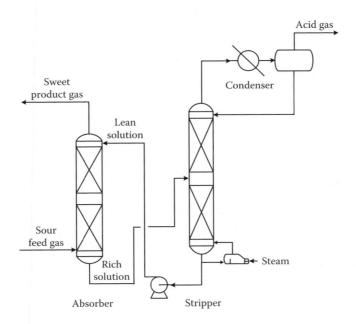

FIGURE 5.7 Process flow diagram for hot potassium carbonate process. (Adapted from Engineering Data Book, 2004b.)

at 240°F (115°C). The rich carbonate pressure is reduced to approximately 5 psig (0.3 barg) as it enters the stripper. Approximately one third to two thirds of the absorbed CO_2 is released by the pressure reduction, reducing the amount of steam required for stripping (Kohl and Nielsen, 1997). The lean carbonate solution leaves the stripper at the same temperature as it enters the contactor, and eliminates the need for heat exchange between the rich and lean streams. The heat of solution for absorption of CO_2 in potassium carbonate is small, approximately 32 Btu/cu ft of CO_2 (Benson et al., 1954), and consequently the temperature rise in the contactor is small and less energy is required for regeneration. Kohl and Nielson (1997) give a thorough discussion of the process and its many modifications.

5.3 PHYSICAL ABSORPTION

In the amine and alkali salt processes, the acid gases are removed in two steps: physical absorption followed by chemical reaction. In processes such as Selexol® or Rectisol®, no chemical reaction occurs and acid gas removal depends entirely on physical absorption. Some of the inherent advantages and disadvantages of physical absorption processes are summarized below:

- Absorption processes are generally most efficient when the partial pressures of the acid gases are relatively high, because partial pressure is the driving force for the absorption.
- Heavy hydrocarbons are strongly absorbed by the solvents used, and consequently acid gas removal is most efficient in natural gases with low concentrations of heavier hydrocarbons.
- Solvents can be chosen for selective removal of sulfur compounds, which allows CO_2 to be slipped into the residue gas stream and reduce separation costs.
- Energy requirements for regeneration of the solvent are lower than in systems that involve chemical reactions.
- Separation can be carried out at near-ambient temperature.
- Partial dehydration occurs along with acid gas removal, whereas amine processes produce a water saturated product stream that must be dried in most applications.

Selexol® is a typical application of physical absorption and a number of open literature articles describe the process. Consequently, it was selected as an example to describe the absorption process. The handbook of Kohl and Nielsen (1997) provides information on other absorption processes.

5.3.1 SOLVENT PROPERTIES

Selexol® is a polyethylene glycol and has the general formula:

$$CH_3-O-CH_2-(CH_2-O-CH_2)_N-CH_2-O-CH_3$$

TABLE 5.6
Typical Relative Ratio of K-values

Component	R_K	Component	R_K
CH_4	1	H_2S	134
C_2H_6	6.4	C_6H_{14}	165
CO_2	15	CH_3SH	340
C_3H_8	15.3	C_7H_{16}	360
Isobutane	28	CS_2	360
n-Butane	35	SO_2	1400
COS	35	C_6H_6	3800
Isopentane	67	C_4H_4S	8100
n-Pentane	83	H_2O	11000

$R_K = (K_{CH4}/K_{Component})$ in Selexol® solvent for various solutes, where $K_{CH4} = 1$.

Source: Sweny (1980).

Because Selexol® is a mixture of homologues (the value of N in the general formula for Selexol® varies), the physical properties discussed below are approximate.

Table 5.6 (Sweny, 1980) presents R_K, the ratio of the K-value* for methane, K_{CH4} (arbitrarily assigned a value of 1), to the K-values of the other component, $R_K = K_{CH4}/K_{component}$. Table 5.7 lists some physical properties of Selexol®. An R_K value greater than unity indicates that the solubility of the component in Selexol® is greater than that of methane, whereas a value less than unity indicates the opposite. The values should be regarded as only representative because pressure and temperature are not specified and, as previously noted, the composition of Selexol® is variable.

Because R_K for CO_2 and H_2S are 15 and 134, respectively, these gases are preferentially absorbed (relative to CH_4), and, consequently, physical absorption is an effective technique for acid gas removal. The process can reduce H_2S to 4 ppmv, reduce CO_2 to levels below 50 ppmv, and essentially remove all mercaptans, CS_2, and COS. Two additional features of Table 5.6 are worth mentioning. Because the R_K values for hydrocarbons heavier than CH_4 are fairly high (6.4 for C_2H_6, 15.3 for C_3H_8, and 35 for n-C_4H_{10}), Selexol® will remove substantial quantities of these hydrocarbons, a feature that can be either positive or negative, depending on the composition of the gas being processed and the desired products. Finally, the R_K value of H_2O is extremely high and consequently, Selexol® provides some dehydration.

Bucklin and Schendel (1985) reported gas solubility data in the form of absorption coefficients (volume of gas absorbed per volume of liquid). We converted values for some of the compounds into scf gas absorbed/gal Selexol® and

---

* The K-value is the ratio of the mole fraction of the component in the vapor phase (y) to its mole fraction in the liquid phase (x), $K = y/x$. High K-values indicate the material is predominately in the vapor phase, whereas low K-values indicate a higher concentration in the liquid phase (x).

TABLE 5.7
Representative Property Data for Selexol®

Molar mass (approximate) = 280

Flash point (Cleveland open cup) = 304°F (151°C)

Freezing point[a] = −8 to −20°F (−22 to −28°C)

Vapor pressure at 25°C = <0.01 mm Hg

Specific heat at 77°F (25°C) = 0.49 Btu/lbm °F
= 2.0 kJ/kg K

Density at 77°F (25°C) = 1030 kg/m3 (8.60 lb/gal)

Viscosity at 77°F (25°C) = 5.8 cp (5.8 × 10⁻³ Ns/m²)

Thermal conductivity at 77°F (25°C) = 0.11 Btu/(h)(ft)(°F)
=0.19 W/m K

Surface tension at 77°F (25°C) = 34.3 dynes/cm

Heat of solution at 77°F, Btu/lb of solute (kJ/kg)

CO_2 = 160 (372)

H_2S = 190 (442)

CH_4 = 75 (174)

Odor = very mild

Toxicity = nil

---

[a] Slush appears at −8°F complete solidification at −20°F

Sources: Sweny and Valentine (1970) and Clare and Valentine (1975).

---

plotted them in Figure 5.8. The figure assumes a Henry's law relationship,* which provides approximate solubility at higher pressures. The lines also ignore probable interaction between solutes. Epps (1994) discusses the limited solubility of hydrocarbons in the liquid (when the liquid is saturated, a second liquid hydrocarbon phase forms above the Selexol® phase) and the effect of absorbed water on the solubility of hydrocarbons.

5.3.2 REPRESENTATIVE PROCESS CONDITIONS

Applications of Selexol® are varied and, consequently, no common process flow diagrams are available. Sweny (1980) presents flow diagrams for nine different applications, and Epps (1994) discusses plants for the dehydration of natural gas and hydrocarbon dew point control.

One plant discussed by Epps (1994), designated only as a European distribution plant, is shown in Figure 5.9. This plant was selected for discussion because it is a modern application, and both inlet and outlet gas compositions were reported (Table 5.8). The plant pretreats the gas to reduce CO_2, ethane,

---

* For an ideal system, Henry's law assumes a linear relation between the solubility of gas component i and its partial pressure, $y_iP = k_ix_i$ where k_i is the Henry's constant.

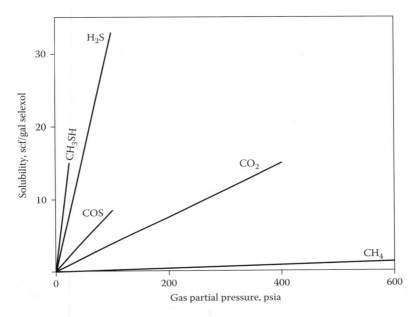

FIGURE 5.8 Solubility of various gases in Selexol® solvent at 70°F (21°C) as a function of partial pressure. (Based on data by Sweny and Valentine, 1970.)

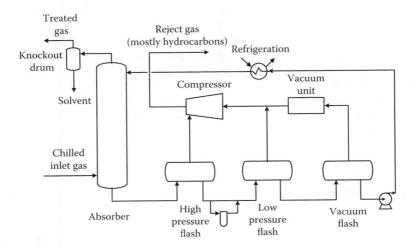

FIGURE 5.9 Process schematic for a Selexol® gas treating facility. (Adapted from Epps, 1994.)

TABLE 5.8
Composition of Inlet and Outlet Gas in a Selexol® Unit

| Component | Gas Composition (Mol%) | |
	Inlet	Outlet
Hydrogen sulfide	0.0002	—
Methyl mercaptan	0.00050	0.00012
Carbon dioxide	2.44	0.29
Nitrogen	0.785	0.88
Methane	86.317	93.02
Ethane	7.539	5.33
Propane	2.403	0.35
Heavier hydrocarbons	0.515	0.130

Source: Epps (1994).

and heavier hydrocarbon levels before final purification in molecular sieve units and subsequent liquefaction. The plant is designed to process 26 MMscfd (0.74 MMSm^3/d) entering the Selexol® unit at 603 psia (41.6 bar) and 32°F (0°C). The lean solvent, cooled to 25°F (−3.9°C) with propane refrigerant, enters the absorber where it absorbs CO_2 and some of the ethane and heavier hydrocarbons. The rich solvent from the absorber is regenerated by reduction of the pressure in three flash drums, from 603 to 106 psia (41.6 to 7.3 bar) in the high-pressure drum, from 106 to 16 psia (7.3 to 1.1 bar) in the medium-pressure drum, and from 16 to 3 psia (1.1 to 0.21 bar) in the vacuum drum. Lean Selexol® from the vacuum drum is recompressed and sent to the propane chiller. The treated gas that leaves the absorber passes through a knockout drum and filter separator to remove entrained Selexol® and condensed hydrocarbons. Table 5.8 shows that the treated gas meets the specifications of a maximum of 0.50% CO_2 and a maximum of 6.5% ethane and heavier hydrocarbons. In addition, the water content of the gas is reduced from 75 ppmv to 12 ppmv, H_2S is reduced from 2 ppmv to essentially nothing, and methyl mercaptan is reduced from 5 ppmv to 1 ppmv. Unlike the amine systems, no irreversible products are generated in the process, which thus eliminates the need for reclaiming.

Other detailed descriptions of acid gas removal with Selexol® are given by Hegwar and Harris (1970), Johnson (1984), Judd (1978), Mortko (1984), Oberding et al. (2004), Raney (1976), and Shah (1989).

5.3.3 HYBRID PROCESSES

Table 5.3 showed the strengths and weaknesses of amine and physical solvent systems. To take advantage of the strengths of each type, a number of hybrid processes commercially used, and under development, combine physical solvents with amines (Engineering Data Book, 2004b). Depending upon the solvent–amine combination, nearly complete removal of H_2S, CO_2, and COS is possible. Other hybrid systems provide high H_2S and COS removal while slipping CO_2. Sulfinol® currently is one of the more commonly used processes. The process uses a combination of a physical solvent (sulfolane) with DIPA or MDEA. The selected amine depends upon the acid gases in the feed and whether CO_2 removal is required. Like the physical solvent processes, the hybrid systems may absorb more hydrocarbons, including BTEX, but that property can be adjusted by varying water content.

5.4 ADSORPTION

Acid gases, as well as water, can be effectively removed by physical adsorption on synthetic zeolites. Applications are limited because water displaces acid gases on the adsorbent bed. (Chapter 6 provides more details on adsorption and its use in dehydration.) Figure 5.10, which shows typical isotherms for CO_2 and H_2S on molecular sieve, indicates that at ambient temperatures substantial quantities of both gases are adsorbed even at low partial pressures.

Molecular sieve can reduce H_2S levels to the 0.25 gr/100 scf (6 mg/m³) specification. However, this reduction requires regeneration of the bed at 600°F (315°C) for extended time (Engineering Data Book, 2004b) with the potential for COS formation if 4A* is used.

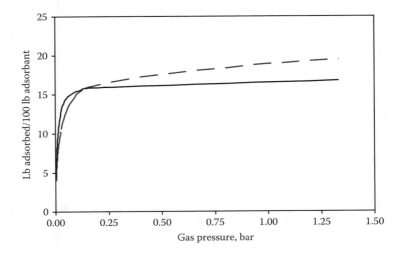

FIGURE 5.10 Adsorption of H_2S (—) and CO_2 (- -) on type 5A molecular sieve at 77°F (25°C). (adapted from Linde, undated).

---

* Molecular sieves are classified by their nominal pore diameter in Angstroms. See Chapter 6 for details.

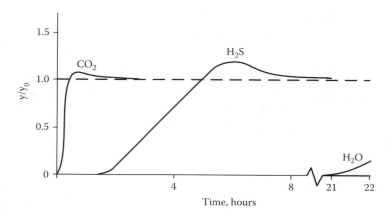

FIGURE 5.11 Effluent H_2S and CO_2 concentration from adsorption bed as a function of time (adapted from Chi and Lee, 1973). See text for details.

Chi and Lee (1973) studied the coadsorption of H_2S, CO_2, and H_2O on a 5A molecular sieve from a natural gas mixture under a variety of conditions. Figure 5.11, from their paper, shows a typical concentration versus time curve. In the figure, y is the concentration in the exit stream and y_o is the concentration in the inlet to the bed. The gas that entered the bed was saturated with H_2O and contained both CO_2 (1.14 mol%) and H_2S (0.073 mol%.). Because the CO_2 content of the gas was 15.6 times that of the H_2S, the bed quickly saturated with CO_2, and its breakthrough was almost instantaneous. As the H_2S was adsorbed and moved down the column, it displaced the CO_2 and, consequently, after approximately 30 minutes, the CO_2 exit concentration peaked at a value greater than its inlet concentration.

The same phenomenon occurs when the H_2S is displaced by the water. Because the H_2S must be removed to extremely low levels, the bed is effectively exhausted from the perspective of H_2S purification shortly after H_2S breakthrough occurs and, thus, well before the bed is totally saturated. Chi and Lee (1973) present empirical correlations for prediction of dynamic saturation capacity and breakthrough times for H_2S under a variety of operating conditions.

Detailed information on adsorber design calculations is available in the paper by Cummings and Chi (1977) and the handbook by Kohl and Nielsen (1997). The Engineering Data Book (2004b) notes that a key point in design is to properly design for treatment of the regeneration gas because the peak H_2S concentration may be 30 times the H_2S concentration in the feed.

Figure 5.12 shows a typical flow diagram for removal of H_2S from natural gas. The configuration is similar to that for dehydration but with the significant difference that the regeneration gas contains high quantities of H_2S as well as water as it leaves the adsorbent bed and, thus, must be treated. (Chapter 6 presents

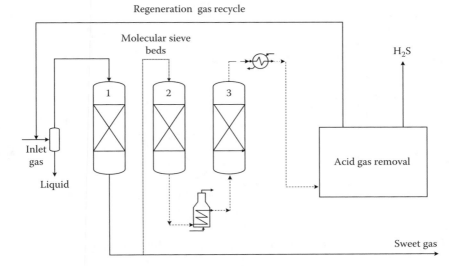

FIGURE 5.12 Schematic of integrated natural gas desulfurization plant. Dashed line denotes regeneration gas stream. (Adapted from Engineering Data Book, 2004b.)

a thorough discussion of adsorption.) The flow configuration shows the first bed in the adsorption cycle, the second bed cooling down after regeneration, and the third bed undergoing regeneration with hot gas.

5.5 CRYOGENIC FRACTIONATION

Distillation is the most widely used process to separate liquid mixtures, and at first glance it seems a good prospect for removing CO_2 and H_2S from natural gas, because the vapor pressures of the principal components are considerably different (Figure 5.13). However, problems are associated with the separation of CO_2 from methane, CO_2 from ethane, and CO_2 from H_2S. These difficulties are summarized below (Engineering Data Book, 2004 b).

CO_2 from methane: Relative volatilities (K_{C1}/K_{CO2}) at typical distillation conditions are about 5 to 1. Therefore one would expect simple fractionation to work. However, because the liquid CO_2 phase freezes when it becomes concentrated, the practical maximum-vapor concentration of methane is only 85 to 90 mol%.

CO_2 from ethane: In addition to solidification problems, CO_2 and ethane form an azeotrope (liquid and vapor compositions are equal) and consequently, a complete separation of these two by simple distillation is impossible. Figure 5.14 shows an isotherm at $-64°F$ ($-53°C$) for the system, and the azeotrope appears at approximately 0.6 mole fraction of CO_2.

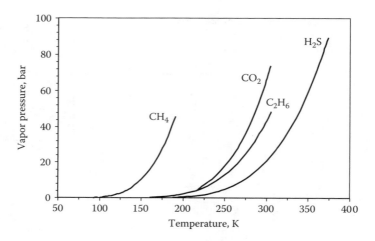

FIGURE 5.13 Vapor pressure of light natural gas components (NIST, 2005).

CO_2 from H_2S: This distillation is difficult because, as Figure 5.15 shows, the mixture forms a pinch at high CO_2 concentrations. This separation by conventional distillation is complicated by the need to have an overhead product that has roughly 100 ppmv H_2S if the stream is vented. The bottoms product should contain less than two-thirds CO_2, assuming the stream is feed to a Claus unit.

A number of techniques are available for solving these problems, but the Ryan/Holmes process (Holmes et al., 1982) is probably the most widely used.

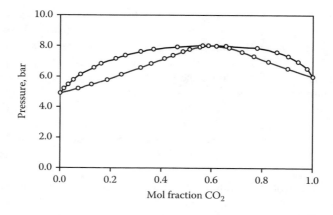

FIGURE 5.14 Vapor–liquid equilibrium curve for CO_2 and C_2H_6 at −64°F (−53°C). Azeotropic composition is 0.60 mol fraction CO_2. (Data from Brown et al., 1988.)

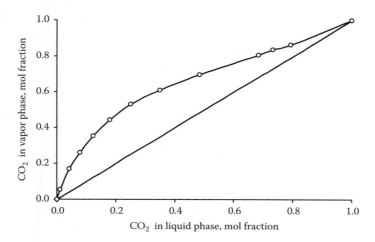

FIGURE 5.15 Vapor–liquid equilibrium curve for CO_2 and H_2S at 20 atm. (Data from Bierlein and Kay, 1953.)

This process is an extractive distillation process* that uses hydrocarbons to significantly alter the behavior of the system and thus, effectively eliminate the distillation problems. The hydrocarbons are normally mixtures of propane and heavier hydrocarbons obtained from the feed mixture. As a result, no additional separations are required. Readers are referred to the original papers for details.

5.6 MEMBRANES

Membranes are used in natural gas processing for dehydration, fuel-gas conditioning, and bulk CO_2 removal, but presently CO_2 removal is by far the most important application. We first present the fundamentals of membrane separations and then discuss their application to CO_2 removal.

5.6.1 MEMBRANE FUNDAMENTALS

From thermodynamics, the driving force for movement through the membrane is the difference in chemical potential, μ, for a given component on the two sides of the membrane. If subscript i is the diffusing component, then

$$\mu_{i,\text{feed}} > \mu_{i,\text{permeate}}$$

---

* Extractive distillation makes distillation of close boiling components possible by addition of a solvent to the mixture to alter the relative volatility of the two key components. The products from the distillation include one of the keys at high purity and a mix of the other key plus the solvent. This mixture is fractionated in another column for recovery of the solvent and production of the pure second key.

where the material is moving from feed side to the permeate side. To make the equation physically comprehensible in gas systems, fugacity, f, which is proportional to the chemical potential is used and gives

$$f_{i,feed} > f_{i,permeate} \tag{5.5}$$

If ideal behavior is assumed for the diffusing gas, then the fugacities can be replaced by the partial pressures and

$$y_{i,feed}\, P_{i,feed} > y_{i,permeate}\, P_{i,permeate}, \tag{5.6}$$

where y is the mole fraction and P is the total pressure. This equation shows what terms affect the driving force across the membrane. The equation can be rearranged to obtain

$$\left(\frac{y_{i,permeate}}{y_{i,feed}} \right) \le \left(\frac{P_{feed}}{P_{permeate}} \right). \tag{5.7}$$

This relation tells us that the separation achieved ($y_{i,permeate}/y_{i,feed}$) can never exceed the pressure ratio ($P_{feed}/P_{permeate}$).

Because the discussion above was based on classical thermodynamics, it tells us nothing about the rate at which the diffusion processes take place. A discussion of diffusion rate requires use of Fick's law. Fick's law for solution-diffusion membranes in rectangular coordinates is (e.g., Echt et al., 2002)

$$J_i = (S_i\, D_i\, p_i)/L, \tag{5.8}$$

where J is the flux of component i, that is, the molar flow of component i through the membrane per unit area of membrane, S_i is the solubility term, D_i is the diffusion coefficient, Δp_i is the partial pressure difference across the membrane, and L is the thickness of the membrane. Customarily, S_i, and D_i are combined into a single term, the permeability, P_i, and thus divides Fick's law into two parts, P_i/L, which is membrane dependent and Δp_i, which is process dependent. Note that P_i/L is not only dependent on the membrane but also dependent on operating conditions, because S_i and D_i depend on both temperature and pressure. P_i also depends weakly upon the composition of the gases present.

All the mixture components have a finite permeability, and the separation is based upon differences in them. Customarily, selectivity, α_{1-2}, is used, which is the ratio of two permeabilities, P_1/P_2, a term important in process design and evaluation. An α of 20 for CO_2/CH_4 means that CO_2 moves through the membrane 20 times faster than does methane.

The above discussion introduces the basic principles regarding membrane separation. Readers desiring more depth should consult the engineering handbook of Kohl and Nielsen (1997).

5.6.2 CARBON DIOXIDE REMOVAL FROM NATURAL GAS

Many different types of membranes have been developed or are under development for industrial separations (Echt et al., 2002; Baker, 2002), but for CO_2 removal, the industry standard is presently cellulose acetate. These membranes are of the solution-diffusion type, in which a thin layer (0.1 to 0.5 μm) of cellulose acetate is on top of a thicker layer of a porous support material. Permeable compounds dissolve into the membrane, diffuse across it, and then travel through the inactive support material. The membranes are thin to maximize mass transfer and, thus, minimize surface area and cost, so the support layer is necessary to provide the needed mechanical strength.

Commercial membrane configurations are either hollow fiber elements or flat sheets wrapped into spirally wound elements. Presently, about 80% of gas-separation membranes are formed into hollow fiber modules (Baker, 2002), like those shown in Figure 5.16. The low-pressure, bore-feed configuration is a

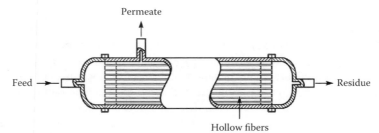

Low pressure, bore-side gas feed module

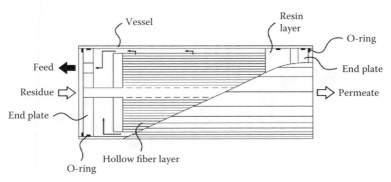

High pressure, shell-side gas feed module

FIGURE 5.16 Cutaway view of the two module configurations used with hollow fiber membranes. (Courtesy of Membrane Research and Technology, Inc.)

countercurrent flow configuration similar to a shell-tube heat exchanger with the gas entering on the tube side. It has the advantage of being more resistant to fouling because the inlet gas flows through the inside of the hollow fibers. However, the mechanical strength of the membrane limits the pressure drop across the membrane. The configuration is only used in low-pressure applications, such as air separation and air dehydration (Baker, 2002).

To handle high pressures, the permeate flows into the hollow fiber from the shell side. This feature makes the membrane much more susceptible to plugging, and gas pretreatment is usually required (Baker, 2002). The gas flow is cross current and provides good feed distribution in the module. This configuration is widely used to remove CO_2 from natural gas.

In the spiral wound element shown in Figure 5.17, two membrane sheets are separated by a permeate spacer and glued shut at three ends to form an envelope or leaf. Many of these leaves, separated by feed spacers, are wrapped around the permeate tube, with the open end of the leaves facing the tube. Feed gas travels along the feed spacers, the permeating species diffuse through the membranes and down the permeate spacers into the permeate tube, and the residue gas exits at the end. The gas flow is cross flow in this configuration.

The spiral configuration is inherently more resistant than the hollow fiber membranes to trace components that would alter the polymer permeability. It also allows a wider range of membrane materials to be used. However, the hollow fiber membranes are cheaper to fabricate, and thus dominate the field (Baker, 2002).

Once the elements have been manufactured, they are grouped into modules, as shown in Figure 5.18. These modules are then mounted on a skid to make a complete unit, as shown in Figure 5.19. The unit on the left of the picture composed of long cylindrical tubes contains the membranes, and the separate unit on the right of the picture is the feed-pretreatment unit, described in more detail in Section 5.6.3.5, Feed Gas Pretreatment. Echt et al. (2002) and Kohl and Nielsen (1997) provide more complete details of the construction of elements and modules.

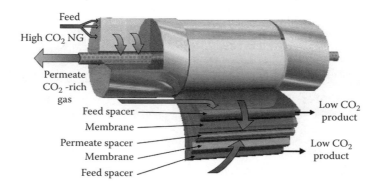

FIGURE 5.17 Cutaway view of spiral wound membrane element. (Copyright 2002 UOP LLC. All rights reserved. Used with permission.)

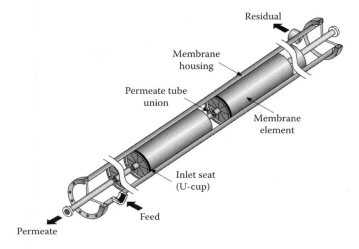

FIGURE 5.18 Cutaway view of spiral wound membrane module. (Copyright 2002 UOP LLC. All rights reserved. Used with permission.)

5.6.3 OPERATING CONSIDERATIONS

5.6.3.1 Flow Patterns

A number of different flow patterns can be used with membranes. Figures 5.20 and 5.21 show simplified examples for the CH_4/CO_2 separation. In the single-stage unit, the overall methane recovery is only 90.2%, but the process requires flow

FIGURE 5.19 Carbon dioxide membrane skid installed in gas plant.

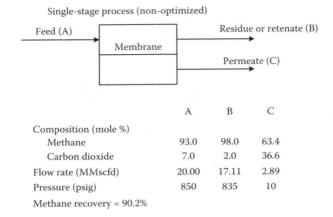

FIGURE 5.20 A non-optimized single-stage CO_2/CH_4 membrane separation process (Spillman and Cooley, 1989).

through only one membrane, and no recompression is needed. To increase methane recovery to 98.7%, a two-stage unit requires recompression of the first-stage permeate. Greater levels of methane recovery are obviously possible by application of three or more stages, but additional elements can quickly become uneconomical because of both membrane cost and recompression energy required.

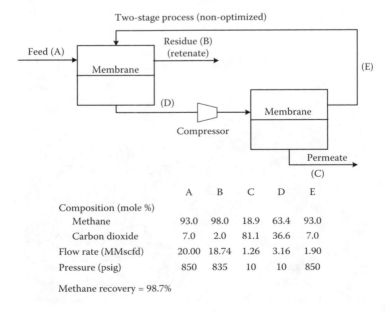

FIGURE 5.21 A non-optimized two stage CO_2/CH_4 membrane separation process (Spillman and Cooley, 1989).

The second law of thermodynamics dictates that energy is required for a separation. For membrane processes, this law translates into loss of pressure. However, in many processes, the cost of recompression, if needed, still makes membranes an attractive separation process.

5.6.3.2   Flow Rate

A maximum acceptable feed gas rate per unit area applies to the membrane, and required membrane area is directly proportional to the flow rate. Membrane units perform well at reduced feed rates, but their performance drops when design flow rates are exceeded. Additional modules are added in parallel to accept higher flow rates.

5.6.3.3   Operating Temperature

Increased operating temperature increases permeability but decreases selectivity. Because membranes are organic polymers, they have a maximum operating temperature that depends upon the polymer used. Exceeding this temperature will degrade membrane material and shorten the useful life of the unit.

5.6.3.4   Operating Pressures

Increased feed pressure decreases both the permeability and selectivity, but at the same time, the pressure difference across the membrane (the driving force in the membrane flux equation) is increased, which results in a net increase in flow through the membrane. The same effect of increasing the driving force is achieved by lowering the permeate pressure. Echt et al. (2002) point out that the pressure ratio across the membrane is an important design parameter, and thus, design engineers try to maintain the lowest possible permeate pressure (see Equation 5.7).

5.6.3.5   Feed Gas Pretreatment

Because membranes are susceptible to degradation from impurities, pretreatment is usually required. The impurities possibly present in natural gas that may cause damage to the membrane (Echt et al., 2002) include:

- Liquids. The liquids may be entrained in the feed to the unit or formed by condensation within the unit. Liquids can cause the membrane to swell, which results in decreased flux rates and possible membrane damage. Liquids can form internally by two mechanisms: (1) because of condensation of higher molar mass compounds caused by the cooling that occurs (Joule-Thomson effect) as the gas expands to a lower pressure through the membrane, and (2) because CO_2 and the lighter hydrocarbons diffuse more quickly than the heavier hydrocarbons, the dew point of the nondiffusing gas may increase to the point where condensation occurs.

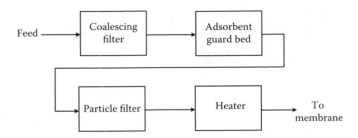

FIGURE 5.22 Schematic of membrane-pretreating equipment. (Adapted from Echt et al., 2002.)

- High-molar-mass hydrocarbons ($C_{15}+$) such as compressor lube oils. These compounds coat the membrane surface and result in a loss of performance. The concentrations are low but the effect is cumulative.
- Particulates. These materials block the small flow passages in the membrane element. Erosion of the membrane could also be a problem.
- Corrosion inhibitors and well additives. Certain of these compounds are destructive to membrane material.

A common method for pretreating the feed gas to a membrane system is shown in Figure 5.22. The coalescing filter removes any entrained liquids; the adsorbent bed takes out trace contaminants such as volatile organic compounds (VOC); the particulate filter removes any dust from the adsorbent bed; and the heater superheats the gas to prevent liquid formation in the membrane unit. The system shown has the following disadvantages (Echt et al., 2002):

- The adsorbent bed is the only unit that removes heavy hydrocarbons. Consequently, if the gas contains more heavy hydrocarbons than anticipated, or in the event of a surge of these materials, the adsorbent bed may become saturated in a relatively short time, and thus allow heavy hydrocarbons to contact the membrane.
- Only the heater provides superheat, and, consequently, if this unit fails, the entire membrane system must be shut down. Echt et al (2002) discuss other pretreatment methods that address the disadvantages discussed above.

5.6.4 ADVANTAGES AND DISADVANTAGES OF MEMBRANE SYSTEMS

The pros and cons of commercial membrane systems are discussed by Kohl and Nielsen (1997) and Echt et al. (2002). The following summarizes their conclusions:

Advantages
- Low capital investment when compared with solvent systems
- Ease of operation: process can run unattended

- Ease of installation: Units are normally skid mounted
- Simplicity: No moving parts for single-stage units
- High turndown: The modular nature of the system means very high turndown ratios can be achieved
- High reliability and on-stream time
- No chemicals needed
- Good weight and space efficiency

Disadvantages

- Economy of scale: Because of their modular nature, they offer little economy of scale
- Clean feed: Pretreatment of the feed to the membrane to remove particulates and liquids is generally required
- Gas compression: Because pressure difference is the driving force for membrane separation, considerable recompression may be required for either or both the residue and permeate streams

For natural gas systems, the following disadvantages also apply:

- Generally higher hydrocarbon losses than solvent systems
- H_2S removal: H_2S and CO_2 permeation rates are roughly the same, so H_2S specifications may be difficult to meet
- Bulk removal: Best for bulk removal of acid gases; membranes alone cannot be used to meet ppmv specifications

For offshore applications, membranes are attractive because of unattended operation, no chemicals needed or generated, and good weight and space efficiency.

5.7 NONREGENERABLE HYDROGEN SULFIDE SCAVENGERS

When the quantity of sulfur to be recovered is small, on the order of 400 lb/d (180 kg/d) or less, the processes discussed previously are uneconomical, and small-scale batch processes are used for H_2S removal. These processes generally use a nonregenerable scavenger. Many such technologies are in general use, and a summary is shown in Table 5.9.

According to Houghton and Bucklin (1994), the most widely used of the scavengers are those that utilize iron oxides, especially iron sponge. In addition to the references mentioned above, detailed discussions of all the above processes may be found in Veroba and Stewart (2003), Engineering Data Book (2004b), and Kohl and Nielsen (1997). Detailed economic comparisons of the most popular scavenger processes may be found in Houghton and Bucklin (1994). The Gas Technology Institute markets a PC program (GRI Scavenger Calcbase™, Gas Research Institute) that makes approximate cost comparisons for a variety of scavengers.

TABLE 5.9
H₂S Scavenger Processes

Solid-based processes:	Iron oxides
	Zinc oxides
Liquid-based processes:	Nitrites
	Amine-aldehyde condensates
	Caustic scrubbing
	Aldehydes
	Oxidizers
	Metal-oxide slurries

Source: Adapted from Holub and Sheilan (2000).

Although the iron sponge is a commonly used solid, nonregenerative scavenger, liquid oxidation–reduction (redox) processes use iron as an oxidizing agent in solution (Engineering Data Book, 2004b). The process involves four basic steps:

1. Removal of H_2S from the gas by absorption into a caustic solution
2. Oxidation of the HS^- ion to elemental sulfur via the oxidizing agent
3. Separation and removal of sulfur from the solution
4. Regeneration (i.e., oxidation) of the oxidizing agent by use of air

While the oxidizing agent forms the desired reaction, it also reacts with the sulfur species to form metal sulfides that precipitate from solution. To avoid the metal sulfide reaction, the solution contains chelating agents (organic compounds that bind with the metal ion to restrict its reactivity but still permit electron transfer for oxidation–reduction reactions). A common chelating agent for iron is EDTA (ethylenediaminetetraacetic acid). Several processes, including Lo-Cat II®, SulF-erox®, and Sulfint-HP, use iron and chelating agents.

5.8 BIOLOGICAL PROCESSES

A biological process for removing H_2S from natural gas has been reported by Cline et al. (2003). In this process, the gas stream that contains the H_2S is first absorbed into a mild alkaline solution, and the absorbed sulfide is oxidized to elemental sulfur by naturally occurring microorganisms.

Cline et al. (2003) reported the successful startup and operation of a commercial plant in Canada. The plant is designed to treat a high-pressure natural gas and produce a product that meets H_2S specifications (4 ppmv or lower). The sulfur plant capacity is approximately 1 ton per day, with a sulfur removal efficiency of 99.5% or higher. Cline et al. (2003) offer the following claims regarding the process:

• Cost effective up to 50 tons/d (50 tonnes/d) of sulfur
• H_2S concentrations in the sour gas from 50 ppmv to 100 vol%

- Sour gas pressures to 75 barg (1,100 psig)
- H_2S concentrations in the sweet gas 1 ppmv
- Formation of hydrophilic sulfur that prevents equipment fouling or blocking

5.9 SAFETY AND ENVIRONMENTAL CONSIDERATIONS

5.9.1 AMINES

The obvious safety concern with amine treating is the potential for H_2S leaks in the plant, even from spilled rich amine. In addition, some sections operate at high temperatures. Caustic handling is another hazard if MEA reclaiming is performed on-site. Reclaimer waste products are toxic and must be handled with care (Engineering Data Book, 2004b).

From an environmental perspective, in addition to the remote chance of hydrogen sulfide release, amines have an affinity for BTEX (benzene, toluene, ethylbenzene, and xylenes), which may be vented during amine regeneration if the sulfur is not recovered. Chapter 9 provides more details. If MEA or DGA are used with reclaimers, the reclaimer solids present a disposal problem, especially with MEA because caustic or soda ash is added to help reverse the reactions.

5.9.2 PHYSICAL ABSORPTION

When H_2S or other sulfur compounds are removed from a gas stream that contains high levels of these materials, the potential always exists for a leak. Depending upon the process, the solvent may be hazardous or toxic.

5.9.3 ADSORPTION

Safety and environmental problems associated with adsorbents such as molecular sieve are relatively minor. Thorough regeneration and purging must be done before the adsorbent can be replaced. However, it should be nonhazardous and disposable in a land fill.

Most solid scavengers are respiratory and eye irritants. Spent iron sponge is pyrophoric, and great care must be taken in the removal and disposal of reacted iron-sponge material (Veroba and Stewart, 2003). The manufacturer's recommendations for this material must be carefully followed to prevent a serious incident.

5.9.4 MEMBRANES

Membranes are probably the safest and most environmentally friendly of the processes for gas treating. No chemicals are used, no waste disposal by-products are generated, and membranes operate at low pressures and generally ambient temperatures.

REFERENCES

Baker, R.W., Future directions of membrane gas separation technology, *Ind. Eng. Chem. Res.* 41, 1391, 2002.

Baker, R.W., Wijmans, J.G., and Kaschemekat, J.H., The design of vapor-gas separation systems, *J. Membrane Sci.* 151, 55, 1998.

Benson, H.E., Field, J.H., and Jimeson, R.M., CO_2 absorption employing hot potassium carbonate solutions, *Chem. Eng. Progress*, 50, 356, 1954.

Benson, H.E., Field, J.H., and Haynes, W.P., Improved process for CO_2 absorption uses hot carbonate solutions, *Chem. Eng. Progress*, 52, 433, 1956.

Bierlein, J.A. and Kay, W.B., Phase equilibrium properties of system carbon dioxide-hydrogen sulfide, *Ind. Eng. Chem.*, 45, 618, 1953.

Brown, T.S., Kidnay, A.J., and Sloan, E.D, Vapor-liquid equilibria in the carbon dioxide + ethane system, *Fluid Phase Equil.*, 40, 169, 1988.

Bucklin, R.W. and Schendel, R.L., Comparison of physical solvents used for gas processing, in *Acid and Sour Gas Treating Processes*, Newman, S.A., Ed., Gulf Publishing Company, Houston, TX, 1985. (As quoted in Kohl, A. and Nielsen, R., *Gas Purification*, 5th ed., Gulf Publishing, Houston, TX, 1997, 1197.

Carson, J.K., Marsh, K.N., and Mather, A.E., Enthalpy of solution of carbon dioxide in (water + monoethanolamine, or diethanolamine, or N-methyldiethanolamine) and (water + monoethanolamine + N-methyldiethanolamine) at $T = 298.15K$, *J. Chem. Thermo.*, 32, 1285, 2000.

Chi, C. W. and Lee, H., Natural gas purification by 5A molecular sieves and its design method, *AIChE Symp. Ser.*, 69, 95, 1973.

Clare, R.T. and Valentine, J.P., Acid Gas Removal Using the Selexol® Process, Proceeding of the Second Quarterly Meeting of the Canadian Natural Gas Processors Association, Edmonton, Alberta, Canada, June 5, 1975.

Cline, C., Hoksberg, A., Abrey, R., and Janssen, A., Biological Process for Removal from Gas Streams, The Shell-Paques/THIOPAQ Gas Desulfurization Process, Proceedings of the Laurance Reid Gas Conditioning Conference, Norman, OK, 2003, 1.

Cummings, W. and Chi, C.W., Natural Gas Purification with Molecular Sieves, Canadian Chemical Engineering Conference, Calgary, Alberta, Canada, October 23–26, 1977.

Echt, W.I., Dortmund, D.D., and Malino, H.M., Fundamentals of Membrane Technology for CO_2 Removal from Natural Gas, Proceedings of the Laurance Reid Gas Conditioning Conference, Norman, OK, 2002, 1.

Echterhoff, L.W., State of the Art of Natural Gas Processing Technologies, Gas Research Institute Report GRI - 91/0094, 1991.

Engineering Data Book, 12th ed., Sec. 2, Product Specifications, Gas Processors Supply Association, Tulsa, OK, 2004a.

Engineering Data Book, 12th ed., Sec. 21, Hydrocarbon Treating, Gas Processors Supply Association, Tulsa, OK, 2004b.

Epps, R, Use of Selexol® solvent for Hydrocarbon Dewpoint Control and Dehydration of Natural Gas, Proceedings of the Laurance Reid Gas Conditioning Conference, Norman, OK, 1994, 26.

Gas Research Institute, www.gastechnology.org, Retrieved July 2005.

Hegwar, A.M. and Harris, R.A, Selexol® solves high H_2S/CO_2 problem, *Hydrocarbon Process.*, 49 (4) 103, 1970.

Holmes, A.S., Ryan, J.M., Price, B.C., and Styring, R.E., Pilot Tests Prove Ryan/Holmes Cryogenic Acid Gas/Hydrocarbon Separations, Proceedings of the Sixty-First Annual Convention of the Gas Processors Association, Tulsa, OK, 1982, 75.

Holub, P.E. and Sheilan, M., Fundamentals of Gas Sweetening, Proceedings of the Laurance Reid Gas Conditioning Conference, Norman, OK, 2000.

Houghton, J.E. and Bucklin, R.W., Nonregenerable H_2S Scavenger Update, Proceedings of the Laurance Reid Gas Conditioning Conference, Norman, OK, 1994, 110.

Johnson, J.E., Selexol® solvent process reduces lean, high-CO_2 natural gas treating costs, *Energy Progr.*, 4 (4) 241, 1984.

Jou, F.-Y, Otto, F.D., and. Mather, A.E, Vapor liquid equilibrium of carbon dioxide in aqueous mixtures of monoethanolamine and methyldiethanolamine, *Ind. Eng. Chem. Res.* 33, 2002, 1994.

Judd, D.K., Selexol® unit saves energy, *Hydrocarbon Process.*, 57 (4) 122, 1978.

Kent, R.L. and Eisenberg, B., Better data for amine treating, *Hydrocarbon Process.*, 55 (2) 87, 1976.

Klinkenbijl, J.M., Dillon, M.L., and Heyman, E.C., Gas Pre-Treatment and Their Impact on Liquefaction Processes, Proceedings of the Seventy-Eighth Annual Convention of the Gas Processors Association, Tulsa, OK, 1999, 299.

Kohl, A. and Nielsen, R., *Gas Purification*, 5th ed., Gulf Publishing, Houston, TX, 1997.

Kopperson, D., Horne, S., Kohn, G., Romansky, D., Chan, C., and Duckworth, G.L., Acid gas disposal 1: Injecting acid gas with water creates new disposal option, *Oil Gas J.*, 96 (31) 33, 1998.

Kopperson, D., Horne, S., Kohn, G., Romansky, D., Chan, C., and Duckworth, G.L., Two gases illustrate acid gas/water injection scheme, *Oil Gas J.*, 96 (32) 64, 1998.

Linde Molecular Sieves Non-Hydrocarbon Materials Data Sheets, Form 9691-F.

Lyddon, L. and Nguyen, H., Analysis of Various Flow Schemes for Sweetening with Amines, Proceedings of the Seventy-Eighth Annual Convention of the Gas Processors Association, Tulsa, OK, 1999, 177.

McCartney, D.G., private communication, 2005.

Meyer, H.S. Volume and Distribution of Subquality Natural Gas in the United States, GasTIPS, Gas Research Institute, Winter 2000, p.10.

Moritis, G., New companies, infrastructure, projects reshape landscape for CO_2 EOR in US, *Oil Gas J.*, 99 (20) 68, 2001.

Mortko, R.A., Remove H_2S selectively, *Hydrocarbon Process.*, 63 (6) 78, 1984.

NIST, National Institute of Standards and Technology, U.S. Department of Commerce, Thermophysical properties of fluid systems, 2005, webbook.nist.gov/chemistry/fluid/, Retrieved August 2005.

Oberding, W., Goff, R., Townsend, M., Chapin, D., Naulty, D., and Worah, V., Lessons Learned and Technology Improvements at the Lost Cabin Gas Plant, Proceedings of the Laurance Reid Gas Conditioning Conference, Norman, OK, 2004, 273.

Raney, D.R., Remove carbon dioxide with Selexol®, *Hydrocarbon Process.*, 55 (4) 73, 1976.

Shah, V.A., Integrated Gas Treating and Hydrocarbon Recovery Process Using Selexol® Solvent Technology, Proceedings of the Sixty-Eighth Annual Convention of the Gas Processors Association, Tulsa, OK, 1989, 197.

Simmons, C.V., Reclaiming Used Amine and Glycol Solutions, Proceedings of the Laurance Reid Gas Conditioning Conference, Norman, OK, 1991, 335.

Sigmund, P.W., Butwell, K.F., and Wussler, A.J., The H_2S Process, an Advanced Process for Selective H_2S Removal, Proceedings of the Sixtieth Annual Convention of the Gas Processors Association, Tulsa, OK, 1981.

Smith, W.A., Good Operating Practices for Amine Treating Systems, Proceedings of the Laurance Reid Gas Conditioning Conference, Norman, OK, 1996, 303.

Spillman, R.W. and Cooley, T.E., Membrane Gas Treating, Proceedings of the Eighty-Eighth Annual Convention of the Gas Processors Association, Tulsa, OK, Tulsa, Oklahoma, 1989, 186.

Sweny, J.W., High CO_2-H_2S Removal with Selexol® Solvent, Proceedings of the Sixty-Ninth Annual Convention of the Gas Processors Association, Tulsa, OK, 1980,163.

Sweny, J.W. and Valentine, J.P., Physical solvent stars in gas treatment/purification, *Chemical Eng.*, 54 (19) 77 1970.

Tennyson, R.N. and Schaaf, R.P., Guidelines can help choose proper process for gas-treating plants, *Oil Gas J.*, 75 (2) 78, 1977.

Veroba, R. and Stewart, E., Fundamentals of Gas Sweetening, Proceedings of the Laurance Reid Gas Conditioning Conference, Norman, OK. 2003, 1.

Wichert, E. and Royan, T., Acid gas injection eliminates sulfur recovery expense, *Oil Gas J.*, 95 (17) 1997.

6 Gas Dehydration

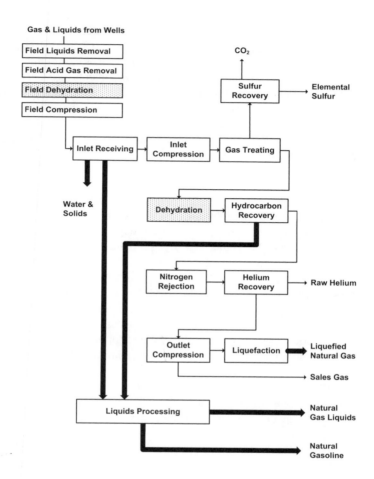

6.1 INTRODUCTION

Typically, dehydration is important in three areas:

- Gas gathering. Water needs to be removed to reduce pipeline corrosion and eliminate line blockage caused by hydrate formation. The water dew point should be below the lowest pipeline temperature to prevent free water formation. Chapter 3 discusses field operations and gas hydrates.

- Product dehydration. Both gas and liquid products have specifications on water content. Sales gas that leaves a plant is usually dry if cryogenic hydrocarbon liquid recovery is used. Liquid and gas streams may be water saturated after amine treatment or coming from underground storage. Most product specifications, except for propane, require that no free water be present (Engineering Data Book, 2004a). This requirement puts the maximum water in sales gas at 4 to 7 lb/MMscf (60 to 110 mg/Sm^3). For liquids, the water content is 10 to 20 ppmw. Commercial grades of propane require lower water contents (see Chapter 1).
- Hydrocarbon recovery. Most plants use cryogenic processes to recover the C_2+ fraction from inlet gas. If acid gases are removed by use of amine processes, the exit gas leaves water saturated. To prevent hydrate formation in the cryogenic section of hydrocarbon recovery, the water concentration should be 0.1 ppmv or less. Chapter 7 discusses hydrocarbon recovery processes.

The first two areas are less demanding and a variety of processes are available to meet the need. However, the third application, as well as gas liquefaction and cryogenic nitrogen rejection units, requires water contents an order of magnitude lower than the other two. Molecular sieves can provide dehydration to this level.

Water content is stated in a number of ways:

- Mass of water per volume of gas, lb/MMscf (mg/Sm^3)
- Dew point temperature, °F (°C), which is the point that liquid water, real or hypothetically subcooled, will condense out of the hydrocarbon phase
- Concentration, parts per million by volume (ppmv)
- Concentration, parts per million by mass (ppmw)

The first three water-content values are commonly used with gases, and the second and fourth values are used with liquids. The first and third values are convenient, as they are easily related (21 ppmv per lb/MMscf [18 mg/Sm^3]). Use of ppmw requires knowledge of the hydrocarbon molecular weight for conversion to other concentrations. Unfortunately, the dew point is commonly used because it makes practical sense. However, no simple conversion method exists between dew point and the other concentration units.

This chapter provides ways to estimate the water content of gases and then discusses the common processes used for its removal. The same basic processes apply to drying liquids, but the liquid, which has a higher density, alters the processes slightly. Chapter 10 discusses dehydration of liquids.

6.2 WATER CONTENT OF HYDROCARBONS

Determining the saturation water content of a gas (the dew point) is a standard but complex problem in thermodynamics, and excellent discussions of phase equilibria calculations are given in many thermodynamics texts (e.g., Smith et al. [2001] and Prausnitz et al. [1999]). We first assume ideal behavior, and then present a commonly used engineering correlation on the basis of experimental measurements.

In any mixture, where both the gas and liquid phases are in equilibrium, each component, i, in the mixture obeys the relationship

$$x_i \gamma_i P_i^{Sat} = y_i \varphi_i P \tag{6.1}$$

where x_i is the mole fraction in the liquid phase, γ_i is the activity coefficient, P^{Sat} is the saturation or vapor pressure, y_i is the vapor phase concentration, φ_i the vapor phase fugacity coefficient, and P the total pressure. The vapor pressure is temperature dependent, whereas the activity and fugacity coefficients are temperature, pressure, and composition dependent. If three of the four variables, temperature, pressure, liquid composition, and vapor composition, are known, the fourth variable can be calculated (flash calculation). The activity and fugacity coefficients can be calculated by use of equations of state and empirical equations. However, present-day equations of state have difficulty modeling the strong polar nature of water in hydrocarbon–water mixtures; calculations from different simulators can give different results, especially in prediction of water content in the liquid phase. If the mixture contains methanol or brine, the calculations become more uncertain.

However, reasonably good estimates of the concentration of water in the vapor phase in equilibrium with liquid water can be made at pressures below 500 psia (35 bar). If we make the good assumption of negligible hydrocarbon in the liquid water phase, which, thus makes both x_i and γ_i unity for water and assume the gas phase to be ideal, which makes φ_i unity, we obtain

$$y_{H_2O} = \frac{P_{H_2O}^{Sat}}{P} \tag{6.2}$$

Equation 6.2 provides reasonably good values for gas-phase water content, provided that the gas contains less than a few mol% of either CO_2 or H_2S. Appendix B tabulates water vapor pressure as a function of temperature.

A more accurate way to determine the water content is to use Figure 6.1. Figure 6.1(a) gives the water content for a hydrocarbon gas as a function of temperature and pressure. Figure 6.1(b) provides corrections for gas gravity and for salinity if the water phase contains brine. The chart is widely used in the gas industry and gives good results, provided the gas contains less than about 5 mol% CO_2 plus H_2S. The Engineering Data Book (2004b) provides an empirical method to predict water concentrations when the gas contains higher concentrations of the acid gases. Note that the dashed lines in Figure 6.1(a) represent the water phase in equilibrium with either ice or gas hydrate, depending upon the gas composition

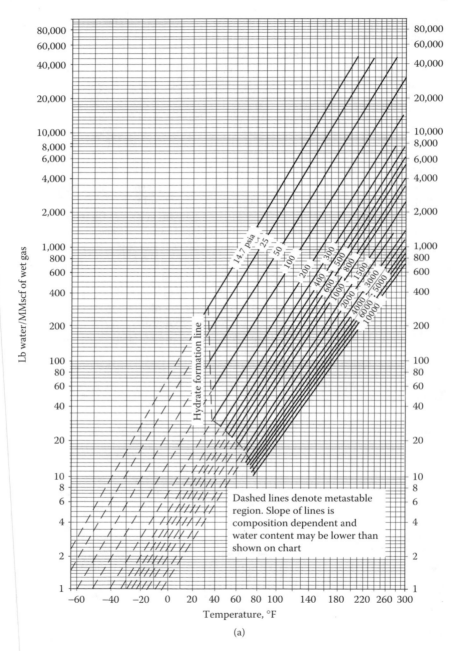

FIGURE 6.1(a) Water content of hydrocarbon gases as a function of temperature and pressure.

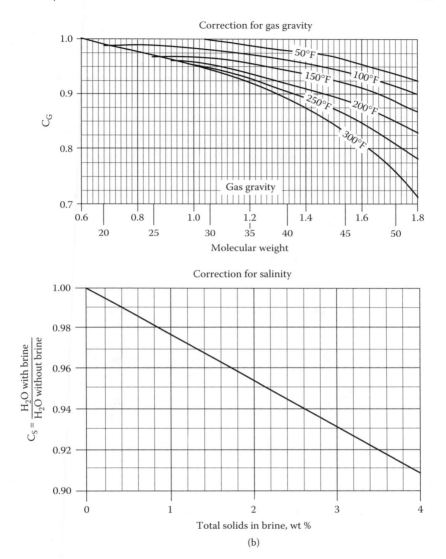

FIGURE 6.1(b) Gas gravity and salinity corrections to water content in hydrocarbon gases. (Adapted from Engineering Data Book, 2004b.)

and pressure. Actual gas-phase water contents can be lower than that obtained from the chart.

Example 6.1 Calculate the water content of the sweet natural gas shown in Table 6.1 at 300 psia (20.7 bar) and 80°F (26.7°C) by use of Equation 6.2 and Figure 6.1.

TABLE 6.1
Gas Composition Data for Example 6.1

Component	Mole Fraction	Molar Mass	Mol Fraction × Molar Mass
CH_4	0.90	16.043	14.44
C_2H_6	0.04	30.070	1.20
C_3H_8	0.03	44.097	1.32
n-C_4H_{10}	0.02	58.123	1.16
N_2	0.01	28.013	0.28
Totals	1.00		18.41

Equation 6.2 using vapor-pressure data from Appendix B gives

$$y_{H_2O} = \frac{P_{H_2O}^{Sat}}{P} = \frac{0.507\,psia}{300\,psia} = 0.0017 \quad \text{mol fraction}$$

Next convert to lb Water/MMscf:

$$W = \left(0.0017\,\frac{mol\,H_2O}{mol\,gas}\right)\left(18\,\frac{lb\,H_2O}{lb\text{-}mol}\right)\left(\frac{1\,lb\text{-}mol}{379.5\,scf}\right)(10)^6 = 81\,\frac{lb\,H_2O}{MMscf}$$

$$= \left(1,300\,\frac{mg}{m^3}\right)$$

From Table 6.1, the MW of the gas mixture is 18.41 and the specific gravity is

$$SpGr = \rho_{gas}/\rho_{air} = 18.41/28.96 = 0.636$$

From Figure 6.1(a) W^{Sat} = 85 lb/MMscf (1,400 mg/Sm³). Correct for specific gravity by obtaining C_G from Figure 6.1b (C_G = 0.99), and multiplication gives

$$W^{Sat} = (0.99)(85) = 84 \text{ lb/MMscf (1,400 mg/Sm}^3)$$

The values differ by 4%. Increasing the pressure to 1,000 psia (69 bar). Equation 6.2 is 27% below the value obtained from Figure 6.1a.

6.3 GAS DEHYDRATION PROCESSES

This section discusses the conventional methods for drying natural gas, and then briefly describes some less-conventional methods. Two processes, absorption and adsorption, are the most common and are discussed in more detail.

6.3.1 Absorption Processes

6.3.1.1 Overview of Absorption Process

Water levels in natural gas can be reduced to the 10 pmmv range in a physical absorption process in which the gas is contacted with a liquid that preferentially absorbs the water vapor. The solvent used for the absorption should have the following properties:

- A high affinity for water and a low affinity for hydrocarbons
- A low volatility at the absorption temperature to reduce vaporization losses
- A low viscosity for ease of pumping and good contact between the gas and liquid phases
- A good thermal stability to prevent decomposition during regeneration
- A low potential for corrosion

In practice, the glycols, ethylene glycol (EG), diethylene glycol (DEG), triethylene glycol (TEG), tetraethylene glycol (TREG) and propylene glycol are the most commonly used absorbents; triethylene glycol is the glycol of choice in most instances. For operations in which frequent brine carryover into the contactor occurs, operators use EG because it can hold more salt than the other glycols. The solubility of sodium chloride in EG water mixtures is around 20 wt% (Masaoudi, 2004; Parrish, 2000; Trimble, 1931), whereas it is only around 5 wt% in TEG (Kruka, 2005).

Appendix B gives some of the more important properties of the commonly used glycols. Table 6.2 shows their formulas, along with the maximum recommended

TABLE 6.2
Glycols Used in Dehydration

Name	Formula		Maximum Recommended Regeneration Temperature, °F (°C)[a]
Ethylene glycol (EG)	HO—(CH$_2$)$_2$—OH	C$_2$H$_6$O$_2$	
Diethylene glycol (DEG)	HO—((CH$_2$)$_2$—O)—(CH$_2$)$_2$—OH	C$_4$H$_{10}$O$_3$	325 (160)
Triethylene glycol (TEG)	HO—((CH$_2$)$_2$—O)$_2$—(CH$_2$)$_2$— OH	C$_6$H$_{14}$O$_4$	360 (180)
Tetraethylene glycol (TREG)	HO—((CH$_2$)$_2$—O)$_3$—(CH$_2$)$_2$— OH	C$_8$H$_{18}$O$_5$	400 (200)
Propylene glycol	HO—(CH$_2$)$_3$—OH	C$_3$H$_8$O$_2$	

[a] *Source*: Kohl and Nielsen (1997).

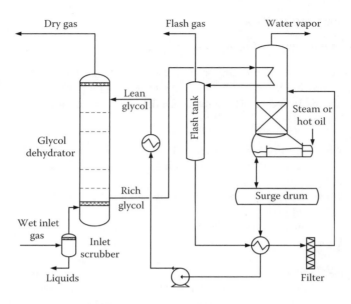

FIGURE 6.2 Schematic of typical glycol dehydrator unit. (Adapted from Engineering Data Book, 2004b.)

regeneration temperature. Smith (2004) reports temperatures for onset of slow thermal degradation as 328°F (164°C) and 385°F (196°C) for DEG and TEG, respectively. However, the author gives typical maximum regenerator temperatures of about 350°F (175°C) and 400°F (204°C) for DEG and TEG, respectively. Manufacturer's literature (Dow Chemical, 2003) gives a decomposition temperature of 464°F (240°C) for TEG.

Figure 6.2 shows a typical, simplified flow sheet for a glycol absorption unit. The wet gas passes through an inlet scrubber to remove solids and free liquids, and then enters the bottom of the glycol contactor. Gas flows upward in the contactor, while lean glycol solution (glycol with little or no water) flows down over the trays. Rich glycol absorbs water and leaves at the bottom of the column while dry gas exits at the top. The rich glycol flows through a heat exchanger at the top of the still where it is heated and provides the coolant for the still condenser. Then the warm solution goes to a flash tank, where dissolved gas is removed. The rich glycol from the flash tank is further heated by heat exchange with the still bottoms, and then becomes the feed to the still. The still produces water at the top and a lean glycol at the bottom, which goes to a surge tank before being returned to the contactor.

Operating conditions for glycol units are governed principally by the degree of dehydration required, the physical properties of the glycol solutions, and the inlet pressure of the gas to the processing unit. Some typical operating conditions for TEG absorbers are given in the next section. The articles of Hernandez-Valencia et al (1992), Parrish et al. (1986), and Wieninger (1991), along with the

Engineering Data Book (2004b), provide detailed information on absorber design and operation. Although many design references refer to trays, most modern TEG absorbers use structured packing instead of trays (McCartney, 2005).

6.3.1.2 Representative Operating Conditions for TEG Absorbers

As noted above, triethylene glycol is the most common absorbent for dehydration, although some operations in colder climates (e.g., the North Sea) will use diethylene glycol because of its lower viscosity. Tetraethylene glycol has a lower vapor pressure and withstands higher regeneration temperatures than does TEG, but the additional cost outweighs the marginal increased benefits. Smith (2004) reports about 20,000 TEG dehydration units operating in the United States.

Gas flow rates, compositions, inlet pressures and temperatures, and the required degree of dehydration vary widely. However, Table 6.3 provides some useful guidelines for typical service.

Figure 6.3 shows the equilibrium dew points that can be achieved with TEG solutions of different concentrations as a function of contactor temperature. For example, if an equilibrium dew point of −15°F (−26°C) is required and the contactor operates at 80°F (27°C), then a TEG solution of 99.5 wt% is required. However, the assumption is that the dry gas exiting is in equilibrium with the incoming lean glycol (i.e., infinite number of contactor trays). To account for nonequilibrium concentrations, the Engineering Data Book (2004b) suggests use of an equilibrium temperature that is 10 to 20°F (5 to 10°C) below the desired dew point temperature. Therefore, to obtain the −15°F (−26°C) exit dew-point temperature, the lean glycol concentration should be 99.8 to 99.9 wt%.

The Engineering Data Book (2004b) notes that Figure 6.3 is based upon equilibrium between water vapor and a liquid water phase. At lower temperatures, the true equilibrium condensed phase is gas hydrate, which will form at higher temperatures than does subcooled water. Therefore, solids formation could be as much as 15 to 20°F (8 to 10°C) higher than the dew point value obtained from Figure 6.3. The actual error depends upon temperature, pressure, and gas composition. However, the pressure effect is minimal, and the chart can be used to 1,500 psia (100 bara).

The required lean-glycol concentration dictates still reboiler operating conditions. Higher reboiler temperatures yield leaner glycol. At a 400°F (204°C), the typical maximum regeneration temperature, TEG yields a lean-glycol concentration of 98.6 wt% (Engineering Data Book, 2004b) at sea level. Higher purity requires reduction of the partial pressure of water in the reboiler vapor space. The most common way to achieve this pressure reduction is to use a stripping gas* or vacuum distillation, which yields lean glycol concentrations of 99.95 and 99.98, respectively. The Engineering Data Book (2004b), Smith

---

* A stripping gas is a noncondensing gas at operating conditions that is used to strip the more volatile solutes from a liquid.

TABLE 6.3
Typical Operating Conditions of TEG Dehydrators

Contactor	
Inlet pressures	<2000 psig (139 bar)
Inlet temperatures	60°F to 100°F (16°C to 38°C)
	(Lower temperatures enhance absorption capacity but can lead to hydrate formation at high pressure.)
Pressure drop	5 to 10 psi (34 to 69 kPa)
Glycol circulation rate	2 to 5 gal/lb H_2O removed, with 3 common. (17 to 42 L/kg)
Tray efficiencies	25 to 30%.
Dew points	> −25°F (−32°C) (Enhanced regeneration required for lower dew point temperatures.)
Glycol losses	
Vaporization	~ 0.012 gal/MMscf (1.6 L/MM Sm³)
Total	0.025 gal/MMscf (3.3 L/MM Sm³)
Regenerator (reboiler and still)	
Column internals	Packed equivalent to 3 or 4 trays
Reboiler temperatures	375°F to 400°F (191°C to 204°C)
Flash tank	
Pressure	50 to 75 psig (446 to 618 kPa)
Temperature	150°F (66°C)
Retention times[a]	
C4+ Lean Gas	~10 minutes.
C4+ Rich Gas[a]	~20 minutes (Use three-phase separator.) TEG absorbs about 1 scf gas/gal TEG at 1,000 psig and 100°F (0.0076 Sm³/L at 70 barg and 38°C)

[a] For treatment of gas streams that contain high concentrations of C4+, adequate time needs to be allowed for removal of the less-volatile components from the glycol to minimize hydrocarbon losses in the still overhead.

Source: Engineering Data Book, (2004b).

(1993), and Smith and Humphrey (1995) describe these methods, along with two proprietary methods, one of which claims to obtain a 99.999 wt% glycol.

Figure 6.3 shows that lower absorption temperatures improve dehydration. As a general guideline, absorber inlet temperatures can be as high as 150°F (66°C), although temperatures above 100°F (38°C) may result in unacceptable vaporization losses for the glycol solutions (Engineering Data Book, 2004b). At inlet temperatures below approximately 50°F (10°C), the viscosities of the glycol solutions (see Appendix B) are sufficiently high to reduce plate efficiencies and increase pumping costs. Typical maximum operating pressures are around 2,000 psig (140 barg).

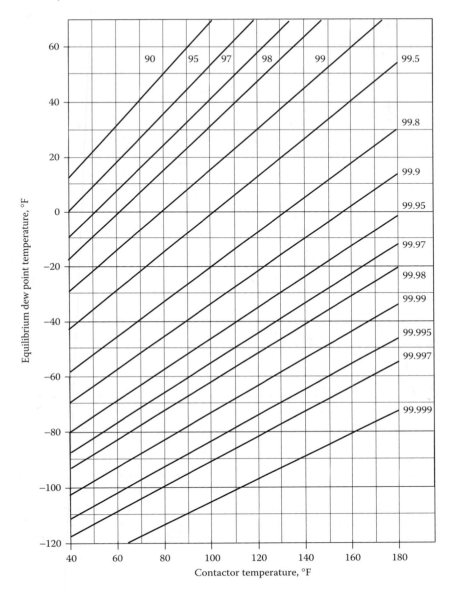

FIGURE 6.3 Equilibrium water dew point as a function of contactor temperature and TEG concentration in wt%. (Adapted from Engineering Data Book. Used with permission of Fluor Enterprises.)

Triethylene glycol losses from a properly operating plant, excluding spillage, should be minimal. The Engineering Data Book (2004b) estimates about 0.1 gal/MMscf (13 L/10⁶Sm³) from carryover if a standard mist eliminator is used. Other losses range between 0.05 gal/MMscf (7 L/MMSm³) for high-pressure,

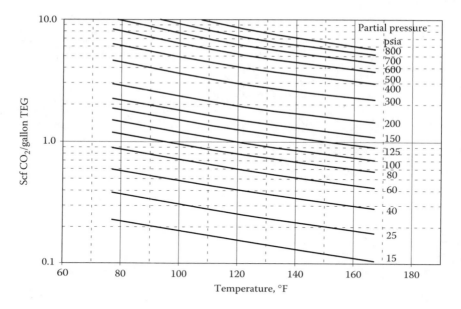

FIGURE 6.4 Solubility of CO_2 in lean triethylene glycol as a function of temperature and partial pressure (Wichert, 2005).

low-temperature dehydration and 0.3 gal/MMscf (40 L/MMSm³) for low-pressure, high-temperature dehydration. Higher losses suggest foaming problems in the absorber or contactor.

If the dehydrator is used to dry high-pressure CO_2 streams, the glycol losses can be much higher because glycol is more soluble in dense CO_2 than in natural gas. The Engineering Data Book (2004b) refers to the paper by Wallace (1985) on this subject.

Sour gases can be dried with glycol solutions if appropriate anticorrosion measures are taken. Figures 6.4 and 6.5 show the solubility of the acid gases in TEG as a function of temperature and pressure in lean TEG. These figures are used to estimate the vapor volumes, and this is added to the estimated 1 scf hydrocarbon gas/gal TEG at 1,000 psig and 100°F (0.0076 Sm³/L at 70 barg and 38°C).

6.3.1.3 Other Factors That Affect Glycol Dehydrator Performance

Oxygen reacts with the glycols to form corrosive acidic compounds. The products also increase the potential for foaming and glycol carryover (Wieninger, 1991). A dry natural gas blanket is often put over the storage and surge tanks to minimize air intrusion. The Engineering Data Book (2004b) suggests that precautions be taken if oxygen is present in the gas but offers no options.

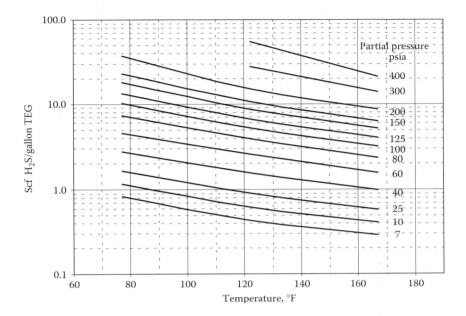

FIGURE 6.5 Solubility of H_2S in lean triethylene glycol as a function of temperature and partial pressure (Wichert, 2005).

The Engineering Data Book (2004b) notes that low pH accelerates glycol decomposition. The book suggests addition of trace amounts of basic hydrocarbons (e.g., alkanolamines). For glycol units downstream of amine treaters, the trace amount of amine carryover tends to mitigate the acidity problem.

Glycol dehydration units in the typical gas plant are downstream of the amine treaters; consequently, they process a relatively clean gas stream that contains no H_2S and reduced CO_2 levels. Amine carryover is the main concern in these situations. Dehydrators in gathering systems can be in more demanding situations where they may process streams that contain acid gases and methanol or other hydrate inhibitors. They also may be subjected to oil and liquid water slugs, as well as compressor oil mist. Water-soluble components increase the regenerator duty, and hydrocarbon impurities can cause foaming in the system. Proper sizing of inlet scrubbers is critical in these situations. In some cases, filter separators or coalescing filters are required to prevent fine mists from entering the absorber.

As noted above, offshore producers sometimes use (EG) instead of TEG when produced water is carried over to the contactor. In these situations, the purpose of dehydration is hydrate prevention. To meet the 4- to 7-lb H_2O/MMscf (60 to 110 mg/Sm^3) specification, either stripping gas or running the still at a vacuum is required. The higher (EG) evaporation losses are considered acceptable. In some cases, reclaimers, which distill the (EG) from the brine, are on the platform.

The Engineering Data Book (2004b) reports that glycol removes 40 to 60% of methanol from the feed. The increased liquid loading places an additional heat load on the reboiler and an added vapor load on the regenerator. Even with adverse conditions, many unmanned glycol systems function well in the field.

6.3.2 ADSORPTION PROCESSES

6.3.2.1 Overview of Adsorption

The two types of adsorption are physical adsorption and chemisorption. In physical adsorption, the bonding between the adsorbed species and the solid phase is called van der Waals forces, the attractive and repulsive forces that hold liquids and solids together and give them their structure. In chemisorption, a much stronger chemical bonding occurs between the surface and the adsorbed molecules. This chapter considers only physical adsorption, and all references to adsorption mean physical adsorption.

Physical adsorption is an equilibrium process like vapor–liquid equilibria and equations analogous to Equation 6.1 apply. Thus, for a given vapor-phase concentration (partial pressure) and temperature, an equilibrium concentration exists on the adsorbent surface that is the maximum concentration of the condensed component (adsorbate) on the surface. Figure 6.6 shows the equilibrium conditions for water on a commercial molecular sieve. Such curves are called isotherms. The figure is based upon a water–air mixture but is applicable to natural gas systems. The important parameter is the partial pressure of water; total pressure has only a minor effect on the adsorption equilibrium.

Because adsorbate concentrations are usually low, generally only a few layers of molecules will build up on the surface. Thus, adsorption processes use solids with extremely high surface-to-volume ratios. Commercially used synthetic zeolites (i.e, molecular sieves) have surface-to-volume ratios in the range of 750 cm^2/cm^3, with most of the surface for adsorption inside of the adsorbent. In the case of molecular sieves, the adsorbent consists of extremely fine zeolite particles held together by a binder. Therefore, adsorbing species travel through the macropores of the binder into the micropores of the zeolite. Adsorbents such as silica gel and alumina are formed in larger particles and require no binder. Pore openings that lead to the *inside* of commercial adsorbents are of molecular size; they normally range from approximately 4 Å (1 Å = 10^{-8} cm) to 100 Å. Molecular sieves have an extremely narrow pore distribution, whereas silica gel and alumina have wide distributions. However, a molecular sieve binder, which is usually about 20% of the weight of the total adsorbent, has large pores capable of adsorbing heavier components.

Two steps are involved in adsorbing a trace gas component. The first step is to have the component contact the surface and the second step is to have it travel through the pathways inside the adsorbent. Because this process is a two-step process and the second step is relatively slow, solid adsorbents take longer to come to equilibrium with the gas phase than in absorption processes.

In addition to concentration (i.e., partial pressure for gases), two properties of the adsorbate dictate its concentration on the absorbent surface: polarity and size. Unless the adsorbent is nonpolar, which is not the case for those used in

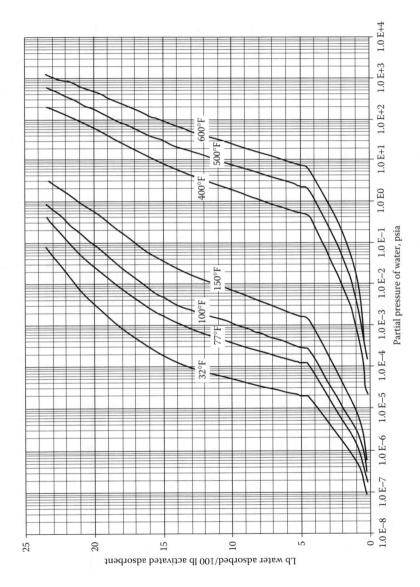

FIGURE 6.6 Water loading on UOP Adsorbent 4A-DG MOLSIV Pellets. Activation conditions for the adsorbent were 662°F (350°C) and less than 10 microns Hg. (Adapted from Engineering Data Book 2004 b. Used with permission of UOP LLC.)

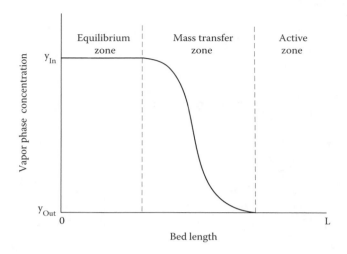

FIGURE 6.7 Vapor-phase concentration profile of an adsorbate in the three zones of an adsorption bed.

gas plants, polar molecules, like water, will be more strongly adsorbed than weakly polar or nonpolar compounds. Thus, methane is displaced by the weakly polar acid gases that are displaced by the strongly polar water.

How size affects adsorption depends upon the pore size of the adsorbent. An adsorbate too large to fit into the pores adsorbs only on the outer surface of adsorbent, which is a trivial amount of surface area compared with the pore area. If the pores are sufficiently large to hold different adsorbates, the less volatile, which usually correlates with size, adsorbates will displace the more volatile ones. Therefore, ethane is displaced by propane.

In commercial practice, adsorption is carried out in a vertical, fixed bed of adsorbent, with the feed gas flowing down through the bed. As noted above, the process is not instantaneous, which leads to the formation of a mass transfer zone (MTZ) in the bed. Figure 6.7 shows the three zones in an adsorbent bed:

1. The equilibrium zone, where the adsorbate on the adsorbent is in equilibrium with the adsorbate in the inlet gas phase and no additional adsorption occurs
2. The mass transfer zone (MTZ), the volume where mass transfer and adsorption take place
3. The active zone, where no adsorption has yet taken place

In the mass transfer zone (MTZ), the concentration drops from the inlet value, y_{in}, to the outlet value, y_{out}, in a smooth S-shaped curve. If the mass transfer rate were infinite, the MTZ would have zero thickness. The MTZ is usually assumed to form quickly in the adsorption bed and to have a constant length as it moves through the bed, unless particle size or shape is changed. The value of y_{in} is dictated by upstream processes; the y_{out} value is determined by the regeneration gas adsorbate content.

The length of the MTZ is usually 0.5 to 6 ft (0.2 to 1.8 m), and the gas is in the zone for 0.5 to 2 seconds (Trent, 2004). To maximize bed capacity, the MTZ needs to be as small as possible because the zone nominally holds only 50% of the adsorbate held by a comparable length of adsorbent at equilibrium. Both tall, slender beds, which reduce the percentage of the bed in the MTZ, and smaller particles make more of the bed effective. However, smaller particle size, deeper beds, and increased gas velocity will increase pressure drop.

For a point in the MTZ, the gas phase adsorbate content increases in time from y_{in} to y_{out} in an S-shaped curve that mirrors the curve shown in Figure 6.7. In principle, beds can be run until the first sign of breakthrough. This practice maximizes cycle time, which extends bed life because temperature cycling is a major source of bed degeneration, and minimizes regeneration costs. However, most plants operate on a set time cycle to ensure no adsorbate breakthrough.

Trent (2004) presents data that show a change in the L/D from 0.8 to 2.7 in the bed increases the useful adsorption capacity from 8.7 to 10.0 wt% in useful water capacity for an equal amount of gas dried. However, the pressure drop increases from 0.4 to 4.3 psi (0.020 to 0.20 kPa).

When used as a purification process, adsorption has two major disadvantages:

- It is a fixed-bed process that requires two or more adsorption beds for continuous operation.
- It has limited capacity and is usually impractical for removing large amounts of impurity.

However, adsorption is very effective in the dehydration of natural gas because water is much more strongly adsorbed than any of the alkanes, carbon dioxide, or hydrogen sulfide. Generally, a higher degree of dehydration can be achieved with adsorbents than with absorption processes.

6.3.2.2   Properties of Industrial Adsorbents for Dehydration

Three types of commercial adsorbents are in common use in gas processing plants:

- Silica gel, which is made of pure SiO_2
- Activated alumina, which is made of Al_2O_3
- Molecular sieves, which are made of alkali aluminosilicates and can be altered to affect adsorption characteristics

Table 6.4 lists the more important properties of three adsorbents compiled primarily from commercial literature. The properties are representative and vary between manufacturers.

TABLE 6.4

Representative Properties of Commercial Silica Gels, Activated Alumina, and Molecular Sieve 4A

	Silica Gel	Activated Alumina	Molecular Sieve 4A
Shape	Spherical	Spherical	Pellets (extruded cylinders) and beads
Bulk density lb/ft³ (kg/m³)	49 (785)	48 (769)	40 –45 (640 – 720)
Particle size	4 – 8 mesh 5 –2 mm	7–14 mesh, 1/8-inch, 3/16-inch, 1/4-inch diameter (3-mm, 5-mm, 6-mm)	1/16-inch, 1/8-inch, 1/4-inch diameter cylinders (1.6-mm, 3.2-mm, 6-mm)
Packed bed % voids	35	35	35
Specific heat Btu/lb-°F (kJ/kg-K)	0.25 (1.05)	0.24 (1.00)	0.24 (1.00)
Surface area m²/g	650 – 750	325 – 360	600 – 800
Pore volume cm³/g	0.36	0.5	0.28
Regeneration temperature, °F (°C)	375 (190)	320 to 430 (160 to 220)	400 to 600 (200 to 315)
Average pore diameter (Å)	22	NA	3,4,5,10
Minimum dew point temperature of effluent, °F (°C)[a]	−80 (−60)	−100 (−75)	−150 (−100)
Average minimum moisture content of effluent gas, ppmv	5 –10	10 – 20	0.1

[a] As reported by Blachman and McHugh (2000).

Silica gels are used mostly where a high concentration of water (>1 mol%) vapor is present in the feed, and low levels of water in the dehydrated gas are not needed. They are relatively noncatalytic compounds. Aluminas are very polar and strongly attract water and acid gases. They are used for moderate levels of water in the feed when low levels of water in the product are not required. They have the highest mechanical strength of the adsorbents considered here. However, for gas going into cryogenic processing, the only adsorbent that can obtain the required dehydration is a molecular sieve. Of these, 4A is the most common, but the smaller pore 3A is sometimes used. It has the advantage of being a poorer catalyst for generation of COS if both H_2S and CO_2 are present* because a

---

* Carbonyl sulfide (COS) is formed in the following reaction: $H_2S + CO_2 \leftrightarrow H_2O + COS$. The equilibrium constant for the reaction is of the order of magnitude of 10^{-6} at adsorption temperatures but increases to 10^{-4} at regeneration temperatures (Trent et al., 1993). Its concentration in feed gas is normally extremely low.

portion of the more active sodium cations in 4A has been replaced with potassium. If both oxygen and H_2S are present 3A reduces the production of elemental sulfur that can block adsorbent pores. However, plant operators usually have little incentive to use 3A for dehydrating gas going to hydrocarbon recovery. Blachman and McHugh (2000) discuss use of multiple adsorbents in the same bed for applications in which both higher water concentrations and acid gases are present.

6.3.2.3 Adsorption Process

Although this discussion uses molecular sieve as the example of an adsorbent to remove water, with the exception of regeneration temperatures, the basic process is the same for all gas adsorption processes. Figure 6.8 shows a schematic of a two-bed adsorber system. One bed, adsorber #1 in Figure 6.8, dries gas while the other bed, adsorber #2, goes through a regeneration cycle. The wet feed goes through an inlet separator that will catch any entrained liquids before the gas enters the top of the active bed. Flow is top-down to avoid bed fluidization. The dried gas then goes through a dust filter that will catch fines

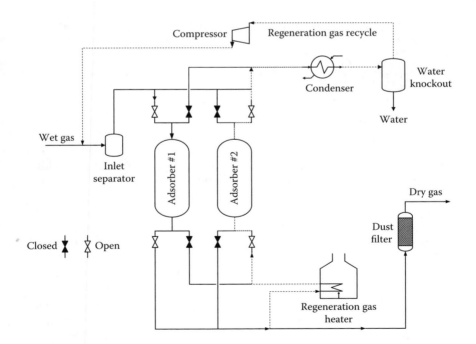

FIGURE 6.8 Schematic of a two-bed adsorption unit. Valving is set to have absorber #1 in drying cycle and absorber #2 in regeneration cycle. (Adapted from Engineering Data Book, 2004b.)

before the gas exits the unit. This filter must be kept working properly, especially if the gas goes on to a cryogenic section with plate-fin heat exchangers, as dust can collect in the exchangers and reduce heat transfer and dramatically increase pressure drop.

Figure 6.8 shows a slip stream of dry gas returning to the bed that is being regenerated. (Sales gas is sometimes used instead of a slip stream. The sales gas stream has the advantage of being free of heavier hydrocarbons that can cause coking.) This gas is usually about 5 to 10% of gas throughput. Regeneration involves heating the bed, removing the water, and cooling. For the first two steps, the regeneration gas is heated to about 600°F (315°C) to both heat the bed and remove adsorbed water from the adsorbent. If COS formation is a problem, it can be mitigated by lowering regeneration temperatures to 400 to 450°F (200 to 230°C) or lower, provided sufficient time for regeneration is available, or by switching to 3A. Regeneration gas enters at the bottom of the bed (countercurrent to flow during adsorption) to ensure that the lower part of the bed is the driest and that any contaminants trapped in the upper section of the bed stay out of the lower section. The high temperature required makes this step energy intensive and in addition to furnaces, other heat sources (e.g., waste heat from gas turbines that drive compressors) are used when possible. The hot, wet regeneration gas then goes through a cooler and inlet separator to remove the water before being recompressed and mixed with incoming wet feed. To complete the regeneration, unheated regeneration gas passes through the bed to cool before it is placed in drying service. Gas flow during this step can be concurrent or countercurrent.

The Engineering Data Book (2004b) recommends that the bed pressure not be changed more than 50 psi/min (6 kPa/s). Therefore, if the adsorption process operates at high pressure, regeneration should take place at as high a pressure as possible to reduce the time needed for changing the pressure. However, as Malino (2004) points out, higher pressures increase the amount of water and hydrocarbons that condense at the top of the bed and fall back onto the adsorption bed. This unavoidable refluxing is a major cause of bed aging, as it leads to adsorbent breakdown and subsequent fines agglomeration (Richman, 2005). The caking leads to higher pressure drop. Condensation at the bed walls can also occur, which can cause bed channeling.

Table 6.5 lists design parameters that are guidelines for typical molecular sieve dehydrators (Chi and Lee, 1973; Cummings, 1977; Engineering Data Book, 2004b; Kohl and Nielsen,1997; Lukchis [date unknown]; Petty, 1976; UOP,1991; Trent, 2001). Both the Engineering Data Book (2004b) and Trent (2001) give extensive details on designing adsorber systems.

The combination of feed rate, pressure drop, and adsorbent crush strength dictates the adsorption bed geometry. As noted in the above discussion regarding minimizing MTZ thickness, the bed diameter should be kept small. This feature also reduces the wall thickness of the high-pressure vessels and increases the superficial velocity, which improves mass transfer in the gas phase. However, it does not affect intraparticle mass transfer, which is the slower of the two processes.

TABLE 6.5
Typical Operating Conditions for Molecular Sieve Dehydration Units

Feed rate	10 to 1500 MMscfd (0.3 to 42 MMSm³/d)
Superficial velocity	Approximately 30 to 35 ft/min (9 to 11 m/min)
Pressure drop	Approximately 5 psi (35 kPa), not to exceed 10 psi (69 kPa)
Cycle time	Four to 24 hours; 8 or a multiple thereof is common
Temperatures and pressures	
Adsorption	
	Temperatures: 50 to 115°F (10 to 45°C)
	Pressures: to 1500 psig (100 barg),
Regeneration	
	Temperatures: 400 to 600°F (200 to 315°C)
	Pressures: Adsorption pressure or lower.

Source: See text for references.

Thus, higher velocities increase the MTZ thickness. Accurate calculation of the MTZ thickness is complex. Trent (2004) suggests the following dimensioned equation for estimating the thickness of the MTZ, L_{MTZ}, in feet of 4×8 (1/8-inch [3 mm] diameter) mesh beads:

$$L_{MTZ} \text{ (ft)} = 2.5 + 0.025 \ V_S \text{ (ft/min)}, \tag{6.3}$$

where V_S is the superficial gas velocity in ft/min. For 8×12-mesh (1/16-inch) beads, the length is about 70% of the values calculated by Equation 6.3. (Multiply m/s by 197 to obtain ft/min and feet by 0.30 to obtain meters.) The Engineering Data Book (2004b) suggests the following equation:

$$L_{MTZ} \text{ (ft)} = F \ [V_S(\text{ft/min})/35]^{0.3}, \tag{6.4}$$

where the factor $F = 1.70$ ft for a 1/8-inch (3-mm) sieve and 0.85 for 1/16-inch (1.5-mm) material. Over the typical gas flow ranges, Equation 6.3 gives an L_{MTZ} double that of Equation 6.4.

As noted above, higher velocities increase pressure drop through the bed. This pressure drop has two adverse effects:

- Higher inlet compression discharge pressures to maintain the same refrigeration requirements and outlet pressure
- Increased mechanical load on the adsorbent, which leads to particle breakdown and causes further increases in pressure drop

The adsorbent beds typically have a 6-inch (15-cm) deep layer of inert 1/2 - to 1-inch (13- to 25-mm) diameter alumina or ceramic balls (density of about

80 lb/ft³ [1,200 kg/m³]) resting on a floating screen at the top of the bed. This layer is on top of another floating screen-supported layer of 1/8-inch (3 mm) diameter beads. These layers help distribute the incoming gas flow but, more importantly, help keep the bed from shifting. The screens float to account for thermal expansion during regeneration. The bottom of the bed has a similar layering, with the smaller beads on top of the larger ones, which are supported by a fixed screen. Some manufacturers offer molecular sieves that are more resistant to the attrition caused by refluxing at the top of the bed (Richman, 2005).

The most common reasons for replacing a bed are loss of adsorbent capacity and unacceptable pressure drop, which usually occur simultaneously. Values for the loss of capacity with time vary considerably, but common values used for molecular sieves in dehydration service are a 35% capacity loss over a 3 to 5 year period or a 50% loss in approximately 1,600 cycles. Typically, a rapid loss occurs in the beginning and a gradual loss thereafter. The adsorbent decays primarily because of carbon and sulfur fouling and caking caused by instability in the clay binder. These effects occur during bed regeneration. de Bruijn et al. (2001) provides a complete discussion of these phenomena.

Increased pressure drop is usually caused by breakdown of adsorbent into finer particles and by caking at the top of the bed because of refluxing. Attrition can occur when the pressure is increased or decreased after or before regeneration. Monitoring the pressure drop is important, as it provides a good diagnostic to bed health. The Engineering Data Book (2004b) recommends a modified form of the Ergun equation to compute pressure drop:

$$\Delta P/L \text{ (psi/ft)} = B\mu V_S + C\rho V_S^2, \tag{6.5}$$

with viscosity, μ, in centipoises, density, ρ, in lb/ft³, and superficial velocity, V_S, in ft/min (multiply psi/ft by 22.62 to obtain kPa/m and m/s by 197 to obtain ft/min). The coefficient values for typical adsorbents are given in Table 6.6.

A number of factors affect bed capacity. For the commonly used 4A molecular sieve, the Engineering Data Book (2004b) suggests that the design water content of a molecular sieve when at equilibrium with saturated gas at 75°F

TABLE 6.6
Coefficient Values for Typical Adsorbents

Particle Type	B	C
1/8-inch (3-mm) bead	0.0560	0.0000889
1/8-inch (3-mm) pellets	0.0722	0.000124
1/16-inch (1.5-mm) bead	0.152	0.000136
1/16-inch (1.5-mm) pellets	0.238	0.000210

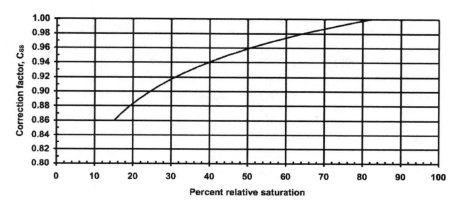

FIGURE 6.9 Molecular-sieve capacity correction for unsaturated inlet gas. (Adapted from Engineering Data Book, 2004b.)

(24°C) will be 13 lb H_2O/100 lb sieve compared with a new molecular sieve, which holds about 20 lb H_2O/100 lb sieve. Two factors affect this number: water content of entering gas and adsorption temperature. Figure 6.9 (Engineering Data Book 2004b). Provides a correction for the capacity as a function of percent relative saturation. Figure 6.10 (Engineering Data Book 2004b) gives a correction for the decrease in bed capacity at temperatures greater than 75°F (24°C). The effect of unsaturated gas and higher temperature can be computed by

$$C_{SS} = 0.636 + 0.0826 \ln(\text{Sat}) \qquad (6.6)$$

and

$$C_T = 1.20 - 0.0026 \; t(°F) \qquad (6.7a)$$
$$C_T = 1.11 - 0.0047 \; t(°C), \qquad (6.7b)$$

where C_{SS} and C_T are correction factors for subsaturation and temperature, respectively. Sat is the percent of saturation.

Example 6.2 An existing 4A molecular sieve bed has been processing 80 MMscfd on a 12-hour cycle with two beds. Exit gas goes to a cryogenic turboexpander section. Gas flow is increased to 100 Mscfd. Estimate the increased pressure drop and determine whether the bed capacity allows continued operation on a 12-hour cycle or the cycle time should be changed. The gas enters the bed at 120°F and 950 psig. Water content is 60% of saturation at 120°F. The molar mass of the gas is 18.5, with a viscosity of 0.014 cP and a compressibility factor of 0.84.

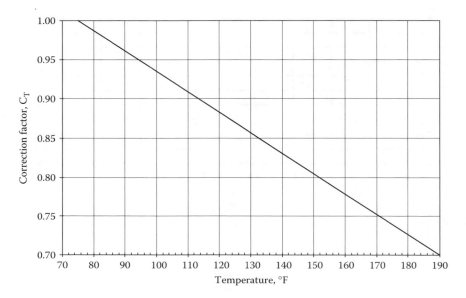

FIGURE 6.10 Correction for decrease in adsorption capacity caused by inlet-gas temperature in molecular sieve. (Adapted from Engineering Data Book, 2004b.)

The adsorption bed contains 41,000 lbs of 1/8-inch diameter beads with a bulk density of 44 lb/ft³. The inside wall diameter of the bed is 7.5 ft. The absorbent was installed 2 years ago.

Pressure-Drop Calculation

Determine the height of the adsorption bed. The volume of adsorbent V = mass/adsorbent density = 41,000/44 = 932 ft³. The cross-sectional area of the bed $A = \pi D^2/4 = \pi(7.5)^2/4 = 44.2$ ft². The bed height then is $V/A = 932/44.2 = 21.1$ ft.

Gas density,

$\rho = MW\ P/(zRT) = 18.5 \times (950 + 14.7)/[\ 0.84 \times 10.73\ (460 + 120)] = 3.41$ lb/ft³.

To obtain superficial velocity, we first need the increased actual volumetric flow rate, Q (Equation 4.18a):

$$Q = scfm\left(\frac{14.7}{P_1(psia)}\right)\left(\frac{T_1(°R)}{520}\right)\left(\frac{z_1}{z_R}\right)$$

$$= \frac{100 \times 10^6}{1440}\left(\frac{14.7}{950+14.7}\right)\left(\frac{460+120}{520}\right)\left(\frac{0.84}{1}\right) = 992\ \text{ft}^3/\text{min}.$$

The superficial velocity is then $V_S = Q/A = 992/44.2 = 22.4$ ft/min.

Use of Equation 6.5, $\Delta P/L$ (psi/ft) $= B\mu V_S + C\rho V_S^2$, with $B = 0.0560$ and $C = 0.0000889$ gives

$$\Delta P/L = (0.0560)(0.014)(22.4) + (0.000089)(3.41)(22.4)^2 = 0.170 \text{ psi/ft.}$$

Total bed pressure-drop is 0.170 psi/ft × 21.1 ft = 3.58 psi, which is in a good operating range.

Bed Capacity Calculation

To determine the capacity, first calculate the capacity per pound of absorbent. The bed has aged, so at 75°F the absorbent should hold 13 lb water/100 lb of sieve if the entering gas is saturated with water vapor. However, the gas enters at 60% of saturation. With either Figure 6.9 or Equation 6.6, the absorbent holds only 97% of capacity or 12.6 lb water/100 lb sieve. The temperature effect from Figure 6.10 or Equation 6.7 is to reduce capacity to 88%, so the bed should hold 12.6 × 0.88 = 11.1 lb water/100 lb sieve in the equilibrium zone.

The length of the MTZ is calculated by either Equation 6.3 or Equation 6.4, but use the former because it is more conservative

$$L_{MTZ} \text{ (ft)} = 2.5 + (0.025 \times 22.4) = 3.1 \text{ ft.}$$

This value represents 15% (3/21) of the bed height. Assume the MTZ holds 50% of the equilibrium loading; the bed should hold

$$11.1/100 \times 41,000(0.85 + 0.15 \times 0.5) = 4,210 \text{ lb of water.}$$

From Figure 6.1a, the water content at saturation for a gas at 950 psig and 120°F is 100 lb/MMscf. Essentially, all water into the bed must be removed. The gas enters at 60% of saturation, which is 0.6 × 100 = 60 lb/MMscf. For a 12-hour cycle, the water adsorbed is 60 × 100 × 0.5 = 3,000 lb water. The bed is slightly oversized but can remain on a 12-hour cycle.

The major operating costs of adsorption are the energy required for regeneration and the compression power required to overcome bed pressure drop. To minimize the heat load, the adsorption beds are insulated. Insulation may be external or internal. Internal insulation saves energy during bed regeneration because it eliminates heating of the vessel walls and reduces the regeneration time. However, insulation imperfections and cracks may cause wet gas to bypass the adsorbent. Internal insulation also requires a larger diameter pressure vessel, which adds to capital costs.

Regeneration usually takes 8 or more hours with 50 to 60% of the time involved in heating and driving off water. The balance of time is in cooling down the bed and having about 30 minutes of standby and switching.

The Engineering Data Book (2004b) recommends the following simple and conservative technique for estimating the heat required for regeneration, q.

$$q_W = (1800 \text{ Btu/lb})(\text{lbs of water on the bed}) \qquad (6.8a)$$
$$q_W = (4200 \text{ kJ/kg})(\text{kg of water on the bed}) \qquad (6.8b)$$

$$q_{si} = \text{(lb of sieve)}(0.24 \text{ Btu/lb } °\text{F})(T_{rg} - T_1) \qquad (6.9\text{a})$$
$$q_{si} = \text{(kg of sieve)}(1.0 \text{ kJ/kg K})(T_{rg} - T_1) \qquad (6.9\text{b})$$
$$q_{st} = \text{(lb of steel)}(0.12 \text{ Btu/lb } °\text{F})(T_{rg} - T_1) \qquad (6.10\text{a})$$
$$q_{st} = \text{(kg steel)}(0.5 \text{ kJ/kg K})(T_{rg} - T_1), \qquad (6.10\text{b})$$

where q_w, q_{si}, and q_{st} represent the heat required to desorb the water, heat the adsorbent, and heat the vessel walls, assuming external insulation, respectively. T_{rg} is the regeneration temperature of the gas that enters the bed minus 50°F (28°C), and T_1 is the inlet gas temperature. The reduction in T_{rg} accounts for incomplete heating of the bed because the regeneration is stopped before the entire bed reaches the regeneration gas temperature. The total estimated heat of regeneration is then

$$q = 2.5 \, (q_w + q_{si} + q_{st}) \, (1.10) \qquad (6.11)$$

where the factor of 2.5 corrects for the change in temperature difference (in − out) across the bed with time during the regeneration cycle; 40% of the heat is actually transferred to the bed and walls and the rest is lost in the exiting gas. The 1.10 assumes a 10% heat loss from the system to the surroundings.

Example 6.3 Estimate the heat required to regenerate an adsorption bed that holds 40,000 lb of 4A molecular sieve and 4,400 lb of water. The vessel contains 55,000 lb of steel, and the regeneration temperature is 600°F. The bed operates at 100°F. Also estimate the gas flow rate under the assumption that the C_P for the gas is 0.68 Btu/lb-°F, that 60% of the regeneration time involves heating the bed, and that the bed is on an 8-hour cycle. The gas leaves the regeneration gas heater at 650°F.

Use Equations 6.8 through 6.11 to compute the heat load. The temperature $T_{rg} = 600 − 50 = 550°F$.

Water: $q_w = (1,800 \text{ Btu/lb})(\text{lbs of water on the bed}) = 1,800 \times 4,400 = 7,920 \text{ Btu}$

Sieve: $q_{si} = \text{(lb of sieve)}(0.24 \text{ Btu/lb-°F})(T_{rg} - T_1) = 40,000 \times 0.24 \, (550 - 100)$
$\qquad = 4,320 \text{ Btu}$

Steel: $q_{st} = \text{(lb of steel)}(0.12 \text{ Btu/lb-°F})(T_{rg} - T_1) = 55,000 \times 0.12 \, (550 - 100)$
$\qquad = 2,970 \text{ Btu}$

Use Equation 6.11 to obtain the total heat requirement.

$q = 2.5 \, (q_w + q_{si} + q_{st}) \, (1.10) = 2.5 \, (7,920 + 4,320 + 2,970)(1.10) = 41,800 \text{ Btu.}$

Regeneration Gas Flow Rate Calculation

Time allocated for heating, Θ, is $0.6 \times 8 = 4.8$ hours. Then the gas flow rate is

$$\dot{m} = q/[C_P \; \Theta \; (t_{Hot} - t_B)] \; = \; 41,800/[0.68 \times 4.8 \, (650 - 100)] = 23,300 \text{ lb/h.}$$

Occasionally the adsorbent beds produce increased fines in the exit gas. The fines usually result from attrition of the bottom support balls and indicate failure of the mechanical supports at the bottom of the bed (Richman, 2005).

The most common bed configuration is the two-bed system shown in Figure 6.8. Trent (2004) notes that many three-bed systems were set up such that two beds would be run in series, and the second bed would be used as a guard bed. However, this arrangement is not used that much now, and the primary advantage of the third bed is that regeneration gas requirements can be reduced by regenerating beds placed in series. Also, three beds provide reduced heating requirements because regeneration gas can be used to transfer heat from a cooling bed to a warming bed. Trent (2004) suggests the best approach for four-bed systems is to run two beds in parallel. For a given gas throughput, the pressure drop is reduced to one eighth of the two-bed system.

6.3.2.4 Other Factors That Affect the Adsorption Process

Trace amounts of oxygen affect bed life and performance in a variety of ways. At the normal regeneration gas temperature of 600°F for molecular sieves, 2 moles of oxygen react with methane to form 2 moles of water and 1 mole of CO_2. As this reaction is exothermic, higher amounts of oxygen in the gas can lead to temperatures above the design temperature of the molecular sieve vessel. When oxygen is present, the temperature of the beds during regeneration must be monitored for safety reasons (McCartney, 2005). Clark, et al. (1975) found that oxygen undergoes partial oxidation reactions with heavier hydrocarbons, which are adsorbed in the binder, to form alcohols and carbocylic acids that ultimately turn to water and CO_2. If H_2S is present, it undergoes oxidation to elemental sulfur, sulfur dioxide, and water. Trent (2001) points out that oxygen concentrations greater than 20 ppmv also generate olefins that become coke in the bed. These reactions reduce molecular sieve capacity by forming solid deposits and by causing incomplete removal of water during regeneration because the partial pressure of water is higher (see Figure 6.6).

To avoid the above reactions as well as COS formation, regeneration temperatures are lowered to the 300 to 375°F range. However, this range increases the required regeneration time and the amount of regeneration gas used, which increases recompression cost.

As in all processes, ensuring that the beds are protected from entrained water and hydrocarbons is important. Even trace amounts of entrained water load the bed quickly and increase the regeneration heat load. If the gas comes to the dehydration unit fully saturated, which is often the case, cooling the gas and removing the condensed water before the gas enters the bed lowers water loading and potentially increases the drying-cycle time. Trent (2001) notes that cooling a water-saturated gas by 20°F (10°C) drops the saturated gas-phase water content by 50%.

Example 6.4 Gas, with a molar mass of 23, leaves an amine treater water saturated at 100°F at 1,000 psia and enters a molecular-sieve dehydrator. How much would the water load on the adsorption bed be decreased if the gas were cooled to 80°F before entering the adsorber? Assume that all of the condensed water is removed.

Figure 6.1a shows the water content of gas at 1,000 psia and at 100 and 80°F to be 64 and 33 lb/MMscf, respectively. The molar mass and temperature are low enough that the gravity correction is only 1% and can be neglected. The reduction in water content is $(64–33)/64 = 0.48$. Thus, if all mist is removed, a 48% reduction of the water load would occur on the adsorption bed. Recall from Figure 6.10 or Equation 6.7 that the cooler gas also increases the bed capacity as the temperature-correction factor increases from 0.94 to 0.99. Thus, cooling the gas by 20°F (1°C) would essentially double the cycle time.

6.3.3 DESICCANT PROCESSES

In some situations, such as remote gas wells, use of a consumable salt desiccant, such as $CaCl_2$, may be economically feasible. The system can reduce the water content down to 20 ppmv. Typical salt capacities are 0.3 lb $CaCl_2$ per lb H_2O. The Engineering Data Book (2004b) provides more details.

6.3.4 MEMBRANE PROCESSES

Membranes offer an attractive option for cases in which drying is required to meet pipeline specifications. Their modular nature, light weight, large turndown ratio, and low maintenance make them competitive with glycol units in some situations.

Feed pretreatment is a critical component of a membrane process (see Chapter 5). The inlet gas must be free of solids and droplets larger than 3 microns. Inlet gas temperature should be at least 20°F (10°C) above the dew point of water to avoid condensation in the membrane.

Units operate at pressures up to 700 to 1,000 psig (50 – 70 barg) with feed gases containing 500 to 2,000 ppmv of water. They produce a product gas stream of 20 to 100 ppmv and 700 to 990 psig (48 to 68 barg). The low-pressure (7 to 60 psig [0.5 to 4 barg]) permeate gas volume is about 3 to 5% of the feed gas volume. This gas must be recompressed or used in a low-pressure system such as fuel gas.

Smith (2004) suggests that membranes used for natural gas dehydration are economically viable only when dehydration is combined with acid-gas removal. On the basis of commercial units installed and several studies (Binci et al. [undated]; Bikin et al., 2003), membranes are economically attractive for dehydration of gas when flow rates are less than 10 MMscfd (0.3 MMSm³/d). Binci et al.(undated) claim that membrane units are competitive with TEG dehydrators on offshore platforms at flows below 56 MMscfd (1.6 MMSm³/d). Certainly, the reliability and simplicity of membranes make them attractive for offshore and remote-site applications, provided the low-pressure permeate gas is used effectively. An added benefit compared with TEG units is the absence of BTEX emissions with membranes.

6.3.5 Other Processes

Three relatively new processes are worth mentioning. The first process is a refrigeration process that mixes methanol with the gas and cools the gas to very low temperatures. The water–methanol mixture drops out and the methanol is recovered in a stripper column. The process has several major advantages:

- It can obtain dew points in the −100 to −150°F (−70 to −100°C) range.
- It requires no heat input other than to the methanol regenerator.
- It requires no venting of hydrocarbon-containing vapors.

However, it requires external refrigeration to cool the gas, and minimal methanol losses occur in the stripper. The Engineering Data Book (2004b) provides more details.

The second process is the Twister technology, which is discussed in Chapter 7. It has been considered attractive in offshore applications (Wilson and Yuvancic, 2004) for dehydration because of its simplicity (no moving parts) along with its small size and weight. Brouwer et al. (2004) discuss the successful implementation on an offshore platform. Some offshore field pressures are greater than 2,000 psi (140 bar), so recompression is not needed with the unit where overall pressure drop is 20 to 30%.

The third process is the vortex tube technology, which also is discussed in Chapter 7. It also has no moving parts. According to vendor information, it is used in Europe in conjunction with TEG addition to remove water from gas stored underground. We found no examples of its use in gas plants.

6.3.6 Comparison of Dehydration Processes

A number of factors should be considered in the evaluation of a dehydration process or combination of processes. If the gas must be dried for cryogenic liquids recovery, molecular sieve is the only long-term, proven technology available. It has the added advantage that it can remove CO_2 at the same time. If CO_2 is being simultaneously removed, because water displaces CO_2, the bed must be switched before the CO_2 breaks through, which is before any water breakthrough. Enhanced TEG regeneration systems may begin to compete with molecular sieve. Skiff et al. (2003) claim to have obtained less than 0.1 ppmv water by use of TEG with a modified regeneration system that uses about 70% of the energy required for molecular sieves.

High inlet water-vapor concentrations make molecular sieve dehydration expensive because of the energy consumption in regeneration. Two approaches are used to reduce the amount of water going to the molecular sieve bed. First, another dehydration process, (e.g., glycol dehydration) is put in front of the molecular sieve bed. The second option is to have combined beds with silica gel or activated alumina in front of the molecular sieve. The bulk of the water is removed with the first adsorbent, and the molecular sieve removes the remaining water. This configuration reduces the overall energy required for regeneration.

If dehydration is required only to avoid free-water formation or hydrate formation or to meet the pipeline specification of 4 to 7 lb/MMscf (60 to 110 mg/Sm³), any of the above-mentioned processes may be viable. Traditionally, glycol dehydration has been the process of choice. System constraints dictate which technology is the best to use. Smith (2004) provides an overview of natural gas dehydration technology, with an emphasis on glycol dehydration.

When considering susceptibility to inlet feed contamination, one should keep in mind that replacing a solvent is much easier and cheaper than changing out an adsorbent bed. However, prevention of contamination by use of properly designed inlet scrubbers and coalescing filters, if required, is the best solution. In a conventional gas plant, where inlet fluctuations are handled in inlet receiving, feed contamination is generally limited to possible carryover from the sweetening unit. However, in field dehydration the possibility exists of produced water, solids, oil, and well-treating chemicals entering the dehydrator. Wieninger (1991) discusses how these components detrimentally affect glycol dehydrators.

6.4 SAFETY AND ENVIRONMENTAL CONSIDERATIONS

Dehydration processes offer few safety considerations outside of having high-temperature and high-pressure operations. Probably the most unique safety consideration is when adsorbent beds are being changed. The bed must be thoroughly purged, preferably with nitrogen, to remove adsorbed hydrocarbons before the adsorbent is dumped. A potential exists for hydrocarbons on the adsorbent to ignite when exposed to air because the adsorbent heats as it adsorbs moisture from the air. Either a highly trained company expert or an adsorbent company representative should be present to help ensure safe dumping and filling operations. The dumping process produces dust, and operators must wear protective clothing and dust masks.

A major environmental concern in dehydration with glycol solutions is BTEX emissions. Chapter 9 discusses this issue in more detail. Disposal of spent $CaCl_2$ brine must be in accordance with local environmental regulations. Spent adsorbents, if properly regenerated, may be discarded into normal land fills. Ethylene glycol is toxic to humans and must be handled properly.

REFERENCES

Bikin, B., Giglia, S, and Hao, J., Novel Composite Membranes and Process for Natural Gas Upgrading, Annual Report to Department of Energy, DE- FC26-99FT40497, March 2003, www.osti.gov/bridge/servlets/purl/823967-PH6Fq7/native/, Retrieved July 2005.

Binci, F., Ciarapica, F.E., and Giacchetta, G., Natural Gas Dehydration in Offshore Rigs. Comparison Between Traditional Glycol Plants and Innovative Membrane Systems, www.membrane.unsw.edu.au/imstec03/content/papers/IND/imstec033.pdf, Retrieved July 2005.

Blachman, M. and McHugh, T., Sour Gas Dehydration Technology and Alternatives, Proceedings of the Laurance Reid Gas Conditioning Conference, Norman, OK, 2000.

Brouwer, J.M., Bakker,G., Verschoof, H.-J., and Epsom, H.D., Supersonic Gas Conditioning First Commercial Offshore Experience, Proceedings of the Eighty-Third Annual Convention of the Gas Processors Association, Tulsa, OK, 2004.

Chi, C.W. and Lee, H., Natural gas purification by 5A molecular sieves and its design method, *AIChE Symposium Series,* **69** (134) 95, 1973.

Clark, K.R., Corvini, G., and Bancroft, W.G., Molecular Sieve Treating of Natural Gas Containing Oxygen, Proceedings of the Laurance Reid Gas Conditioning Conference, Norman OK, 1975.

Cummings, W. and Chi, C.W., Natural Gas purification with molecular sieves, *presented at the Canadian Chemical Engineering Conf.,* Calgary, Alberta, Canada, October 23–26, 1977.

de Bruijn, J.N.H. and van Grinsven, P.F.A., Otimizing the on stream times of a mol sieve dehydration unit, *Proceedings of the Laurance Reid Gas Conditioning Conference,* Norman OK, 2001.

Dow Chemical Company, Triethylene Glycol, 2003, www.dow.com/PublishedLiterature/dh_0451/09002f13804518f1.pdf, Retrieved September 2005.

Engineering Data Book, 12th ed., Sec. 2, Product Specifications, Gas Processors Supply Association, Tulsa, OK, 2004a.

Engineering Data Book, 12th ed., Sec. 20, Dehydration, Gas Processors Supply Assocition, Tulsa, OK, 2004b.

Hernandez-Valencia, V.N., Hlavinka, M.W., and Bullin, J.A., Design Glycol Units for Maximum Efficiency, Proceedings of the Seventy-First Annual Convention of the Gas Processors Association, Tulsa, OK, 1992, 310.

Kruka, V.R., private communication, 2005.

Kohl, A. and Nielsen, R., *Gas Purification*, 5th ed., Gulf Publishing, Houston, TX, 1997.

Lukchis, G.M. Adsorption systems, Part 1: Design by Mass-Transfer Zone concept, Part II, Equipment Design, and Part III Adsorption Regeneration. *UOP Brochure XF04A.*

Malino, H.M., Fundamentals of Adsorptive Dehydration, Proceedings of the Laurance Reid Gas Conditioning Conference, Norman OK, 2004, 61.

Masaoudi, R., Tohidi, B., Anderson, R., Burgass, R.W., and Yang, J., Experimental measurements and thermodynamic modeling of clathrate hydrate equilibria and salt solubility in aqueous ethylene glycol and electrolyte solutions, *Fluid Phase Equil.*, 31, 219, 2004.

Parrish, W.R., Won, K.W., and Baltatu, M.E., Phase Behavior of the Triethylene Glycol-Water System and Dehydration/Regeneration Design for Extremely Low Dew Point Requirements, Proceedings of the Sixty-Fifth Annual Conversion Gas Processors Association, Tulsa, OK, 1986, 202.

Parrish, W.R., Thermodynamic Inhibitors in Brines, Final Report, DeepStar CTR 4210, 2000. (Contact Adminstrator@DeepStar.org for information.)

Petty, L.E., Practical aspects of molecular sieve unit design and operation. *Proceedings of the Fifty-fifth annual convention of the Gas Processors association,* Tulsa OK, 1976.

Prausnitz, J.M., Lichtenthaler, R.N., Azevedo, E.G., *Molecular Thermodynamics of Fluid-Phase Equilibria*, Prentice-Hall, Englewood Cliffs, NJ, 1999.

Richman, P., private communication, 2005.

Skiff, T., Szuts, A., Szujo, V., and Toth, A., Drizo Unit Competes with Solid Bed Desiccant Hydration, Proceedings of the Laurance Reid Gas Conditioning Conference, Norman OK, 2003, 213.

Smith, J.M., Van Ness, H.C., and Abbott, M.M., *Introduction to Chemical Engineering Thermodynamics*, 6th ed., McGraw-Hill, New York, 2001.

Smith, R.S., Custom Glycol Units Extend Operating Limits, Proceedings of the Laurance Reid Gas Conditioning Conference, Norman OK, 1993, 101.

Smith, R.S., Fundamentals of Gas Dehydration Inhibition/Absorption Section, Proceedings of the Laurance Reid Gas Conditioning Conference, Norman OK, 2004, 17.

Smith, R.S. and Humphrey, S.E., High Purity Glycol Design Parameters And Operating Experience, Proceedings of the Laurance Reid Gas Conditioning Conference, Norman OK, 1995, 142.

Trent , R.E., Craig, D.F., and Coleman, R.L., The Practical Application of Special Molecular Sieves to Minimize the Formation of Carbonyl Sulfide During Natural Gas Dehydration, Proceedings of the Laurance Reid Gas Conditioning Conference, Norman OK, 1993, 239.

Trent, R.E., Dehydration with Molecular Sieves, Proceedings of the Laurance Reid Gas Conditioning Conference, Norman OK, 2001.

Trent, R.E., Dehydration with Molecular Sieves, Proceedings of the Laurance Reid Gas Conditioning Confereence, Norman OK, 2004, 75.

Trimble, H.M., Solubilities of salts in ethylene glycol and in its mixtures with water, *Ind. Eng. Chem.* 23, 165, 1931.

Wallace, C.B., Dehydration of supercritical CO2, Proceedings of the Laurance Reid Gas Conditioning Conference, Norman OK, 1985.

Wichert, E., private communication, 2005.

Wieninger, P., Operating Glycol Dehydration Systems, Proceedings of the Laurance Reid Gas Conditioning Conference, Norman OK, 1991, 23.

Wilson, J.L. and Yuvancic, J., Process Selection for Dehydrating Gulf of Mexico Offshore Platform Gas, Proceedings of the Laurance Reid Gas Conditioning Conference, Norman OK, 2004, 125.

7 Hydrocarbon Recovery

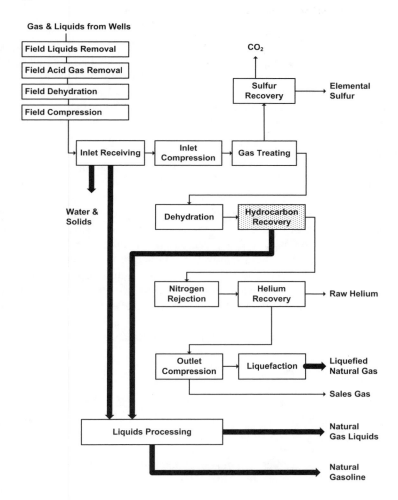

7.1 INTRODUCTION

Pipeline quality natural gas specifications include limits on sulfur and water content, along with higher heating value, which must be about 950 to 1,150 Btu/scf (35,400 to 42,800 kJ/Sm^3) (Engineering Data Book, 2004a). Exact limits are set by negotiation between the processor and the purchaser. The previous two chapters addressed the first two specifications. This chapter addresses the heating value.

165

Unless the treated gas contains high concentrations of inerts (N_2, CO_2), the heating value may be too high because of the C_2+ fraction present. This chapter discusses hydrocarbon recovery methods to both lower the heating value and create, simultaneously, valuable NGL liquid hydrocarbon products.

Processors have additional reasons for reducing the C_2+ fraction. Hydrocarbon recovery frequently is required in field operations for fuel conditioning or dew point control. As noted in Chapter 3, booster stations in gathering systems may fuel the compressors with raw gas taken directly from the pipeline. Raw gas usually is far too rich, and simple systems are used to lower the heating value (i.e., condition the fuel) by removing heavier hydrocarbons.

Dew point control (or "dew pointing") is necessary when raw gas lines are constrained in liquid content as the liquid reduces gas throughput, causes slugging, and interferes with gas metering. This situation can be a problem for offshore production, where several operators share a common gas line to shore. These lines may operate at 1,800 psig (125 barg) or higher and be at seafloor temperatures of 36 to 40°F (2 to 4°C). Dew point control is also necessary if a potential for condensation is present in a process because of temperature or pressure drops. The latter happens when the gas is in the retrograde condensation region. However, effective dew point control is much less demanding than C_2+ recovery, as it can be accomplished without removal of a large portion of the C_3+ fraction.

After a brief discussion of retrograde condensation, this chapter discusses some common process elements used in conventional hydrocarbon recovery and then describes some common processes that utilize these elements. A myriad of process variations are available, and this chapter illustrates only a few simple process configurations. The chapter also includes brief descriptions of some new technologies.

7.1.1 RETROGRADE CONDENSATION

A major reason for dew point control is the fact that rich natural gas mixtures that contain heavier hydrocarbons exhibit a nonintuitive behavior called retrograde condensation. Figure 7.1 shows the pressure–temperature relationship for a hypothetical (but typical) natural gas mixture that contains 85 mol% methane with 4.8 mol% of C_3+. The envelope is the bubble point–dew point line of the mixture. At any temperature and pressure combination outside the envelope, the mixture is single phase. At temperatures and pressures inside the envelope, two phases exist.

Three points on the envelope are important:

- The cricondentherm, the maximum temperature at which two phases can exist
- The cricondenbar, the maximum pressure at which two phases can exist
- The critical point, the temperature and pressure where the liquid and vapor phases have the same concentration

The retrograde condensation effect can be seen by following the vertical dashed line in Figure 7.1. For the mixture at the temperature and pressure at the top of the

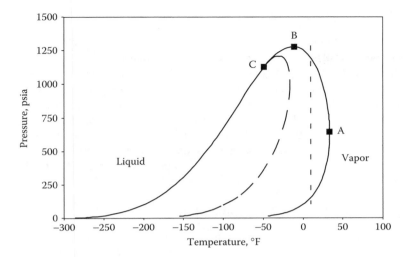

FIGURE 7.1 Pressure–temperature diagram for a hypothetical raw natural gas that contains predominately methane, with trace components up to heptane. The dashed curve represents the vapor-phase line at 95% quality. Points A, B, and C denote the cricondentherm, cricondenbar, and critical point of the mixture, respectively.

line, a single phase exists. Dropping the pressure causes a liquid phase to form (retrograde condensation), which will be present until the pressure is below the envelope. The dotted-line path is similar to what happens in a pipeline because of line pressure drop if pipeline temperature is constant.

The dashed curve inside the envelope denotes the pressure and temperature of the mixture when the vapor quality is 95 mol%. This curve shows the dramatic effect on the phase behavior from condensation of only 5 mol% of the vapor. The cricondentherm of this vapor phase is about 50°F (30°C) lower than the original mixture, and condensation at typical pipeline temperatures would not be possible.

As can be seen from the previous paragraph, the cricondentherm of a mixture strongly depends on the molecular weight of the heavy components. The cricondenbar increases with increased molecular weight. Concentration of the heavy components present is relatively less important than their molecular weight. On the basis of simple flash calculations, a mixture that contains methane and 10 mol% propane has a cricondentherm comparable to a methane–heptane mixture with only 0.06 mol% heptane.

7.2 PROCESS COMPONENTS

The process elements involved in hydrocarbon recovery vary, depending upon the desired products and gas volume being processed as well as inlet composition and pressure. Not all of the elements listed here will be found in all plants, and often it is a choice of one component or another.

7.2.1 EXTERNAL REFRIGERATION

External refrigeration plays a major role in many hydrocarbon recovery processes, as it is used to cool the gas stream to recover a significant amount of C_3+ and to lower gas temperatures as the gas goes into other stages of hydrocarbon recovery. It may be the only source of refrigeration when inlet pressure is low. Adsorption and vapor-compression refrigeration are used in special situations. However, vapor compression using propane as the working fluid is the most common in gas plants and will be discussed here. (LNG facilities also use ethane or ethylene as well as hydrocarbon mixtures as refrigerants. See Chapter 13 for details.) The basic refrigeration cycle is discussed first and is then followed by a look at process variations. The Engineering Data Book (2004d) provides valuable design guidelines that are omitted here.

7.2.1.1 Basic Propane Refrigeration Process

As described in any basic thermodynamics book (e.g., Smith et al., 2001), the refrigeration cycle consists of four steps that are depicted on the pressure–enthalpy chart in Figure 7.2:

1. Compression of saturated refrigerant vapor at point A to a pressure well above its vapor pressure at ambient temperature at point B
2. Condensation to point C by heat exchange with a cooling fluid, usually air
3. Expansion through a valve (Joule-Thomson expansion) to cool and condense the refrigerant to point D
4. Heat exchange with the fluid to be cooled by evaporation of the refrigerant back to point A

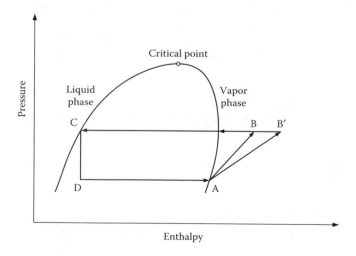

FIGURE 7.2 Schematic of refrigeration cycle on a pressure–enthalpy chart.

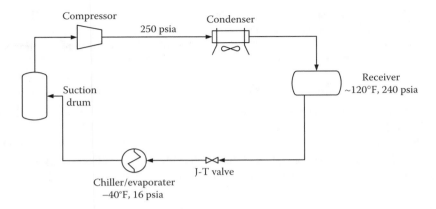

FIGURE 7.3 Single-stage propane refrigeration system. (Adapted from Engineering Data Book, 2004d.)

Figure 7.3 shows the flow diagram for a single-stage propane refrigeration system, with typical operating conditions. Each of the steps is described below.

Compression Step—Cycle analysis begins with propane vapor entering the compressor as a vapor at 14.5 psia (1 bar) and approximately −40°F (−40°C), where it is compressed to 250 psia (17 bar) (point A to point B in Figure 7.2). The power required and compressor discharge temperature depends upon compressor efficiency (see Chapter 4 for details). Plants once used multistaged reciprocating compressors, but oil-injected screw compressors are now preferred because they can complete the compression in one stage. (Large refrigeration units such as those used in LNG plants use centrifugal compressors.)

The work of compression is simply

$$w_S = (h_B - h_A)/\eta_{IS}, \tag{7.1}$$

where η_{IS} is the adiabatic efficiency of the compressor (see Chapter 4). Taking into account compressor nonideality, the actual enthalpy at the end of the expansion is

$$h_{B'} = h_A + w_S. \tag{7.2}$$

Compressor power to the refrigeration system is the product of the mass flow rate and shaft work

$$HP = \dot{m} \cdot w_S. \tag{7.3}$$

Condensation Step—The warm gas goes to an air- or water-cooled condenser, where the propane cools to 100 to 120°F (38 to 50°C), totally condenses, and collects in a receiver (point B′ to point C in Figure 7.2). This step is simply

$$q_{Cond} = h_C - h_{B'}. \tag{7.4}$$

Expansion Step—Propane liquid leaves the receiver and flashes through a J-T valve, where the temperature and pressure drop to −40°F (−40°C) and 16 psia (1 bar) (point C to point D). No change occurs in the enthalpy, but the temperature drops to the saturation temperature of the liquid at the expansion-discharge pressure, and $h_C = h_D$ if there are no heat leaks. If there is a heat leak, q_L, then

$$h_D = h_C + q_L. \tag{7.5}$$

The fraction, f, of propane condensed is computed knowing the initial enthalpy and liquid and vapor enthalpies at the condensation temperature, which for the given case is

$$h_D = h_D^L f + (1 - f) h_D^V .$$

Assuming the vapor leaves the chiller as a saturated vapor,

$$h_D^V = h_A$$

$$f = (h_A - h_D)/(h_A - h_D^L). \tag{7.6}$$

Refrigeration Step—The cold propane then goes to a heat exchanger, the chiller, where it cools the process stream by evaporation (point D to point A in Figure 7.2). Because the propane in the chiller is evaporating, and a minimal heat exchange occurs between cold propane vapor and the inlet gas, the inlet and outlet propane temperature remains constant. The propane returns to the compressor suction slightly above −40°F (−40°C). The heat absorbed by the propane is simply $h_A - h_D$.

In the past, most chillers were the kettle type in which propane is on the shell side and the liquid level is maintained above the tube bundle. Now, other high-performance heat exchangers (e.g., plate-fin) are used (see Section 7.2.3 for more details). The chiller typically has two zones of heat transfer. The first is exchange of boiling propane with gas above its dew point and will involve only sensible heat. The second zone has condensing vapors from the process stream and boiling propane, which gives a much higher overall heat-transfer coefficient (Engineering Data Book, 2004d).

To complete the cycle, the propane vapors leave the chiller and go to the suction drum before being compressed again.

Refrigeration-cycle performance is commonly stated in terms of coefficient of performance (COP), which is the ratio of the refrigeration obtained divided by the work required. On the basis of Figure 7.2, the COP is determined by

$$COP = (h_A - h_D)/(h_B - h_A) = f(h_A - h_D^L)/w_S .\qquad(7.7)$$

Example 7.1 Compute the liquid fraction produced and the COP for the propane refrigeration system on the basis of the conditions given in Figure 7.3. Ignore heat leak into the system and assume the compressor efficiency is 77%. Use the saturation table and PH diagram for propane given in Appendix B for the calculations.

Compression Step—Following the cycle as given above, first calculate the work of compression. To compute compressor work, we use Equation 3.1 and assume that the inlet-gas condition is saturated vapor at –40°F. The work per unit mass (assuming reversible compression) required to compress the propane from 14.5 to 250 psia (from Appendix B) is

$$w_S = (h_B - h_A)/\eta_{IS} = (h_{250\,psia} - h_{14.5\,psia,\,-40°F})/\eta_{IS}$$
$$(240 - 182)/0.77 = 75 \text{ Btu/lb.}$$

Because of the compressor inefficiency, the enthalpy of the vapor that leaves the compressor is

$$h_{B'} = h_{250\,psia} = h_{14.5psia,-40°F} + w_S = 182 + 75 = 257 \text{ Btu/lb,}$$

which corresponds to about 140°F at 250 psia.

Condensation Step—Pressure drop between the condenser inlet and receiver is usually around 8 to 10 psi (0.55 to 0.7 bar). The heat load on the condenser is the change in enthalpy from the heated vapor to condense to all liquid. On the basis of the saturation table, this value is

$$q = h_C - h_{B'} = h_{250\,psia} - h_{242psia,\,120°F} = 257 - 98.7 = 158 \text{ Btu/lb.}$$

To calculate the mass fraction condensed use Equation 7.6

$$f = (h_A - h_D)/(h_A - h_D^L) = (h_{-40°F}^V - h_{240\,psia,120°F})/(h_{-40°F}^V - h_{-40°F}^L)$$
$$(182.2 - 98.7)/(182.2 - 0) = 0.46$$

The COP is, from Equation 7.7:

$$COP = f(h_A - h_D^L)/w_S = f(h_{-40°F}^V - h_{-40°F}^L)/w_S$$
$$= 0.46\,(182.2 - 0)/75 = 1.1$$

7.2.1.2 Alternate Process Configurations

Thermodynamics dictate that to minimize the refrigeration work (i.e., compression required) heat from the chiller should be removed at as high a temperature as possible. One way to reduce compressor duty per unit of refrigeration duty is to multistage the refrigeration process by removal of process heat at more than one temperature. Figure 7.4 shows a two-stage system, with representative operating conditions. In this system, the condensed propane stream expands to about 62 psia (4.3 bar) and 25°F (−4°C) in two parallel J-T valves. One of the expanded streams goes to a chiller before going to the suction surge drum for the second stage of the compressor.

Vapor from the surge drum goes to the second stage of the compressor, while the liquid goes through a second J-T expansion to provide the low-temperature cooling. Pressures are adjusted so that the compression ratio is equal in both compressor stages. Table 7.1 shows the significant savings in going to two stages but less of a benefit when going to three stages.

Multistaging reduces work requirements by removing heat from the process stream at different temperatures. An alternative is removal of heat from the refrigerant before it is expanded. Refrigerant subcooling is sometimes used by exchange of the propane that leaves the receiver with a portion of the cold liquid propane. An Engineering Data Book (2004d) example shows that a 10% reduction in recirculation rate to the second stage of two-stage refrigeration is obtainable when the propane from the condenser is cooled 10°F (5°C).

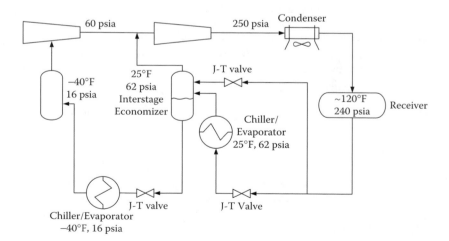

FIGURE 7.4 Two-stage propane refrigeration system, with second heat exchanger and economizer. Units may omit either the first stage heat exchanger or expansion directly to the economizer.

TABLE 7.1
Effect of Multistaging on Condenser and Compression Duty
for Constant Refrigeration Duty with Propane as the Refrigerant

	Number of Stages		
	1	2	3
Change in compressor power (%)	0	−19.2	−23.3
Change in condenser duty (%)	0	−8.2	−9.6

Refrigeration temperature is −40°F (−40°C) and condensing temperature is 100°F (38°C).

Source: Engineering Data Book (2004d).

Although a second heat exchanger provides the most benefit from staging a refrigeration system, savings can be obtained by use of only a single J-T expansion into the economizer or by use of single J-T expansion and the heat exchanger. Gas plants commonly omit the first-stage heat exchanger for cost purposes. However, when energy requirements are critical, such as in an LNG plant, the J-T valve that bypasses the heat exchanger is commonly eliminated and three stages of refrigeration are common. See Chapter 13 for details of more complex heat exchanger systems.

Another way to alter the temperature at which heat is removed is to use refrigerant cascading. In this option, one refrigerant (e.g., propane) is used to remove heat from another refrigerant (e.g., ethane), which then cools the process gas down to as low as −120°F (−85°C). This technique is commonly used in LNG plants and is discussed in Chapter 13. Many gas plant process stream temperatures go well below −120°F (−85°C) but usually rely upon process gas expansion for cooling instead of additional external cooling.

Obviously many combinations are possible for making refrigeration more efficient. However, each must be balanced with the associated additional capital cost, realizable operating cost savings, and operating complexity. Although little incentive exists for more complex systems in gas processing, LNG plants make extensive use of complex refrigeration cycles, as discussed in Chapter 13, to reduce refrigeration (i.e., compression) costs.

7.2.1.3 Effect of Operating Variables on Refrigeration Performance

A number of other process variables affect refrigeration capacity. An obvious one is the condenser outlet temperature, which is limited by the heat removed in the condenser and dictates compressor discharge pressure. If air cooling is used, the temperature there can exhibit wide swings, both seasonal and daily, in ambient and condenser temperatures. Table 7.2 shows the importance the condensing temperature plays on both condenser and compressor duties to maintain a constant

TABLE 7.2
**Effect of Condensed Refrigerant Temperature on Condenser and
Compression Duty for Constant Refrigeration Duty with Propylene
As the Refrigerant**

Condensing Temperature, °F (°C)	60 (15)	80 (27)	100 (38)	120 (49)	140 (60)
Change in Compressor duty (%)	−36.6	−19.8	Base	28.8	66.4
Change in Condenser duty (%)	−16.3	−8.7	Base	13.6	31.5

Source: Engineering Data Book (2004d).

refrigeration capacity when the refrigerant is propylene; results should be comparable for propane.

Another factor that affects performance is refrigerant purity. Propane commonly contains small amounts of ethane. Low concentrations pose no problem. Ethane accumulates in the propane receiver and increases both condensation pressure and temperature. If refrigeration capacity is adversely affected, the propane can be purged into the plant-gas inlet stream for recovery and the system recharged.

Whereas light ends affect performance, heavy components are less of a problem but still must be removed. The most common heavy material is compressor oil. Heavy liquids are removed by periodically taking a slip stream from the bottom of the chiller, which goes to a small heated vessel where the propane is evaporated and returned to the suction of the compressor. The heat source is hot propane vapor from compressor discharge. Butanes may also be present, but unless in large quantities they will not significantly raise the chiller temperature.

7.2.2 TURBOEXPANSION

Until the 1960s, Joule-Thomson expansion was the only way used to cool gas-plant streams by pressure drop. Herrin (1966) describes the first turboexpander plant. The J-T valve, which is essentially a control valve with a variable or fixed orifice, is an extremely simple, inexpensive, and widely used means to reduce gas temperature. Although still extensively used in many applications to produce refrigeration, J-Ts are being widely supplanted by turboexpanders in gas plants for cooling the process stream when it is a gas. Turboexpanders are, in essence, centrifugal compressors that run backwards. Unlike J-T expanders, they perform work during the process. Whereas J-T expansion is essentially an isenthalpic process (therefore, no work is done on or by the gas), an ideal, thermodynamically reversible turboexpander is isentropic. The maximum reversible work required for compression is isentropic, and, conversely, the maximum reversible work recovered by a turboexpander system on expansion is also isentropic. Turboexpansion provides the maximum amount of heat removal from a system for a given pressure drop while generating useful work. The work is used to drive compressors or electrical generators. The equations in Chapter 4 for isentropic compression apply to the turboexpander.

The major breakthrough for turboexpanders came when the design and materials made it possible for condensation to occur inside the expander. The fraction condensed can be up to 50% by weight (Jumonville, 2004). However, the droplets must generally be 20 microns in diameter, or less, as larger droplets cause rapid erosion of internal components.

Most turboexpanders drive centrifugal compressors to provide a portion of the outlet compression. In situations where inlet pressures are very high (e.g., offshore) turboexpanders are used in pressure letdown to provide refrigeration for dew point control and to generate power.

Like compressors, expanders can be positive displacement or dynamic; dynamic can be radial or axial. Reciprocating expanders were used for liquefying gases (see Chapter 13). However, the only type used in gas processing is the radial unit with inward (centripetal) flow, and discussion is restricted to this type.

A cutaway view of a typical turboexpander for gas processing is shown in Figure 7.5. The expander is the unit on the right, and the compressor is the unit on the left. Gas enters the expander through the pipe at the top right, and is guided onto the wheel by the aerodynamically shaped adjustable guide vanes, which completely surround the expander wheel. The swirling high-velocity inlet gas turns the wheel and transfers part of its kinetic energy to the wheel and shaft, and exits to the right through the tapered nozzle. Because part of the energy of the gas has been transferred to the wheel, the exit gas is at a much lower temperature and pressure than the gas entering. The expander wheel, directly coupled to the compressor wheel, provides the work necessary to drive the centrifugal compressor on the left. Low-pressure gas enters in the straight section

FIGURE 7.5 Cutaway view of a turboexpander. (Courtesy of Mafi-Trench Corp.)

on the left, is compressed by the compressor wheel, and exits at the top of the unit. Lubricating oil enters in the top port shown in the center of the unit.

About 50% of the enthalpy change occurs in the turbine (Jumonville, 2004). The increase in velocity over the vanes results in the other 50% of the total pressure and temperature drop across the expander unit. Thus, the inlet guide vanes are a vital part of the energy conversion process in a turboexpander. The high velocity of the gas that exits the vanes and enters the expander wheel greatly improves overall turboexpander efficiency (Peranteaux, 2005).

Figure 7.6 shows a large turboexpander used in a large gas plant. Operating conditions for turboexpanders vary, depending on the process application and the composition of the gas being processed. However, Tables 7.3 and 7.4 provide a general idea of turboexpander operations in a gas plant. Note the size of the wheel in Table 7.3 and the rotating speeds in both tables.

Although simple in principle, turboexpanders are sophisticated, yet extremely reliable pieces of rotating machinery. In addition to the complex aerodynamic design of the expander and compressor components of the system, the unit must be equipped with low drag but 100% effective shaft seals. These seals operate under high pressure and at temperatures that range from cryogenic to well above ambient. By injecting clean, dry "seal gas" into the shaft seals, lube oil is prevented from escaping the bearing housing, and the bearings are protected from the cold process gas. Fluid-film bearings that support the shaft that connects the expander and compressor sections traditionally are supplied with oil by use of a pressurized recirculating system. The Engineering Data Book (2004c) provides more details.

FIGURE 7.6 Turboexpander in large gas plant. The turboexpander processes 30,500 lb/min (830,000 kg/h) and recovers approximately 9.2 MW (12,000 hp) of power. (Courtesy of Mafi-Trench Corp.)

TABLE 7.3
Expander–Compressor Design-Point Conditions

	Expander	Compressor
Inlet gas rate, MMscfd (MMSm³/d)	250 (7.1)	187 (5.3)
Molar mass	22.80	19.69
Inlet pressure, psia (bar)	1,590 (109.7)	855 (58.9)
Inlet temperature, °F (°C)	30 (−1.0)	34 (1.0)
Outlet pressure, psia (bar)	870 (60.1)	1,165 (80.3)
Outlet temperature, °F (°C)	−15 (−26.1)	78 (25.4)
Liquid formation, wt%	36	—
Power, kW	2,004	1,986
As tested efficiency, %	85.0	78.5
Speed, rpm	16,700	16,700
Wheel diameter, in. (mm)	7.75 (197)	10.9 (277)

Source: Lillard and Nicoll (1994).

In principle, both the turboexpander and compressor can be multistaged. However, to date, mechanical sealing problems have made multistaging impractical for both the turbine and expander (Peranteaux, 2005).

Many new units employ magnetic bearings instead of fluid-film bearings. Magnetic bearings keep the shaft in place by constantly adjusting the current in radial and axial electromagnets that support the shaft. Although the initial investment in magnetic bearings is slightly higher than for comparably sized fluid-film bearings, this design provides substantial savings in lifetime utility operating costs

TABLE 7.4
Turboexpander Normal Design Conditions

	Expander	Compressor
Inlet flow rate, lb/h (kg/h)	221,000 (100,243)	208,000 (94,347)
Inlet gas rate, MMscfd (MMSm³/d)	115.6 (3.27)	115.2 (3.26)
Molar mass	17.46	16.5
Inlet pressure, psia (bar)	1,080 (74.5)	470 (32.4)
Inlet temperature, °F (°C)	−50 (−46)	60 (16)
Outlet pressure, psia (bar)	480 (33.1)	576 (39.7)
Outlet temperature, °F (°C)	−113.6 (−80.9)	93.5 (34.2)
Liquid formation, wt%	19.3	—
Efficiency, %	83	74
Speed, rpm	22,000	22,000
Power, bhp (bkW)	1,380 (1,029)	1,350 (1,007)

Source: Agahi and Ershaghi (1992).

when compared with conventional oil-lubricating systems. More importantly, they eliminate the potential for oil contamination in the cold section of the plant (Jumonville, 2004). Nonetheless, magnetic bearing systems still require seal gas. In this case, the seal gas protects the bearings from the cold process gas while simultaneously ensuring adequate cooling for the bearing coils (Peranteaux, 2005).

The Engineering Data Book (2004c) emphasizes some points that should be kept in mind for turboexpanders:

- Entrainment. Gas that enters the turboexpander must be free of both solids and liquids. Fine-mesh screens are used to protect the device, and the pressure drop across the screen should be monitored.
- Seal gas. This gas isolates process gas from the lubricating oil, or isolates process gas from the shaft if magnetic bearings are used, and must be clean and constantly available at the operating pressure. Sales gas is commonly used. Otherwise, a warmed inlet gas stream off of the expander inlet separator is used. (The gas must be warmed to 70°F [20°C] or more to prevent thickening of the lube oil, if used.)
- Lubricant pumps. These pumps must maintain a constant flow to lubricate the bearings if oil is used. A spare pump is mandatory. The Engineering Data Book (2004c) describes the lubrication system.
- Shut-off valves. A quick-closure shut-off valve is used to shut in the inlet for startup and shutdown.

As is the case for centrifugal compressors, turboexpander efficiency diminishes when operating off of the design point. This variance can be about 5 to 7 percentage points when the flow increases or decreases by 50%. However, the turboexpander normally is driving a compressor, which also will suffer loss in efficiency when off of the design point. Therefore, the overall effect on the turboexpander–compressor unit efficiency will be larger. As with centrifugal compressors, surge control is needed.

Example 7.2 Use the inlet conditions and outlet pressure supplied in Table 7.4 to compute the outlet temperature and work generated per lb of gas inlet. For this example, assume the gas is pure methane and use diagrams from Appendix B to do the calculations. Compare this result to the outlet temperature and work done if a J-T expansion occurs over the same pressure drop and inlet temperature.

Turboexpander Calculations Inlet conditions of the gas are −50°F and 1,080 psia. At these conditions, for pure methane, the enthalpy $h = 255$ Btu/lb, and the entropy $s = 0.69$ Btu/lb-°R. The ideal expansion is isentropic, so we follow the constant entropy line to the exit pressure of 480 psia, which is close to the saturation boundary of methane, where the temperature is −138°F and $h = 230$ Btu/lb. The work done is $(255 − 230) = 25$ Btu/lb. However, the expander is only 77% efficient, so the actual work done is $0.77 (25) = 19.2$ Btu/lb. The corrected final enthalpy is then $255 − 19.2 = 236$ Btu/lb at 480 psia, which corresponds to about −135°F. Therefore, the inefficiency did not greatly affect the final temperature in this example. The effect of the efficiency in plants that process large quantities of gas is significant, both in cooling and in lost power for recompression.

J-T Expansion The J-T expansion follows the constant enthalpy line down to the exit pressure of 480 psia, giving an exit temperature of −115°F. The work obtained during expansion is zero.

Therefore, use of a turboexpander with 77% efficiency provides an outlet gas that is 20°F colder and at the same time provides 19.2 Btu/lb of work. The benefit of turboexpansion compared with J-T expansion is noticeable in plants when the turboexpander is bypassed and J-T expansion is used. In such cases, the NGL production can drop by as much as 80%. Although this comparison is slightly unfair because the plant design is optimized for turboexpansion, it points out the value of turboexpanders.

7.2.3 HEAT EXCHANGE

Most heat exchangers in a gas plant operating at or above ambient temperature are conventional shell and tube type and are ideal for steam and hot oil systems where fouling occurs. They are relatively inexpensive and easy to maintain because the tube bundle can be removed and tubes cleaned or replaced as needed. The Engineering Data Book (2004b) provides extensive details of the exchangers for use in gas processing. Where the fluids are clean and fouling does not occur, such as in gas–gas exchangers, compact heat exchangers are ideal. Wadekar (2000) provides a good overview of the more common types. This section briefly discusses two kinds, brazed-aluminum plate-fin heat exchangers and printed circuit heat exchangers, which commonly are used in gas processing. Because the literature for conventional shell and tube heat exchangers is extensive. these exchangers are not discussed.

7.2.3.1 Plate-Fin Exchangers

Cryogenic facilities have made extensive use of brazed-aluminum plate-fin heat exchangers since the 1950s. Instead of a shell and tube configuration, these units consist of channels formed by a thin sheet of aluminum pressed into a corrugated pattern (the fin) sandwiched between two aluminum plates. Each layer resembles the end view of corrugated cardboard. The fin channels may be straight or may have a ruffled or louvered pattern to interrupt the straight flow path.

Advantages of plate-fin exchangers include (Engineering Data Book, 2004b):

- Light weight.
- Excellent mechanical strength at subambient temperatures (used in liquid helium service [−452°F (−268°C)]). Can operate at pressures up to 1,400 psig (96 barg).
- High heat transfer surface area. Up to six times the surface area per unit volume of shell and tube exchanger and 25 times the area per unit mass.

- Complex flow configurations. Can handle more then 10 inlet streams with countercurrent, crossflow, and counter crossflow configurations.
- Close temperature approaches. Temperatures of 3°F (1.7°C) for single-phase fluids compared with 10 to 15°F (6 to 9°C) for shell and tube exchangers and 5°F (2.8°C) for two-phase systems.

Drawbacks and limitations of the exchangers include:

- Single-unit construction. Repair can be more costly and time consuming than with shell and tube exchangers.
- Maximum operating temperature of approximately 150°F (~85°C), although special designs go to 400°F (205°C).
- Narrow channels. More susceptible to plugging, and fine mesh screens are needed where solids may enter. Components that might freeze out, water, CO_2, benzene, and p-xylene, must be in sufficiently low concentrations to avoid plugging. The exchangers can be difficult to clean if plugging occurs.
- Less rugged. Does not accept rough handling or high pipe stress on nozzles.
- Limited to fluids noncorrosive to aluminum. Caustic chemicals are corrosive but not corroded by acid gases, unless free water is present.
- Susceptible to mercury contamination. Mercury amalgamates with aluminum to destroy mechanical strength.
- Susceptible to thermal shock. Maximum rate of temperature change is 4°F/min (2°C/min), and maximum difference between two streams is 55°F (30°C) (Howard, 1998).

The Engineering Data Book (2004b) provides some additional information on plate-fin exchangers, including fin geometries. Lunsford (1996) provides some guidelines on designing plate-fin exchangers. However, because of the complex design, heat exchanger vendors should be contacted for details on a specific application. Most often, heat transfer surface areas will be given for both hot and cold sides. This practice is comparable to adding both the shell and tube heat transfer surface areas together in a conventional shell and tube exchanger.

7.2.3.2   Printed Circuit Heat Exchangers

Another heat exchanger type, the printed circuit heat exchanger (PCHE) is used in clean service (Engineering Data Book, 2004b). This technology is relatively new, commercialized in the 1980s, but hundreds of units are in service (Pua and Rumbold, 2003). Like electronic printed circuits, heat transfer passages are etched in plates, and the plates are bonded together by diffusion bonding. Unlike the brazed-aluminum exchangers, they are rugged and, depending on materials of construction, go to high temperatures and pressures but can still handle complex flow schemes that involve many streams. Heat transfer passage sizes range from "microchannels" (less than 8 mil, 200 microns) to "minichannels" (0.12 in, 3 mm) to provide high heat transfer surface areas. Heat transfer area per unit volume can be 800 compared with 500 for plate-fin exchangers. Like plate-fin exchangers, vendor design is required.

Offshore operations employ PCHEs in many applications because they offer comparable heat transfer at comparable pressure drops at significantly less size and at one fifth the weight (Pua and Rumbold, 2003). One example of a high-temperature onshore application is in gas turbine-driven compressor stations where hot lube oil is used to preheat fuel gas to the gas turbine to avoid condensation problems (Sell et al., 2004). Haynes and Johnson (2002) discuss exchanger performance characteristics.

7.2.4 FRACTIONATION

In addition to conventional distillation columns, two other types of distillation columns are commonly found in gas plants: stabilizers and demethanizers. Stabilizers are stripping columns used to remove light ends from NGL streams. Demethanizers are also stripping columns to remove methane from the NGL bottoms product. Demethanizers also act as the final cold separator, a collector of cold NGL liquids, and source of recovering some refrigeration by cooling warm inlet streams.

7.2.4.1 Stabilizers

The primary focus of dew pointing or fuel conditioning is to obtain a leaner gas. However, the "by-product" is a liquid phase that contains a substantial amount of volatiles. To make the liquid product easier to store and to recover more light ends for fuel or sales gas, many of the systems will "stabilize" the liquid by passing it through a stabilizer column. The stabilizer feed typically enters at the top of a packed or tray column and no reflux occurs. To increase stripping of light ends, the column pressure will be lower than that of the gas separator that feeds the column. In some cases, a stripping gas may be added near the bottom of the column in addition to the externally heated reboiler installed to provide additional vapor flow and enhance light-ends removal. This feature usually comes as an increased operating cost because the gas from the stripper is at low pressure and must be recompressed if put back into the inlet gas stream upstream of the gas treating unit.

7.2.4.2 Demethanizer

A distinguishing feature of gas plants with high ethane-recovery rates is the demethanizer, as shown in Figure 7.7. The column differs from usual distillation columns in the following ways:

- It has an increased diameter at the top to accommodate the predominately vapor feed to the top tray.
- It is typically primarily a stripping column, with no traditional condenser–reflux stream.

FIGURE 7.7 Demethanizer column.

- It may have several liquid feed inlets further down the column that come from low-temperature separators.
- It may have several side reboilers, the primary purpose of which is to cool the gas going through the reboiler to recover some of the refrigeration available in the warming NGL stream.
- It has a large temperature gradient; over 170°F (75°C) is common.

The column serves two main functions: it acts as a flash drum for the top feed, which comes in as a cold, two-phase stream, and it removes methane from the bottoms product. Depending upon the plant configuration, the feed may be from a turboexpander, a J-T valve, or a heat exchanger. In some configurations the columns have reflux, but many demethanizers have no reflux. The NGL bottoms product is usually continuously monitored for methane content, which typically is kept below 0.5 liquid vol% of the ethane, on a C_3+ free basis.

The top of the column usually operates in the −175 to −165°F (−115 to −110°C) range, with pressures in the 200 to 400 psig (14 to 28 barg) range. Operation of the demethanizer at higher pressures reduces compression costs. Shah and Stucky (2004) discuss the advantages and disadvantages of higher pressure operation.

To obtain high ethane recovery, the column must operate near the low end of the temperature range, which translates into a higher compression load. The demethanizer usually has 18 to 26 trays, which operate at a tray efficiency of 45 to 60% (Engineering Data Book, 2004f) or an equivalent packed bed.

7.3 RECOVERY PROCESSES

Many process configurations are used to recover hydrocarbons in the field and in gas plants. The best configuration depends upon many variables, including:

- Product slate
- Gas volumes
- Gas composition
- Pressures, both inlet and outlet

The product slate dictates the required lowest operating temperature of the gas. Both dew point control and fuel conditioning have the same main product— a residue gas with reduced C_3+ fraction. Dew point control is usually a field operation, and stabilization of the produced liquid is site specific. Although gas temperatures in a low-temperature separator (LTS) may go down to $-40°F$ ($-40°C$), only a cold separator is required to separate the light ends from the liquid. Two new technologies, Twister and vortex tube, discussed below make dew point control and fuel conditioning a one-step process. In addition, use of membranes for fuel conditioning is discussed briefly.

If limited ethane recovery (<60% ethane) is desired, the recovery process is essentially a low temperature separator, except that fractionation of the cold liquid is added to increase the recovery. Lean oil absorption is sometimes used for up to approximately 50% ethane recovery. For high ethane recovery, the gas processing temperatures must be as low as $-160°F$ ($-110°C$) and usually require a combination of external refrigeration and expansion. These plants require a demethanizer to increase recovery rates and to strip methane from the NGL.

Gas volumes and gas composition set the optimal plant configuration on an economic basis. This combination makes it difficult to set criteria for establishing the best plant configuration. However, the higher the gas volume and GPM (gal liquid per Mscf [see Chapter 1]), the more attractive are high ethane recoveries.

Inlet gas pressures make a major difference in plant configuration. High pressures permit use of expansion, J-T or turboexpander, to provide all of the cooling if low ethane recovery is desired. For low inlet pressures, either external refrigeration or inlet compression followed by expansion is needed to cool the gas, regardless of extent of ethane recovery. Required outlet pressure helps decide which approach should be taken.

The following three sections discuss the three hydrocarbon-recovery systems:

1. Dew point control and fuel conditioning
2. Low ethane recovery
3. High ethane recovery

The focus is on simple configurations, to give the reader the ability to evaluate and understand the more complex configurations they may see.

7.3.1 Dew Point Control and Fuel Conditioning

Dew point control and fuel conditioning exist to knock out heavy hydrocarbons from the gas stream. These operations are primarily field operations.

7.3.1.1 Low Temperature Separators

Low temperature separators (LTS) (also called low temperature extraction units, or LTX) are used both onshore and offshore. The process consists of cooling and partial condensation of the gas stream, followed by a low temperature separator. When inlet pressures are high enough to meet discharge-pressure requirements to make pressure drop acceptable, cooling is obtained by expansion through a J-T valve or turboexpander. Otherwise, external cooling is required. Water usually is present, and to prevent hydrate formation the separator downstream of the expander is warmed above the hydrate-formation temperature to prevent plugging. An alternative to heating is injection of either ethylene glycol or methanol, which is then recovered and dried for reuse.

Figure 7.8 shows a LTS that uses ethylene glycol injection for hydrate prevention and uses J-T expansion for cooling. The feed initially goes through a water knockout vessel to remove free water. The water-saturated hydrocarbon gas and liquids then mix with ethylene glycol before being precooled. The mixture then passes through a J-T expansion valve and flashes into the low temperature separator to separate the gas, condensate, and glycol–water phases. The condensate goes to the condensate stabilizer for removal of remaining light ends.

Overhead gas from the low temperature separator passes through the pre-cooler before being combined with the stabilizer overhead and put into the pipeline. The low temperature separator is set to maintain the proper dew point of the blended outlet gas.

The C_3+ condensate from the stabilizer goes to product storage. The glycol–water mixture from the low temperature separator goes to the glycol regenerator for removal of the water and then reinjection into the feed.

If inlet pressures are too low for expansion, the stream is cooled by propane refrigeration. The advantage of direct refrigeration is that the pressure drop is kept at a minimum. Hydrate formation must be considered with either feed dehydration

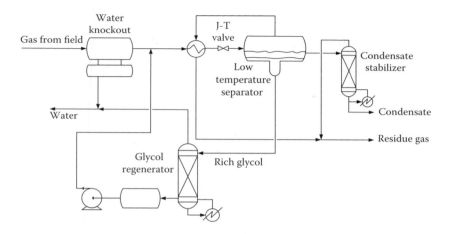

FIGURE 7.8 Low-temperature separator (LTS), with glycol injection and condensate stabilization.

upstream of the unit or inhibitor injection. Glycol injection is usually the more cost effective, but if used, it increases the required refrigeration duty. The Engineering Data Book (2004e) provides additional details on these processes.

7.3.1.2 Twister

Okimoto and Betting (1997) reported use of a relatively new device, Twister, for dew point control and dehydration. It is used in one offshore facility (Brouwer et al, 2004) and has been favorably evaluated for other projects (Wilson and Yuvancic, 2004). Figure 7.9 shows a cutaway view of the device and denotes the salient parts of the unit. Gas enters and expands through a nozzle at sonic velocity, which drops both the temperature and pressure and causes droplet nucleation. The two-phase mixture then contacts a wing that creates a swirl and forces separation of the phases by centrifugal force. The gas and liquid are separated in the diffuser; the liquid is collected at the walls and dry gas exits in the center. Advantages of the system include:

- Simplicity. No moving parts and no utilities required.
- Small size and low weight. A 1-inch (24-mm) throat diameter, 6 feet (2 m) long tube can process 35 MMscfd (1 MMSm³/d) at 1,450 psia (100 bar).
- Driven by pressure ratio, not absolute pressure.
- Relatively low overall pressure drop. System recovers 65 to 80% of original pressure.
- High isentropic efficiency. Efficiency is around 90% compared with 75 to 85% for turboexpanders.

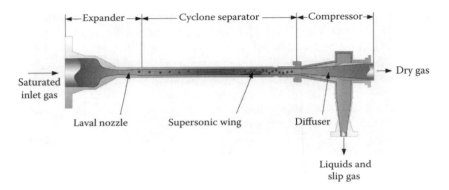

FIGURE 7.9 Cutaway view of Twister device. (Courtesy of Twister BV.)

Drawbacks of the system include:

- Requires a clean feed. Solids erode the tubing and wing, necessitating an inlet filter separator.
- Limited turndown capacity. Flow variability is limited to ±10% of designed flow. This limitation is mitigated by use of multiple tubes in parallel.

Liquids exit with a "slip gas." This mixture is typically 20 to 30% of the total flow volume (Epsom, 2005). The mixture can go to a gas–liquid separator for recovery of the gas, which may require recompression.

7.3.1.3 Vortex Tube

Vortex tubes use pressure drop to cool the gas phase but generate both a cold and warm gas stream. If streams are recombined, the overall effect is comparable to a J-T expansion. The principle of operation is the Ranque-Hilsch tube, developed in the 1940s and commonly marketed as a means to provide cold air from a compressed air stream. Cockerill (1998) provides an excellent general and fundamental discussion of the device. Lorey and Thomas (2005) discuss the principles of the device and provide operating data on its use for gas conditioning.

For dew point control, and dehydration, the device has the vortex tube and a liquid receiver connected to the tube. Gas enters the tube tangentially through several nozzles at one end of the tube, expands, and travels spirally at near sonic velocities to the other end. As it travels down the tube, warm and cool gas separate. The cool gas goes into the center of the tube. Warm gas vents in a radial direction at the end, but the cool gas is reflected back up the tube and exits just beyond the inlet nozzles. Condensation occurs in the cool gas, and the liquid is moved to the walls by centrifugal force, where it collects and drains into the receiver below. The overall cooling effect is comparable to that of a J-T expansion, with a low-temperature separator. However, the vortex tube combines the expansion and separation into a single step.

The working pressure of the tube is 500 to 3,050 psig (36 to 210 barg), and flow rates are 20 to 140 MNm³/h. The turndown ratio is 15% for a single tube but can be increased by use of multiple tubes in parallel; the optimum pressure drop is 25 to 35% (Lorey and Thomas, 2005). The vendor states that liquid condensation must be less than 10 wt%. (FILTAN, undated). Lorey and Thomas (2005) note that the device performs well with up to 5% liquids in the inlet stream. The device has been used to dehydrate gas from underground storage. To prevent hydrate formation in the cold stream, TEG is added.

Like Twister, the vortex tube has the advantage of simplicity and light weight. It could be useful where limited turndown is acceptable. It will be of most value when no compression is required.

7.3.1.4 Membranes

As discussed in Chapter 5, membranes are being used in several areas of gas processing, including dew pointing. Hale and Lokhandwala (2004) discuss use of membranes for fuel conditioning. Membranes are ideal for this application, provided preconditioning is adequate to protect the membrane, and little penalty exists for permeate compression. Figure 7.10 shows the flow configuration. Gas enters the membrane on the discharge side of the compressor, and the residual gas provides fuel to the compressor engine or turbine. The low pressure permeate is recycled to the suction for recompression to recover the permeate. Table 7.5 provides results for one field unit. Gas rates are low because only a slip stream needs to be processed for fuel.

Like the previous two technologies, the process is simple and requires no moving parts. It too has the advantage of being relatively small and light weight. (The membrane unit described by Hale and Lokhandwala [2004] is 6 ft × 6 ft × 6 ft [2 m × 2 m × 2 m]). The authors report this technology is used on several offshore installations. Unlike the Twister and vortex tube, membranes have the advantage of a turndown ratio down to 50%, with no performance penalty. This property may not be an advantage for fuel gas conditioning, where flow rates should be stable.

The reported data are used to compute the permeate composition and fraction of feed removed from the feed. The table points out the selectivity of the membrane

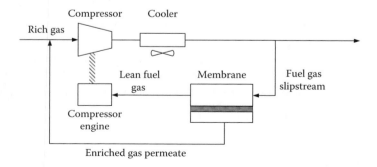

FIGURE 7.10 Schematic for membrane unit used as a fuel conditioner. (Adapted from Hale and Lokhandwala, 2004.)

TABLE 7.5

Operating Conditions and Composition of Natural Gas Stream Using Membrane for Fuel Gas Conditioning

	Membrane Feed	Conditioned Fuel Gas	Permeate
Temperature, °F (°C)	95 (35)	51 (10.5)	—
Pressure psig (barg)	940 (65)	940 (65.0)	—
Total mass flow lb mol/h (kg-mol/h)	110.1 (50)	58.0 (26.3)	52.1 (23.7)
Total volume flow MMscfd (MSm³/d)	0.95 (27)	0.5 (14)	—

Component	Mol%			% Removed from Gas
	Feed	Fuel Gas	Permeate[a]	
Carbon dioxide	1.3	0.6	2.08	76
Methane	72.8	81.2	63.59	41
Ethane	9.6	9.0	10.29	51
Propane	9.9	7.1	13.04	62
i-Butane	2.4	0.8	4.19	82
n-Butane	2.5	0.9	4.29	81
n-Pentane	1.3	0.4	2.30	84
Water	0.11	0.00	0.23	100
Hydrocarbon dew point (°C)	35	3.5	—	

[a] Composition of permeate and fraction removed from gas computed by material balance of normalized reported feed and fuel gas composition.

Source: Hale and Lokhandwala (2004).

and may be poorer than that of the above two technologies. In fuel conditioning, the selectivity is not a major issue because of the relatively small fraction of gas that needs to be recompressed, and the enriched stream is recycled without requiring additional compression.

A membrane that passes heavy hydrocarbons preferentially to smaller molecules is counterintuitive. Chapter 5 points out that membrane permeability is the product of the solubility and diffusion coefficient. For separation of light gases, the primary mode of selectivity is the diffusion coefficient. For dew pointing, solubility drives the selectivity (Baker et al., 1998). These membranes are silicone rubber compounds that preferentially absorb the heavy components.

7.3.2 Low Ethane Recovery

The focus of the previous section was removal of heavy components (C_3+) to avoid condensation or to lower the heating value. This section discusses processes used in conventional gas plants, where the objective is to produce a lean gas and recover up to approximately 60% of the ethane in the feed gas. Two process schemes are used to obtain this level of ethane recovery:

- Cooling by expansion or external refrigeration
- Lean-oil absorption

As noted above, inlet pressure dictates the best means of refrigeration. Lean oil was an early method used for hydrocarbon recovery but is now used on a more limited basis. Many of the refrigerated lean oil absorption plants in operation today are large facilities, where replacing them with a more modern turboexpander plant would be capital cost prohibitive. Both approaches are described briefly below.

7.3.2.1 Cooling by Expansion or External Refrigeration

Figure 7.11 shows the propane recovery obtainable as a function of process temperature for various levels of liquids content. In this case, GPM is based upon the C_3+ fraction. Figure 7.12 shows a similar plot for recovery of ethane and propane, with GPM based upon C_3+ fraction. Both the figures point out how recovery depends upon the relative amount of C_3+ in the gas stream and the lowest gas temperature. A general rule is to assume that recovery increases with increased richness of the gas. This assumption is made because the ethane content in the vapor at the top of

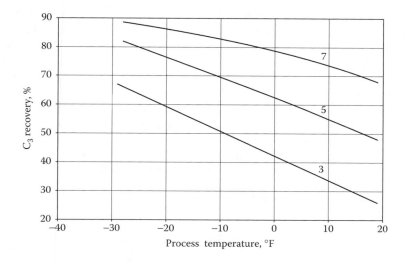

FIGURE 7.11 Recovery of the propane plus fraction as a function of the lowest separation temperature and gas composition of feed. Operating pressure is 600 psig, and numbers given on chart represent GPM based upon C_3+. (Adapted from Engineering Data Book, 2004e.)

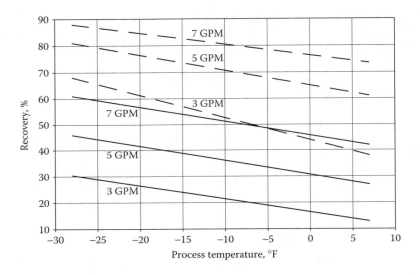

FIGURE 7.12 Recovery of ethane and propane as a function of the lowest separation temperature and gas content of feed. Operating pressure is 600 psig and GPM are based upon C₃+. The solid and dashed lines represent ethane and propane recovery, respectively. (Adapted from Engineering Data Book, 2004e.)

the column is set by column feed composition, along with temperature and pressure. At constant pressure and temperature, the ethane concentration in the liquid decreases with increasing C_3+ fraction, which lowers the ethane concentration in the vapor and, thus, increases the percent ethane recovered. (However, this outcome will not always be the case in plants that use J-T or turboexpanders, because leaner gas puts less of a load on the refrigeration-expander system and may lower column temperatures and increase recovery (McCartney, 2005)).

Figure 7.13 shows one commonly used direct-refrigeration process that employs recycle from a fractionator to maximize liquids recovery. Inlet gas is initially cooled with cold residue gas and cold liquid from the cold separator before going to the propane chiller and to the cold separator. Vapor from the separator is the sales gas, and the liquid goes to a fractionator to strip out light ends and recover liquid product. The column operates at a lower pressure than does the cold separator. Because of system pressure drop and because the fractionator runs at the lower pressure, the recycle stream must be recompressed. Alternatives to the process include:

- Reduction or elimination of the recycle by adding reflux to the fractionator
- Running the fractionator at a higher pressure and use of a pump to feed the column from the cold separator

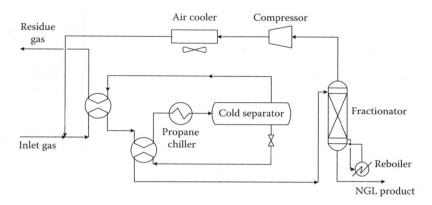

FIGURE 7.13 Schematic of a direct refrigeration process for partial recovery of C_2+ fraction. (Adapted from Engineering Data Book, 2004e.)

The Engineering Data Book (2004e) provides more details of process variations and discusses how processing variables affect compressor and refrigeration requirements. These configurations assume that the gas enters sufficiently dehydrated to prevent hydrate formation. If the water content is higher, ethylene glycol can be added, which increases refrigeration duty. However, temperatures then are limited by glycol viscosity.

Because the unit relies only on external propane refrigeration, the lower temperature limit on the feed to the cold separator is −35°F (−37°C) at best. Unless the feed has a very high GPM, ethane recoveries will be below 60%. Expansion is required to lower temperatures and increase recoveries.

With high inlet gas pressures, replacing the propane system with an expander is an attractive option. However, inlet compression may be necessary to obtain the temperatures required to obtain the desired recoveries. Both J-T and turboexpanders are used. Crum (1981) points out situations where a J-T system may be preferable to turboexpanders, although recent advances in turboexpander technology may temper some of them:

- Low gas rates. J-T is more economically viable at low gas rates. Crum (1981) maintains that at below 10 MMscfd (300 MSm³/d), turboexpanders offer less economic advantage and they lose efficiency below 5 MMScfd (150 MSm³/d).
- Low ethane recovery. For ethane recoveries of 10 to 30%, J-T expansion may be sufficient.
- Variable flow rates. J-T is insensitive to widely varying flow rates, whereas turboexpanders lose efficiency when operating off of design rates.

Crum (1981) also points out that J-T plants are much simpler than turboexpander plants because J-T plants have no need for seal gas and lubricating oil systems. However, because of the inefficiency of J-T valves compared with turboexpanders, if any inlet compression is required, more is required with J-T expansion to obtain the same amount of refrigeration. The Engineering Data Book (2004e) suggests that use of J-T expansion for limited ethane recovery requires inlet pressures around 1,000 psi (70 bar).

Crum (1981) discusses five configurations in which J-T expansion is used to recover C_2+ in plants, ranging from 3 to 10 MMScfd (80 to 300 MSm³/d). Without external refrigeration, recovery rates were up to 39%. A process that utilizes propane refrigeration in combination with J-T expansion obtained 80% recovery. Crum (1981) found that the fuel cost savings for compression failed to justify installation of turboexpanders in these small plants.

7.3.2.2 Lean Oil Absorption

Early gas processing plants used lean oil absorbers to strip NGL from natural gas (Cannon, 1993), and the process is still used in about 70 gas plants today. To improve recoveries, later plants used external refrigeration to cool the feed gas and lean oil. Figure 7.14 shows a representative schematic of a propane-refrigerated lean oil system. The process involves three steps (Engineering Data Book, 2004e):

1. Absorption. An absorber contacts a lean oil to absorb C_2+ plus from raw natural gas.

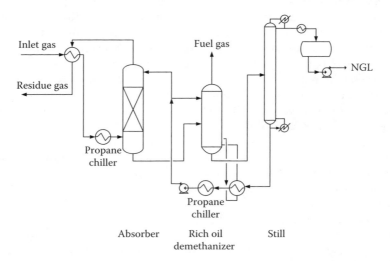

FIGURE 7.14 Refrigerated lean oil absorption process. (Adapted from Engineering Data Book, 2004e.)

2. Stabilization. The rich oil demethanizer (ROD) strips methane and lighter components from the rich oil.
3. Separation. The still separates the recovered NGL components as product from the rich oil, and the lean oil then returns to the absorber.

Gas from the ROD is either blended with the exiting gas stream or used for fuel. Original systems used lean oil with molar mass of 150 to 200, but refrigerated systems use molar masses of 100 to 130.

If no refrigeration is used, and assuming the absorber runs at about 100°F (38°C), over 75% of the butanes and essentially all of the C_5+ fraction are recovered. Using high solvent rates makes possible the recovery of 50% of the ethane and essentially all of the propane and heavier components (McCartney, 2005). With propane refrigeration, typically over 97% of the propane is recovered and up to 50% of the ethane. The Engineering Data Book (2004e) notes that refrigeration of the inlet gas and lean oil, along with heat to the still and ROD, are the key elements of an efficient lean oil system.

An advantage of lean oil absorption is that little pressure drop occurs through the absorber. However, the process is energy intensive and relies on numerous heat exchangers to reduce the energy load. For gas processing, the whole process can be simplified by elimination of the lean oil and use of external refrigeration, as discussed in the previous section. However, many refrigerated lean oil absorption plants remain in operation today with capacities of 1,000 MMscfd (30 MMSm³/d) or more (McCartney, 2005).

One use for lean oil absorbers today is in capturing fugitive hydrocarbons from air streams because refrigeration is unnecessary. Many non gas related industries use this process for pollution control.

7.3.3 HIGH ETHANE RECOVERY

The above processes provided limited recovery of ethane. To obtain 80 to 90% or more ethane recovery requires separation temperatures well below what is obtainable by use of propane refrigeration alone. In principle, direct-refrigeration processes could be used by cascading propane cooling with ethane or ethylene refrigeration or by use of a mixed refrigerant that contains methane, ethane, and propane. The primary motivation for use of only direct refrigeration would be low inlet gas pressures. If significant inlet compression is required to produce refrigeration by expansion, then cascade or mixed-refrigeration cooling, with or without expansion, may be attractive. No matter which option is used, obtaining high ethane recoveries from low inlet-pressure feed streams requires substantial compression, of either the feed stream, the refrigerants, or both.

With recovery of a high ethane fraction, sales gas specifications must be considered. Recovery of too much ethane could reduce the heating value below contract limits. Figure 7.15 shows the maximum recovery attainable to obtain a 1,000 Btu/scf (37,000 kJ/m³) heating value as a function of GPM on the basis of C_2+ and inerts (e.g., N_2, CO_2) in the sales gas. The plot assumes that all inerts in the inlet gas remain

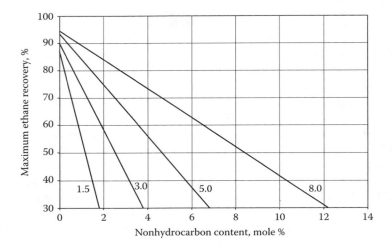

FIGURE 7.15 Maximum ethane recovery as a function of noncombustible concentration in the feed and ethane plus concentration. Numbers denote the GPM (based upon C_2+) in the feed. Lines assume that all inerts remain in the sales gas, and the sales gas heating value is 1,000 Btu/scf. (Adapted from Engineering Data Book, 2004e.)

in the sales gas. Figure 7.16 shows how ethane recovery affects propane and butane recoveries. The Engineering Data Book (2004e) notes that the propane recovery varies, depending upon process configuration.

Figure 7.17 shows a simplified conventional expander plant schematic. It consists of a gas–gas heat exchanger with five gas streams that enter at different

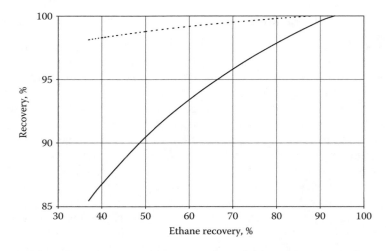

FIGURE 7.16 Recovery of propane and butanes as a function of ethane recovery rate. The solid and dashed lines represent propane and butanes recovery, respectively. (Adapted from Engineering Data Book, 2004e.)

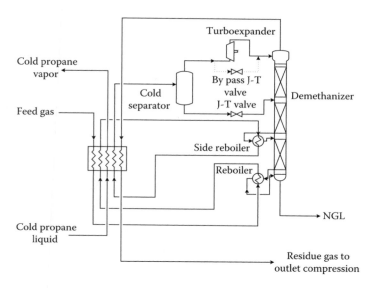

FIGURE 7.17 Schematic of conventional turboexpander process with no recycle to demethanizer. Note that the one heat exchanger represents a network of exchangers. (Adapted from Engineering Data Book, 2004e.)

temperatures, cold separator, turboexpander, and demethanizer. Although the flow sheet shown is schematically simple, in practice most actual designs replace the single exchanger with a more complex and efficient combination of exchangers. The inlet gas stream makes several passes through the gas–gas exchanger before going to the cold separator, where the vapor expands through a turboexpander. Liquid from the cold separator is flashed through a J-T valve and fed to the middle of the demethanizer. The incoming gas provides reboiler heat at the bottom, and then is cooled further in a second reboiler midway up the column.

A J-T valve is always installed parallel to the turboexpander. This configuration helps in plant start-up and in handling excess gas flow. It also is used if the turboexpander goes down.

The maximum ethane recovery with the conventional turboexpander configuration is about 80% (Engineering Data Book, 2004e). Also, the cold separator may be near the critical temperature and pressure of the mixture, which can make the process unstable. Carbon dioxide freezing out can also be a problem. Improvement of C_2+ recovery requires reduction of ethane losses in the top of the demethanizer by addition of reflux. The Engineering Data Book (2004e) discusses a number of configurations used. One that can provide up to 98% recovery, called the cold residue recycle (CRR) process, is shown in Figure 7.18, which gives the maximum ethane recovery with regards to compression requirements of all commonly used processes. It has the added advantage that it can reject ethane and still maximize C_3+ recovery if desired.

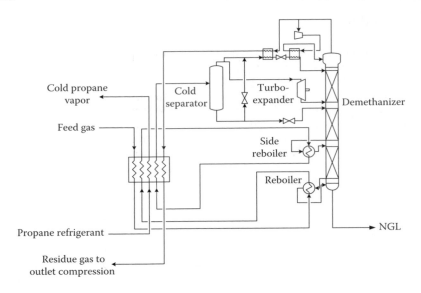

FIGURE 7.18 Cold-residue recycle process for maximizing ethane recovery. All valves in figure are J-T expander valves but are unlabeled for figure clarity and the large heat exchanger represents a network of exchangers. (Adapted from Engineering Data Book, 2004e.) 16-21

In this variation, the cold separator runs at a warmer temperature to avoid the critical point problem. The vapor from the separator splits into two streams. Part goes to the turboexpander and the balance goes through two overhead exchangers, where it is condensed to provide liquid reflux to the column. Turboexpander output then enters further down the column. In addition, part of the overhead is compressed and cooled to provide additional reflux.

The Engineering Data Book (2004e) provides details on a number of other process variations that are commonly used.

7.4 SAFETY AND ENVIRONMENTAL CONSIDERATIONS

Safety considerations primarily exist because of the high pressures and low temperatures involved in hydrocarbon recovery. Plugging from insufficient removal of water, BTEX, and carbon dioxide can be a processing hazard. Properly sized relief systems must be properly functioning in case of plugging.

Environmental concerns are relatively minor; lubricating oils along with methanol or glycol, if used, are the primary potential pollutants from spills. All process fluids are tied directly to the flare in case of emergency shutdown.

REFERENCES

Agahi, R. and Ershaghi, B., Turboexpander Redesign Concepts and Economics, Proceedings of the Seventy-First Annual Convention of the Gas Processors Association, Tulsa, OK, 1992, 29.

Baker, R.W, Wijmans, J.G, and Kaschemekat, J.H., The design of membrane vapor-gas separation systems, *J. Membrane Sci.,* 151, 55, 1998.

Brouwer, J.M., Bakker, G., Verschoof, H.-J., and Epsom, H.D., Supersonic Gas Conditioning First Commercial Offshore Experience, Proceedings of the Eighty-Third Annual Convention of the Gas Processors Association, Tulsa, OK, 2004.

Cannon, R. E., *The Gas Processing Industry, Origins and Evolution*, Gas Processors Association, Tulsa, OK, 1993, 51.

Crum, F.S., Application of J-T Plants for LP-Gas Recovery, Proceedings of the Sixtieth Annual Convention of the Gas Processors Association, Tulsa, OK, 1981, 68.

Cockerill, T.T., Thermodynamics and Fluid Mechanics of a Ranque-Hilsch Vortex Tube. Ph.D. thesis, University of Cambridge, Cambridge, England, 1998, www.southstreet.freeserve.co.uk/rhvtmatl/, Retrieved July 2005.

Engineering Data Book, 12th ed., Sec. 2, Product Specifications, Gas Processors Supply Association, Tulsa, OK, 2004a.

Engineering Data Book, 12th ed., Sec. 9, Heat Exchangers, Gas Processors Supply Association, Tulsa, OK, 2004b.

Engineering Data Book, 12th ed., Sec. 13, Compressors and Expanders, Gas Processors Supply Association, Tulsa, OK, 2004c.

Engineering Data Book, 12th ed., Sec. 14, Refrigeration, Gas Processors Supply Association, Tulsa, OK, 2004d.

Engineering Data Book, 12th ed., Sec. 16, Hydrocarbon Recovery, Gas Processors Supply Association, Tulsa, OK, 2004e.

Engineering Data Book, 12th ed., Sec. 19, Fractionation and Absorption, Gas Processors Supply Association, Tulsa, OK, 2004f.

Epsom, H., private communication, 2005.

FILTRAN GmbH, http://www.filtan.de/ENGLISH/VTS_A.htm, Retrieved June 2005.

Hale, P. and Lokhandwala, K., Advances in Membrane Materials Provide New Solutions in the Gas Business, Proceedings of the Eighty-Third Annual Convention of the Gas Processors Association, Tulsa, OK, 2004.

Haynes, B.S. and Johnson, A.M., High-Effectiveness Micro-Exchanger Performance, presented at AIChE 2002, Spring National Meeting, New Orleans, www.heatric.com/technical_papers.html, Retrieved September 2005.

Herrin, J.P., New process for liquid recovery, *Hydrocarbon Process.*, 45 (6) 144 1966.

Howard, I., Hannibal's Experiences, *Proceedings of the Laurance Reid Gas Conditioning Conference,* Norman, OK, 1998, 194.

Jumonville, J., Tutorial on Cryogenic Turboexpanders, *Proceedings of the Thirty-Third Turbomachinery Symposium*, 2004, http://turbolab.tamu.edu/pubs/Turbo33/T33 pg127.pdf, Retrieved September 2005.

Lillard, J.K., and Nicoll, G., An Operating History of Turboexpanders in the Supercritical Fluid Regime, Proceedings of the Seventy-Third Annual Convention of the Gas Processors Association, Tulsa, OK, 1994, 270.

Lorey, M. and Thomas, K., Gas Conditioning Utilizing the Vortisep Device Application of a Novel Technology on an Industrial Scale, Proceedings of the Eighty-Fourth Annual Convention of the Gas Processors Association, Tulsa, OK, 2005.

Lunsford, K.M., Understand the use of brazed heat exchangers, *Chem. Eng. Progress*, 92 (11) 44, 1996.

McCartney, G.D., private communication, 2005.

Okimoto, F.T, and Betting, M., Twister Supersonic Separator, Proceedings of the Laurance Reid Gas Conditioning Conference, Norman OK, 2001.

Peranteaux, J., private communication, 2005.

Pua, L.M. and Rumbold, S.O., Industrial Microchannel Devices—Where Are We Today? First International Conference on Microchannels and Minichannels, Rochester, New York, 2003.

Sell, K.T., Langston, P.R. and Mitchell, R.H., Compressor Station Fuel Gas Superheating Using Lube Oil Waste Heat, International Pipeline Conference, October 4–8, Calgary, Alberta, Canada, 2004.

Shah, K. and Stucky, B., Operation of High Pressure Cryogenic Demethanizerater El Paso Eunice Gas Plant, Proceedings of the Eighty-Third Annual Convention of the Gas Processors Association, Tulsa, OK, 2004.

Smith, J.M., Van Ness, H.C., and Abbott, M.M., *Introduction to Chemical Engineering Thermodynamics,* 6th ed., McGraw-Hill, New York, 2001.

Wadekar, V.V., Compact heat exchangers, *Chem. Eng. Progress.*, 96 (12) 39, 2000.

Wilson, J. L. and Yuvancic, J., Process Selection for Dehydrating Gulf of Mexico Offshore Platform Gas, Proceedings of the Laurance Reid Gas Conditioning Conference, Norman, OK, 2004, 125.

8 Nitrogen Rejection

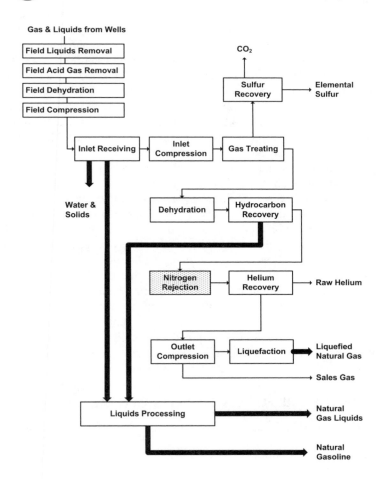

8.1 INTRODUCTION

Three sets of circumstances require nitrogen separation or rejection:

- Processing a gas high in nitrogen to produce a pipeline quality gas.
- Removing nitrogen from a natural gas so that the nitrogen can be used in an enhanced oil recovery (EOR) operation.
- Separating helium from nitrogen in a helium recovery operation.

This chapter discusses nitrogen rejection for EOR and for production of pipeline quality gas. A brief discussion of the helium–nitrogen separation is left to Chapter 9.

As mentioned in Chapter 1, the Gas Research Institute's survey of reserves (Meyer, 2000) found that approximately 24 Tcf (670 Bm^3), 16% of the nonassociated reserves at that time, were subquality in nitrogen and, consequently, will require blending to lower the nitrogen concentration or processing to remove the nitrogen to meet the 3 mol% total inerts specification for pipelines (Engineering Data Book, 2004a).

In 1998 (EPRI, 1999), enhanced oil recovery (EOR) methods contributed about 12% of the total oil production in the United States. Of the EOR production, about 55% was from thermal methods, 28% was from carbon dioxide flooding, 12% was from natural gas flooding, and 4.5% was from nitrogen flooding. Even with the relatively small fraction of production, nitrogen EOR contributed 32,000 barrels of oil per day (5,100 m^3/d).

The inlet nitrogen concentration to a plant upgrading subquality gas is relatively constant, although it may fluctuate if variations exist in nitrogen concentrations in the various gases feeding the plant. This circumstance is not the case in EOR applications. In nitrogen EOR projects, nitrogen is injected into a stimulation well to increase oil production in producing wells. Gas from the producing wells initially contains little or no nitrogen. As the nitrogen gradually breaks through from the injection to the producing wells, the nitrogen concentration gradually rises until it reaches a value so high that the project is terminated. Consequently, in EOR applications the nitrogen rejection unit (NRU) must be designed to accommodate changing inlet feed concentrations. This chapter briefly discusses both upgrading subquality gas and EOR applications.

Tannehill et al. (1999) discuss the processing of high nitrogen gas. They also cover the locations of high nitrogen reserves and the number and types of plants used in the processing, and they summarize some of the economics.

8.2 NITROGEN REJECTION FOR GAS UPGRADING

Three basic methods are used for removal of nitrogen from natural gas:

- Cryogenic distillation
- Adsorption
- Membrane separation

Table 8.1 provides a comparison of the three processes. Cryogenic methods are the most economical and can provide higher nitrogen rejection at high gas throughput. At low gas volumes, membranes and pressure swing adsorption (PSA) by use of molecular sieves are economically feasible. The tabulated flow ranges are guidelines only.

In regard to hydrocarbon recovery, only PSA has heavier hydrocarbons (all C_4+ and part of propane) going with the nitrogen stream. This situation is caused by adsorption in the sieve binder, as the components are too large to enter the sieve pores. The binder also adsorbs water and CO_2. The loss of hydrocarbons may or may not be beneficial.

TABLE 8.1
Comparison of Nitrogen Removal Processes

Process	Flow Range MMscfd (MSm³/d)	Complexity	Heavy Hydrocarbon Recovery	Development Stage
Cryogenic distillation	>15 (400)	Complex	In product gas	Mature
Pressure-swing adsorption (PSA)	2 – 15 (60 – 400)	Simple; batch operation, requires bed switching	In regeneration gas	Early commercialization
Membrane	0.5 – 25 (15 – 700)	Simple continuous operation	In product gas	Early commercialization

Source: Adapted from Hale and Lokhandwala (2004).

8.2.1 CRYOGENIC DISTILLATION

The most common method of removing nitrogen from natural gas is cryogenic distillation. The Engineering Data Book (2004b) notes that for feed concentrations below 20% N_2, a single-column design can be used. For higher concentrations, a dual-column is better. With the addition of a recycle compressor, it can be used at lower N_2 contents. Figure 8.1 shows a flow diagram for a two-column NRU

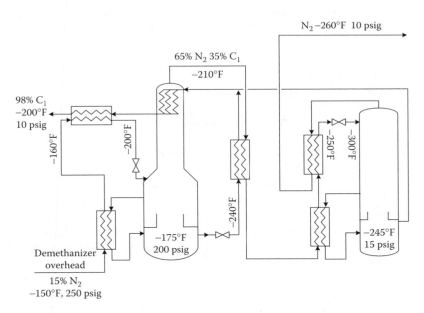

FIGURE 8.1 NRU by use of two-column cryogenic distillation (Handwerk, 1990). Valves are J-T valves.

receiving feed that contains 15% N_2 from a demethanizer in a conventional turboexpander plant. Gas from the demethanizer overhead is cooled by heat exchange and pressure reduction and fed to a distillation column operating at 200 psig (14 barg). The bottoms product from this high-pressure column is reduced in pressure to cool the stream to $-240°F$ ($-151°C$). This stream, combined with the bottoms product from the second low-pressure column, is fed to a heat exchanger in the top of the high-pressure column to provide the necessary reflux. The overhead from the high-pressure column flows through three heat exchangers, is reduced in pressure to approximately 15 psig (1 barg), and enters the low-pressure column at $-300°F$ ($-184°C$). The overhead from this column is 98% N_2, and the bottoms product is approximately 98% CH_4. The Hannibal Gas Plant of British Gas Tunisia (Jones et al., 1999) uses cryogenic distillation to reduce the N_2 content of the feed gas from 16.9% N_2 to the sales-gas specification of 6.5%. (See Chapter 15 for more details.) Complete descriptions of the plant and its operations are given by Jones et al. (1999) and Howard (1998).

8.2.2 PRESSURE SWING ADSORPTION

After cryogenic distillation, pressure swing adsorption (PSA) is probably the most widely used process. At this point, we should briefly discuss the significant differences between the adsorption process used for dehydration (thermal swing adsorption, or TSA) and that used for nitrogen rejection (pressure swing adsorption, or PSA).

As discussed in Chapter 6, differences in adsorbate polarity and size provide the means of separation in adsorption. The amount adsorbed depends on four factors:

1. The adsorbent itself
2. The species being adsorbed (adsorbate)
3. The temperature
4. The partial pressure of the adsorbate

Once the adsorbent and adsorbates are selected, the temperature and partial pressure become the governing variables. All industrial regenerative adsorption separations involve two steps: adsorption to separate the species, followed by desorption and removal of the adsorbate (regeneration) to prepare the adsorbent for further use.

In natural gas systems, if adsorption is used to remove a relatively small amount of material to a very low level, or if the heat of adsorption is very high, TSA is generally used, as discussed in Chapter 6. An example is natural gas dehydration, which meets both criteria. For bulk removal of one component from another (e.g., upgrading natural gas to pipeline specifications by removal of CO_2 or N_2), PSA may be the choice because concentrations of the adsorbate are high and the heat of adsorption is low. We briefly discuss the fundamentals of this process.

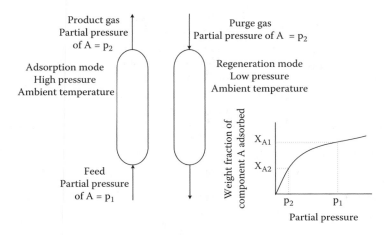

FIGURE 8.2 Simple pressure swing adsorption (PSA) system.

Figure 8.2 shows a very simplified two-bed PSA system (actual plants may have four beds) to separate a binary mixture of compounds A and B. Assume A is strongly adsorbed relative to B. The feed enters the adsorbing bed at ambient temperature and a high pressure with the partial pressure of A = p_1. Because A is more strongly adsorbed than B, the adsorbed phase (adsorbate) is enriched in A, and the A content of the gas leaving the bed is reduced. When the outlet concentration of A begins to increase because of bed breakthrough, it is switched to the regeneration mode, and the feed is switched to the previously regenerated bed to continue the cycle.

Regeneration is accomplished by dropping bed pressure, which causes the gas to desorb, and purging the bed to remove the desorbed gas. The adsorption isotherm in Figure 8.2 (the plot of weight fraction A, X_A, adsorbed versus partial pressure) indicates that the final concentration of A on the regenerated bed drops to what will be in equilibrium with the purge gas A at a partial pressure of p_2. Thus, all of the A cannot be removed from the bed. The residual A that loads when the bed is put back into adsorption service is at equilibrium with p_2. This condition means the gas that leaves the bed during the adsorption mode has a partial pressure of p_2. If the purge gas contains no A, then its partial pressure is reduced to zero, and most of the adsorbed A will be desorbed and purged. Consequently, the bed is capable of reducing the level of A in the product to a very low level.

This brief discussion is very simplified. Readers who desire a more complete treatment should consult the review article of Sherman and Yan (1991).

D'Amico et al. (1993) describe a PSA unit that uses a specially treated carbon molecular sieve (CMS) for nitrogen rejection. The CMS is prepared by processing of coal or wood carbon. Complete details as well as an economic study are presented in the original article. Mitariten (2001) and Mitaritan and Dolan (2001) describe a PSA process, called Molecular Gate™, that uses a new type of molecular

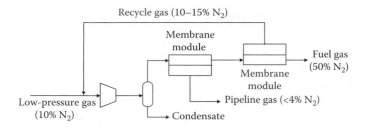

FIGURE 8.3 Separating N_2 from natural gas by use of membranes.

sieve. According to these authors, conventional molecular sieve 3A and 4A have pore openings of 3.2 and slightly less than 4.2 Å, respectively. The new Molecular Gate sieve is a titanium silicate with a pore opening of 3.7 Å, which allows better separation of nitrogen (3.6 in diameter) from methane (3.8 in diameter). Further information on the Molecular Gate process is given by Mitaritan et al (2002).

8.2.3 MEMBRANES

Chapter 5 presented the basic features of membrane separation, and that discussion applies to the present problem of N_2 separation from natural gas. Hale and Lokhandwala (2004) discuss the NitroSep™ process for removal of N_2 from natural gas to produce a pipeline quality gas along with a fuel gas (Figure 8.3). In the process shown, the low-pressure inlet gas is compressed, any condensate formed is removed, and the feed goes to the membrane unit, where pipeline-quality gas and a liquid stream that contains most of the heavier hydrocarbons are produced. The residue gas from the first membrane unit goes to a second unit that produces a gas suitable for use as a compressor fuel and a recycle stream to recover the methane.

8.3 NITROGEN REJECTION FOR ENHANCED OIL RECOVERY

As discussed previously, the primary difference between conventional nitrogen rejection and nitrogen rejection for EOR is the planned increase with time in N_2 feed concentration to the NRU for EOR. The reason for the wide variation lies in the nature of the EOR process. The N_2 is injected into the reservoir through a series of injection wells positioned to force the oil to the producing wells. Initially, little or no N_2 will be in the gas and oil produced, but inevitably, N_2 begins to break through and reach the producing wells. As the field becomes depleted, the N_2 content increases. Eventually, the N_2 level reaches a point at which gas production is no longer economically possible, and the project is stopped.

In the EOR project described by McLeod and Schaak (1986), the anticipated inlet NRU N_2 concentrations ranged from 21 to 75%. As previously noted, feed concentration plays a major role in the design (Engineering Data Book, 2004b). Therefore, these NRU plants must be carefully designed to accommodate a wide range of inlet concentrations. Davis et al. (1983) discuss EOR designs in which the N_2 level of the gas goes from 5 to 80% N_2.

The process of choice for NRUs in EOR service is cryogenic distillation, and the major decision is the type of distillation scheme, single column or double column. Both processes are discussed by Davis et al. (1983); the single column is presented in more detail by Davis et al. (1989). The double-column system shown in Figure 8.1 and described above is described in more detail by Alvarez and Vines (1985). The double-column system can more easily handle large changes in N_2 and is preferred for EOR or where the N_2 feed content can change widely.

McLeod and Schaack (1986) give a detailed account of an enhanced oil recovery project that utilizes N_2 as the injection fluid. Figure 8.4 is a general schematic of the operation, which consists of three separate units:

- An air separation plant to generate nitrogen
- A gas plant to recover liquids from the produced stream
- An NRU to produce a sales gas

The air separation plant produces 30 MMscfd (0.8 MMSm³/d) of N_2 that contains 5 ppmv of O_2, in addition to some liquid nitrogen. The oxygen from the unit is vented to the atmosphere. The N_2 gas is compressed to 4,900 psig (340 barg) in two sets of compressors and injected into the reservoir. The solution gas that comes

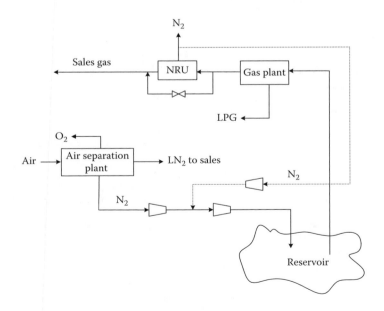

FIGURE 8.4 Schematic of an enhanced oil recovery (EOR) system.

from the reservoir is treated in a conventional gas processing plant to produce a heavier hydrocarbon product (LP gas) and a residue gas that goes to the NRU. The NRU must produce a sales gas with a heating value of at least 975 Btu/scf (36,300 kJ/m^3) from a feed gas that contains from 20 to 75% N_2. The NRU uses double-column cryogenic distillation for the separation, which is described in the paper. At the beginning of the project little N_2 is present in the produced gas, but the concentration continuously increases during the life of the project. In the later stages of the project, N_2 removed in the NRU is recompressed and reinjected into the reservoir. This procedure reduces the need for N_2 from the air separation plant.

The Anschutz Ranch East NGL/NRU plant described by Davis et al. (1989) was designed to process 300 MMscfd (8 MMSm^3/d) of gas with N_2 compositions varying from 2 to 80%. When built, it was the largest NRU plant in the world to process widely varying N_2 feeds. Additional information on NRUs may be found in the papers of Goethe and Mawer (1987) and Young and Maloney (1989).

8.4 SAFETY AND ENVIRONMENTAL CONSIDERATIONS

Safety considerations are centered on the high pressures and extremely low temperatures generally encountered in NRUs. In addition, if the low-temperature nitrogen can be vented, it must be routed to a safe location so that it becomes more buoyant than the ambient air and is not allowed to accumulate in a confined space. This procedure prevents the possibility of an accidental asphyxiation. Environmentally, all of the systems are benign.

REFERENCES

Alvarez, M.A. and Vines, H.L., Nitrogen rejection/NGL recovery for EOR projects, *Energy Prog.*, 5 (2) 67 1985.

Davis, R.A., Vines, H.L., and Pervier, J.W., Cryogenic Schemes for Nitrogen Rejection— an EOR Project, presented at the Rocky Mountain Meeting of the Gas Processors Association, 1983.

Davis, R.A., Gottier, G.M., Blackburn, G.A., and Root, C.R., Integrated Nitrogen Rejection Facility Design and Operation, Proceedings of the Sixty-Eighth Annual Convention of the Gas Processors Association, Tulsa, OK, 1989, 204.

D'Amico, J., Reinhold, H., and Gamez, J., A PSA Process for Nitrogen Rejection from Natural Gas, Proceedings of the Laurance Reid Gas Conditioning Conference, Norman, OK, 1993, 55.

Engineering Data Book, 12th ed., Sec. 2, Product Specifications, Gas Processors Supply Association, Tulsa, OK, 2004a,.

Engineering Data Book, 12th ed., Sec. 16, Hydrocarbon Recovery, Gas Processors Supply Association, Tulsa, OK, 2004b.

EPRI, Enhanced Oil Recovery Scoping Study, TR-113836, Palo Alto, CA. 1999. www.energy.ca.gov/process/pubs/electrotech_opps_tr113836.pdf, Retrieved June 2005.

Goethe, A. and. Mawer, D.J, A New Integrated Nitrogen Rejection Process with NGL Recovery, Proceedings of the Sixty-Sixth Annual Convention of the Gas Processors Association, Tulsa, OK, 1987, 89.

Hale, P. and Lokhandwala, K., Advances in Membrane Materials Provide New Gas Processing Solutions, Proceedings of the Laurance Reid Gas Conditioning Conference, Norman, OK, 2004, 165.

Handwerk, G., private communication, 1990.

Howard, I., Hannibal's Experiences, Proceedings of the Laurance Reid Gas Conditioning Conference Norman, OK, 1998, 194.

Jones, S., Lee, S., Evans, E., and Chen, R., Simultaneous Removal of Water and BTEX From Feed Gas for a Cryogenic Plant, Proceedings of the Seventy-Eighth Annual Convention of the Gas Processors Association, Tulsa, OK, 1999, 108.

McLeod, R.P. and Schaak, J.- P., A Unique Combination of Natural Gas Processing with Cryogenic Technology in an Enhanced Oil Recovery Project, Proceeding of the Sixty-Fifth Annual Convention of the Gas Processors Association, Tulsa, OK, 1986, 71.

Meyer, H.S., Volume and distribution of subquality natural gas in the United States, *Gas TIPS* 6, 10, 2000.

Mitariten, M., New technology improves nitrogen removal economics, *Oil Gas J.*, 99 (17) 42, 2001.

Mitaritan, M., and Dolan, W., Nitrogen Removal from Natural Gas with Molecular Gate Technology, Proceedings of the Laurance Reid Gas Conditioning Conference, Norman, OK, 2001.

Mitaritan, M., Dolan, W., and Maglio, A., Innovative Molecular Gate Systems for Nitrogen Rejection, Carbon Dioxide Removal, and NGL Recovery, Proceedings of the Eighty-First Annual Convention of the Gas Processors Association, Tulsa, OK, 2002.

Sherman, J.D. and Yan, C.M., Adsorption, gas separation, in *Encyclopedia of Chemical Technology*, 4th ed., Wiley-Interscience, New York, 1991.

Tannehill, C.C., Strickland, J.G., and Meyer, H., High N_2 Gas—Snap Shot of the Present Requirements for the Future, Proceedings of the Seventy-Eighth Annual Convention of the Gas Processors Association, Tulsa, OK, 1999, 230.

Young, S.M. and Maloney, J.J., Enhanced Reservoir Value Through Nitrogen Rejection, Proceedings of the Laurance Reid Gas Conditioning Conference, Norman, OK, 1989, 301.

9 Trace Component Recovery or Removal

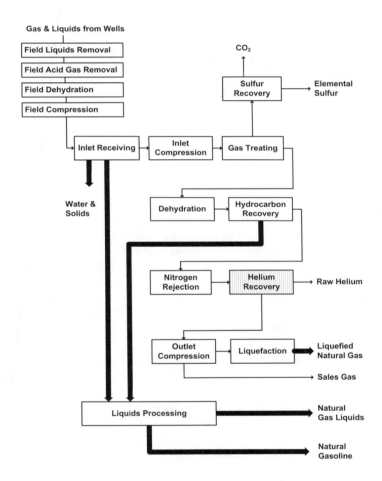

9.1 INTRODUCTION

Chapters 5, 6, 7, and 8 discussed methods to remove the common natural gas impurities. A number of other trace components present processing, product quality, or environmental problems if concentrations are too high. These components include:

209

- Hydrogen
- Oxygen
- Radon (NORM)
- Arsenic
- Helium
- Mercury
- BTEX (benzene, toluene, ethylbenzene, and xylene)

This chapter briefly touches on the first four items in the list and then discusses the last three items in more detail, primarily because they are of such importance that several commercial processes exist to remove them.

9.1.1 HYDROGEN

Hydrogen levels are rarely sufficiently high to cause a problem, and no GPA specifications exist for maximum levels in sales gas. The primary source would be gas streams from refineries, but even there, the concentrations should be low.

9.1.2 OXYGEN

Of the items listed above, oxygen is the only contaminant that is not naturally occurring, and the best approach for treating it is to prevent its introduction into the processing stream. The major source of oxygen is leaking valves and piping in gathering systems that operate below atmospheric pressure. The Engineering Data Book (2004b) gives a 1.0 vol% maximum oxygen concentration in sales gas, but some pipelines have specifications as low as 0.1 vol%. However, it causes problems in gas processing at concentrations of 50 ppmv, such as:

- Enhancing pipeline corrosion if liquid water is present
- Reacting with amines in gas treating, which ultimately leads to heat stable salt formation
- Reacting with glycols to form corrosive acidic compounds
- Reacting with hydrocarbons during the high-temperature regeneration of the adsorption beds to form water, which reduces regeneration effectiveness and, thus, reduces bed capacity

At low levels, the oxygen can be removed by use of nonregenerative scavengers. For higher concentrations, the obvious removal method is to catalytically react it in the gas stream to produce water, which is then removed in the dehydration step. However, gases that contain sulfur compounds poison oxidation catalysts. (The motivation for reduction of sulfur levels in gasoline and diesel fuels.) Thus, combustion upstream of gas treating is infeasible unless the gas is H_2S free. Oxidizing the process stream after treating generates CO_2, which can freeze out in cryogenic sections of the plant. Oxygen is removed in sales gas with nitrogen rejection, but this step follows the processing steps most susceptible to oxygen damage. Also, few gas plants have nitrogen rejection units. Oxygen will become more of a problem as gas fields in the United States are depleted.

9.1.3 RADON (NORM)

Natural gas contains radon, a naturally occurring radioactive material (NORM), at low concentrations, and it rarely poses a health problem because it has a half-life of about 3.8 days (Encyclopedia Americana, 1979). However, radon decays into lead-210, then to bismuth-210 and polonium-210 and finally into stable lead-206. These daughter products of radon, some of which have long half-lives, condense on pipe walls and form a low-level radioactive scale, which may flake off and collect on inlet filters. Because the boiling point of radon is −79.2°F (−61.8°C), it tends to concentrate in propane and ethane–propane mixtures. Storage vessels can accumulate the daughter products as sludge. Discarded piping with the scale generates large quantities of low-level radioactive waste that must be discarded in disposal wells. For further details see, Bland (2002), Gray (1990), Grice (1991), and Railroad Commission of Texas (undated).

9.1.4 ARSENIC

Arsenic is a toxic nonvolatile solid but exists in natural gas predominately as a more volatile trimethylarsine (As $(CH_3)_3$). It usually collects as a fine gray dust. High concentrations tend to be geographically localized. It can be successfully removed from gas by use of a nonregenerative adsorption process. Several facilities reduce arsenic concentrations in sweet raw gas from around 1,000 to less than 1μg/m³ (Rhodes, 2005). Without arsenic removal, the gas streams could not be marketed. The process requires dehydration of the gas to pipeline specifications before it goes to the adsorbers.

9.2 HELIUM

9.2.1 INTRODUCTION

Helium is a difficult diluent to remove in natural gas unless nitrogen rejection is used. Unlike the other trace components discussed here, helium is a valuable product from natural gas processing, which makes high concentrations desirable. Recovery of helium from air is technically feasible but would be very expensive because the helium concentration is so low, 5.24 ± 0.05 ppm (Engineering Data Book, 2004a). The only viable source is from select natural gas fields, and even the best of these gas fields contain relatively small quantities of helium. A helium-rich gas is defined as one that contains more than 0.3% helium, and according to published field concentrations (Deaton and Haynes, 1961; Guccione, 1963; Johnson and Rydjord, 2001), a gas that contains more than 5% is not usually found.

The National Academy of Sciences (2000) presents an excellent overview of helium reserves and usage in the United States. In 2003, the world production of grade-A helium (99.995% purity) was 6,400 MMscf (181 MMm³); the United States produced 84% of the total, and the remainder coming from Algeria, Poland, and Russia. However, two new large helium plants are in the planning and construction stages, one in Darwin, Australia, slated to start up in 2007, and the other in Qatar, which becomes operational in 2005. Both plants will process the offgas from baseload LNG facilities. At full production, the Darwin plant will be capable of producing 150 MMscf (4.2 MMm³) of helium per year (BOC Group, 2005) and the Qatar plant will be capable of producing 300 MMscf (8.5 MMm³) per year (Pacheco, 2003).

The uses of helium are limited but important. Its main applications stem from the fact that it is chemically inert and has the lowest boiling point of any substance. Figure 9.1 shows how helium was used in the United States in 2003.

9.2.2 RECOVERY METHODS

Recovery of helium from natural gas requires refinement of the low concentrations into high-purity helium (99.995%). However, the gas plant may produce only a raw helium stream that contains a roughly equimolar mixture of helium and nitrogen. This stream then goes to another facility for final purification. Early processing schemes (Deaton and Haynes, 1961; Guccione, 1963; Remirez, 1968) were cryogenic and involved cooling and condensing the natural gas and then recovering the helium by distillation as a crude product that was about 65 to 80% helium. The crude helium was then stored or further refined in a series of low-temperature separations, followed by a final adsorption step on activated charcoal to produce the high-purity product.

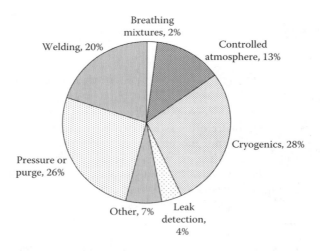

FIGURE 9.1 Helium uses in the United States in 2003 (Pacheco, 2003).

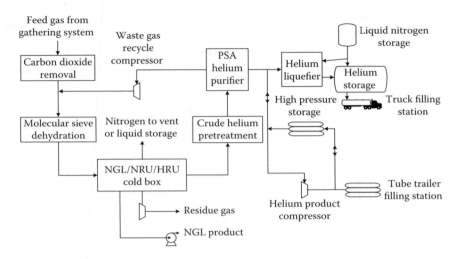

FIGURE 9.2 Schematic of Ladder Creek Helium Recovery Plant (Johnson and Rydjord, 2001).

More recently, interest has increased in alternate technologies such as pressure-swing adsorption (PSA) and membranes for helium recovery. Choe et al. (1988) present a discussion and analysis of hybrid cryogenic and membrane processes.

The newest helium recovery plant in the United States, the Ladder Creek Plant (Johnson and Rydjord, 2001), located in Eastern Colorado, started production in 1998. Figure 9.2 is a simplified process flow diagram for the Ladder Creek Plant. The inlet gas composition varies substantially, but the values in Table 9.1 are representative of the feed gases to the plant.

The plant was designed to process 35 MMscfd (1 MMm³/d), with expansion capability to 50 MMscfd (1.4 MMm³/d). Helium recovery was set at 95%, with the ethane either rejected or recovered. The natural gas liquids (NGL) are recovered while pipeline gas is produced.

In Figure 9.2, the carbon dioxide content is first reduced to less than 10 ppm, and then trace quantities of mercury present are removed before dehydration by use of molecular sieve beds. Details of the cold box are unavailable, but its products are:

- Crude helium stream (nitrogen plus helium)
- Small stream of liquid nitrogen
- Residue gas
- Natural gas liquids

The liquid nitrogen is used to fill the nitrogen shields on the liquid helium trailers and the helium storage. The crude helium is purified to 99.997% by use of pressure swing adsorption (PSA). (See Chapter 8 for details on PSA.) The PSA

TABLE 9.1
Range of Feed Gas Compositions to Ladder Creek
Helium Recovery Plant

Component	Composition (mol%)	
	Rich Helium Gas	Lean Helium Gas
Nitrogen	61.92	31.58
Helium	3.54	1.81
Carbon dioxide	0.98	0.91
Methane	26.65	52.84
Ethane	2.60	6.40
Propane	1.95	3.64
i-Butane	0.37	0.46
n-Butane	0.88	1.09
Pentane+	1.10	1.27
Total	100.00	100.00

Source: Johnson and Rydjord (2001).

adsorbers operate at ambient temperature and 240 to 280 psig (16.5 to 19.3 barg). Adsorbers are on-line for approximately 320 seconds and are then taken off-line, depressurized to approximately 40 psig (3.8 barg), and flushed with pure helium gas for reactivation. All flush gas is recycled to the cold box. The helium produced may be liquefied and stored and shipped by truck as a liquid or compressed and stored and shipped as a high-pressure gas. Details of the patented helium liquefier are not available, but it has a capacity of 1.7 MMscfd (50 MSm³/d).

Example 9.1 Compute the ratio of the minimum work of separation required to obtain pure He from air compared with that from a natural gas that contains 3 mol% He in the feed. Assume the separation takes place at constant temperature and that the gases are ideal.

As shown in any chemical engineering thermodynamics textbook (e.g., Smith et al, 2001), the theoretical minimum amount of work to go from one state to another is $T\Delta S$, where T is the absolute temperature and ΔS is the change in entropy between the initial and final state. The entropy change per mole of ideal gas mixture formed from pure components is

$$\Delta S = -R\Sigma\, x_i \ln x_i,$$

where R is the gas constant and x_i is the mole fraction of component i. Because the He concentration is so low in air, the mole fraction of the other components in air can be assumed to not change, so the change in entropy is assumed to be caused by the change in He concentration only. We are interested in the ratio of work per mole of He produced and base work on 1 mole of initial gas mixture. Thus, the ratio of work is

$$\frac{T\Delta S_{Air}}{T\Delta S_{NG}} = \frac{\ln x_{He,Air}}{\left(\dfrac{x_{He,NG}\ln x_{He,NG} + (1-x_{He,NG})\ln(1-x_{He,NG})}{x_{He,NG}}\right)}$$

$$= \frac{\ln(5\times10^{-6})}{\left(\dfrac{0.03\ln 0.03 + 0.97\ln(0.97)}{0.03}\right)} = 2.9$$

If the two separations are performed at 100% thermodynamic efficiency at a constant temperature, the work required to obtain helium from air would be 2.9 times greater than extracting it from natural gas.

9.3 MERCURY

Two major problems are associated with the presence of mercury in natural gas: amalgam formation with aluminum and environmental pollution. According to Rios et al. (1998) mercury may be present in elemental form or as organometallic compounds, such as dimethylmercury, methylethylmercury, and diethylmercury, or as inorganic compounds such as $HgCl_2$.

Elemental mercury distributes between the gas and liquid phases. Whereas elemental mercury has an extremely low vapor pressure (~4 × 10⁶ psia at 32°F, 3 × 10⁻⁷ bar at 0°C, [Rios, et al. 1998]), vapor phase concentrations are orders of magnitude higher than computed by Equation 6.1 because of the nonideal behavior of mercury at operating conditions. For example, Stiltner (2002) found gas from a high-pressure inlet separator contained about 25% of the mercury that entered the plant. The high-pressure separator liquid was flashed to ambient pressure, and the liquid contained less than 20% of the inlet mercury, whereas the flash gas, which was mostly ethane through pentane, contained over 50% of the inlet mercury. Thus, the majority of mercury goes to treating, dehydration, and then to the cryogenic section. What fraction would reach the cryogenic section was not stated.

The volatility of elemental mercury and the fact that it will accumulate as it condenses makes removal from the gas stream mandatory in cryogenic plants. Mercury corrodes brazed-aluminum heat exchangers as it amalgamates with the aluminum to weaken the material. The first reported instance of this problem occurred in an Algerian LNG facility in 1974 (Rios et al., 1998). Elemental mercury is toxic to humans as well as a poison to many catalysts.

The mercury compounds concentrate in the hydrocarbon liquids, where they potentially can present environmental and safety hazards. The compounds are readily absorbed by most biological systems.

TABLE 9.2

Range of Elemental Mercury Levels in Wellhead
Natural Gas from a Number of Countries

Location	Elemental Mercury Concentration in μg/Nm³ (ppbv)
South America	69 – 119 (8 to 13)
Far East	58 – 93 (6 to 10)
North Africa	0.3 – 130 (0.03 to 15)
Gronigen (Germany)	180 (20)
Middle East	1 – 9 (0.1 to 1)
Eastern US Pipeline	0.019 – 0.44 (0.002 to 0.05)
Midwest US Pipeline	0.001 – 0.10 (0.0001 to 0.01)
North America	0.005 – 0.040 (0.0005 to 0.004)

Source: Abbott and Openshaw (2002).

Although mercury in natural gas is normally at low levels, some gases contain sufficiently high mercury concentrations to cause both safety and health concerns. Table 9.2 shows the range of mercury levels in wellhead gas in a number of countries.

Abbott and Openshaw (2002) state that much higher mercury levels have been detected in individual wells and that the highest known concentration is 4,400 μg/Nm³ (500 ppbv) from a well in Germany. Rios et al. (1998) note that Southeast Asian gases tend to have the higher elemental mercury levels, whereas the United States Gulf Coast gases are usually low, but a wide variation can occur within a given region.

9.3.1 ENVIRONMENTAL CONSIDERATIONS

The environmental considerations of mercury and its compounds fall into two categories: (1) direct harm to individuals from inhalation of mercury vapors or compounds and (2) safe disposal of contaminated materials and equipment.

From a health perspective, ingested mercury, particularly dimethyl mercury, permanently damages the brain and kidneys. Unfortunately it concentrates in the tissue of both the fish and shellfish that are consumed by humans. According to the EPA database on fish advisories (Environmental Protection Agency, 2005a), in 2005 the United States had 2,470 active advisories on eating fish and shellfish because they contained high mercury levels. Stringent emission controls at the federal level on mercury from coal-fired power plants are being considered that will cut mercury emissions from power plants by 70% by 2018 (Environmental Protection Agency, 2005b.).

The mercury concentration in most natural gases is insignificant compared with the coals presently used in power plants (Environmental Protection Agency, 2001). As discussed below, elemental mercury is removed from natural gas to protect processing equipment and catalysts. However, some of these processes do not remove the mercury compounds. The gas industry has not come under any EPA regulations that require mercury removal. However, this situation could change in the future, especially for very large plants that process gases with higher than normal mercury levels.

The disposal of materials and equipment contaminated with mercury is a complex issue, and a discussion is beyond the scope of this book. Rios et al. (1998) and Müssig (1997) present excellent discussions of the problem.

9.3.2 AMALGAM FORMATION

The elemental form of mercury is apparently the cause of low temperature failure in the commonly used plate-fin aluminum heat exchangers (Rios et al., 1998). However, elemental mercury attacks copper, zinc and brass, chromium, iron, and nickel as well. If the natural gas is to undergo cryogenic processing, the mercury must be removed to levels below 0.01 $\mu g/Nm^3$ to avoid damage to aluminum heat exchangers used in low temperature service. A detailed discussion of the mechanism of mercury attack on brazed-aluminum exchangers is given by Nelson (1994).

9.3.3 REMOVAL PROCESSES

Both regenerative and nonregenerative processes are available for the removal of mercury from gas and liquid streams, and both types will be discussed briefly. All of the processes take advantage of the reactivity of elemental mercury for removal. Complete removal of all mercury requires reduction of mercury compounds to elemental mercury. Processes that use chemisorption for removal may simultaneously remove the organic mercury compounds through physical adsorption. Some of the processes remove arsenic and lead as well (Engineering Data Book, 2004d).

9.3.3.1 Nonregenerative Processes

Rios et al. (1998) and Engineering Data Book (2004d) list a number of chemisorption processes for removal of elemental mercury from hydrocarbon streams to $0.01 \mu g/Nm^3$ or lower. Bed capacities of 10% or higher are common. Most of these processes use sulfur impregnated on a support such as activated charcoal or alumina to provide a large surface area. Bourke and Mazzoni (1989) describe a sulfur-impregnated carbon adsorbent for mercury removal. In this process, the mercury reacts with the sulfur to form a stable compound on the adsorbent surface. The use of this type sulfur removal at the Anschutz Ranch East Plant in Wyoming is discussed in some detail by Lund (1996).

Some processes do not adsorb the organic forms of mercury that are inert to metals, and some treat only the gas phase.

9.3.3.2 Regenerative Process

Markovs and Corvini (1996) and Stiltner (2002) describe a regenerative process that utilizes silver on molecular sieve to chemisorb elemental mercury while providing dehydration at the same time. The advantage of silver is that the amalgam it forms with mercury decomposes at typical regeneration temperatures for dehydration (see Chapter 6). The silver-impregnated sieve is added to the standard molecular sieve dehydration bed, and the basic dehydration process remains unchanged. Essentially, all of the mercury condenses with the water on regeneration and forms a separate phase, which easily can be decanted and sold. Stiltner (2002) reports the solubility of water to be about 25 ppbw. Removal of the mercury from the water requires an additional adsorption step. After water and mercury removal, the regeneration gas can be recycled to the front of the bed without additional treatment, or it can be blended into the sales gas. If necessary, the regeneration gas can be further treated with a nonregenerative mercury-removal process to remove the last traces of mercury.

Although the silver-sieve process treats both gas and hydrocarbon liquid, how effective the process is in removing organic mercury compounds is unclear. Like the nonregenerative processes, this process can drop elemental mercury levels to $0.01\mu g/Nm^3$.

9.4 (BTEX) BENZENE, TOLUENE, ETHYLBENZENE, AND XYLENE

Benzene, toluene, ethylbenzene, and xylene (BTEX) present two possible problems in gas processing: environmental concerns and potential freezing in cryogenic units. In most gas processing plants, BTEX is present in the volatile organic compounds (VOC) that must be controlled to meet EPA clean air regulations. Benzene is also classified as a carcinogen under United States regulations and is considered an air toxin. Emission limits for air toxics are lower than those for VOCs. BTEX concentrates in the glycols and amines used to dehydrate and sweeten the gas. Initially, industry interest was in glycol dehydration because it posed the larger problem, but recently the amine sweetening process has been considered as well.

In glycol dehydration, the glycol absorbs some hydrocarbons as well as water. Although the paraffinic compounds are removed in the flash tank, a fraction of the aromatic compounds are not completely removed because of their relatively low vapor pressures at the flash conditions and the slight affinity of glycols for them (see Chapter 6 for process description). Use of triethylene glycol (the most commonly used glycol) to dehydrate a gas stream at 1,000 psia (14.5 bar), 100°F (38°C) and at a typical glycol rate of 3 gal TEG/lb of water absorbed, the TEG will absorb the following percentages of BTEX from the gas stream (Engineering Data Book, 2004c):

Benzene	10%
Toluene	14%
Ethyl benzene	19%
Ortho xylene	28%

When the rich glycol is regenerated, the BTEX exits the still in the vapor stream, and consequently this stream must be processed to recover or eliminate the BTEX before atmospheric venting. The methods presently used to control BTEX emissions to the atmosphere are:

- Adjustment of plant operating conditions to minimize the quantity of BTEX in the glycol absorber off gas
- Burning of the still off gases before venting
- Condensation of the off gases and recovery of the BTEX as a liquid product
- Adsorption of the BTEX on a carbon adsorbent

BTEX removal is thoroughly discussed in the paper by Rueter et al. (1992), and a summary of the main items in that paper is presented in the handbook by Kohl and Nielsen (1997).

Amine treaters absorb BTEX along with the acid gases. Both acid gases and the BTEX are stripped in the regenerator and, traditionally, either vented if the H_2S level is low or sent to a Claus unit for sulfur recovery. Collie et al. (1998) present a comparative study of emissions from glycol dehydrators and amine treaters. They used process simulation programs to model the behavior of both glycol and amine units and, on the basis of simulation results, came to the following conclusions:

- Reduction of the solvent circulation rate in both glycol and amine units and minimization of the lean-amine temperature were the most effective means of reducing BTEX emissions from the flash gas.
- A change of solvent type can also reduce emissions.
- For the cases studied, approximately 25% of the BTEX emissions came from amine units, and the remaining 75% came from glycol dehydrators.

Covington et al. (1998) present a similar paper that discusses how to minimize both operating costs and BTEX emissions from glycol dehydrators by use of ethylene glycol (EG), diethylene glycol (DEG), and triethylene glycol. They conclude that emissions can be reduced by use of either EG or DEG instead of TEG because of the lower solubility of BTEX in the lighter glycols. Operating costs may be reduced as well. However, this option may be attractive only when very low water contents are not required and the cost of glycol losses is not excessive.

Like carbon dioxide, both benzene and p-xylene can cause plugging problems in cryogenic NGL extraction units. If the feed contains significant quantities of BTEX, these compounds must be removed to a level that will prevent freezing in the low-temperature units. (GPA offers a computer program to estimate required concentrations.) Jones et al. (1999) describe a plant that contains a nitrogen-rejection unit (see Chapter 15 for details) where the feed gas contains approximately 0.12% BTEX. The plant uses a glycol dehydrator (TEG) to remove water to a level of 7 lb/MMscf and at the same time remove more than half of the BTEX. The glycol unit is followed by a molecular sieve dehydrator to achieve the level of dehydration necessary for the nitrogen rejection unit.

REFERENCES

Abbott, J. and Oppenshaw, P., Mercury Removal Technology and Its Applications, Proceedings of the Eighty-First Annual Convention of the Gas Processors Association, Tulsa, OK, 2002.

BOC Group, BOC to Build Australian Helium plant, 2005, http://www.boc.com/news/article_866_09mar05.asp, Retrieved September 2005.

Bland, C.J., A Review of NORM in Oil and Natural Gas Extraction, 2002, www.c5plus.com/norm.htm, Retrieved July 2005.

Bourke, M.J. and Mazzoni, A.F., The Roles of Activated Carbon in Gas Conditioning, Proceedings of the Laurance Reid Gas Conditioning Conference, Norman, OK, 1989, 137.

Choe, J.S., et al., Membrane/Cryogenic Hybrid Systems for Helium Purification, Proceedings of the Sixty-Seventh Annual Convention of the Gas Processors Association, Tulsa, OK, 1988, 251.

Collie, J., Hlavinka, M., and Ashworth, A., An Analysis of BTEX Emissions from Amine Sweetening and Glycol Dehydration Facilities, Proceedings of the Laurance Reid Gas Conditioning Conference, Norman, OK, 1998, 175.

Covington, K., Lyddon, L., and Ebeling, H., Reduce Emissions and Operating Costs with Appropriate Glycol Selection, Proceedings of the Seventy-Seventh Annual Convention of the Gas Processors Association, Tulsa, OK, 1998, 42.

Deaton, W.M. and Haynes, R.D., How newest helium plant is working, *Pet. Refiner*, 40, (3) 205, 1961.

Encyclopedia Americana, Danbury, CT, 1979, 23, 130c.

Engineering Data Book, 12th ed., Sec. 1, General Information, Gas Processors Supply Association, Tulsa, OK, 2004a.

Engineering Data Book, 12th ed., Sec. 2, Product Specifications, Gas Processors Supply Association, Tulsa, OK, 2004b.

Engineering Data Book, 12th ed., Sec. 20, Dehydration, Gas Processors Supply Association, Tulsa, OK, 2004c.

Engineering Data Book, 12th ed., Sec. 21, Hydrocarbon Treating, Gas Processors Supply Association, Tulsa, OK, 2004d.

Environmental Protection Agency, Mercury in Petroleum and Natural Gas: Estimation of Emissions from Production, Processing, and Combustion, EPA/600/R-01/066, September 2001, www.epa.gov/ORD/NRMRL/pubs/600r01066/600r01066.htm, Retrieved July 2005.

Environmental Protection Agency, The National Listing of Fish Advisories (NLFA), 2005a, http://epa.gov/waterscience/fish/advisories/index.html, Retrieved October 2005.

Environmental Protection Agency, Fact Sheet—EPA Issues Notice of Data Availability on Proposed Clean Air Mercury Rule, 2005b, http://www.epa.gov/mercury/control_emissions/nodafact.html, Retrieved September 2005.

Gray, P.R., Radioactive materials could pose problems for the gas industry, *Oil Gas J.*, 88 (26) 45, 1990.

Grice, K.J., Naturally Occurring Radioactive Materials (NORM) in the Oil and Gas Industry: A New Management Challenge, Presented at First International Conference on Health, Safety and Environment, The Hague, The Netherlands, Society of Petroleum Engineers Paper SPE 23384, November 1991.

Guccione, E., New Approach to recovery of helium from natural gas, *Chemical Eng.*, 70 (9) 73, 1963.

Johnson, D.J. and Rydjord, K., Unique NGL/NRU/HRU Plant Processes New Gas in Eastern Colorado, Proceedings of the Eightieth Annual Convention of the Gas Processors Assocation, Tulsa, OK, 2001.

Jones, S., Lee, S., Evans, M., and Chen, R., Simultaneous Removal of Water and BTEX from Feed Gas for a Cryogenic Plant, Proceedings of the Seventy-Eighth Annual Convention of the Gas Processors Association, Tulsa, OK, 1999, 108.

Kohl, A. and Nielsen, R., *Gas Purification*, 5th ed., Gulf Publishing, Houston, TX, 1997.

Lund, D.L., Causes and Remedies for Mercury Exposure to Aluminum Coldboxes, Proceedings of the Seventy-Fifth Annual Convention of the Gas Processors Association, Tulsa, OK, 1996, 282.

Markovs, J. and Corvini, J., Mercury Removal from Natural Gas and Liquid Streams, Proceedings of the Laurance Reid Gas Conditioning Conference, Norman, OK, 1996, 133.

Müssig, S., Experience in Removing Mercury from Natural Gas and Subsequent Mercury Decontamination of Process Equipment, Proceedings of the Laurance Reid Gas Conditioning Conference, Norman, OK, 1997, 31.

National Academy of Sciences, *The Impact of Selling the Federal Helium Reserve*, National Academies Press, Washington, D.C, 2000, books.nap.edu/catalog/9860.html, Retrieved July 2005.

Nelson, D.R., Mercury Attack of Brazed Aluminum Heat Exchangers in Cryogenic Gas Service, Proceedings of the Seventy-Third Annual Convention of the Gas Processors Association, Tulsa, OK 1994, 178.

Pacheco, N., *Helium*, U.S. Bureau of Land Management, Amarillo Field Office, Helium Operations, Amarillo, TX, 2003.

Railroad Commission of Texas, NORM—Naturally Occurring Radioactive Material, www.rrc.state.tx.us/divisions/og/key-programs/norm.html, Retrieved July 2005.

Remirez, R., Cold-box design is key in helium recovery unit, *Chemical Eng.*, 75 (27) 80, 1968.

Rueter, C.O., et al., Research on Emissions of BTEX and VOC from Glycol Dehydrators, Proceedings of the Laurance Reid Gas Conditioning Conference, Norman, OK, 1992, 187.

Rhodes, Z., private communication, 2005, www.newpointgas.com.

Rios, J.A., Coyle, D.A. Durr, C.A., and Frankie, B.M., Removal of Trace Mercury Contaminants from Gas and Liquid Streams in the LNG and Gas Processing Industry, Proceedings of the Seventy-Seventh Annual Convention of the Gas Processors Association, Tulsa, OK, 1998, 191.

Stiltner, J., Mercury Removal from Natural Gas and Liquid Streams, Proceedings of the Eighty-First Annual Convention of the Gas Processors Association, Tulsa, OK, 2002.

Smith, J.M., Van Ness, H.C., and Abbott, M.M., Introduction to Chemical Engineering Thermodynamics, 6th ed., McGraw-Hill, New York, 2001.

WEB SITES

Northeast Waste Management Officials Association: www.newmoa.org/Newmoa/htdocs/ This site provides information on a variety of environmental problems, including mercury and hazardous wastes by organization of environmental officials in northeastern United States.

Pollution Probe: www.pollutionprobe.org/Publications/Mercury.htm. This site provides a review of mercury and its health and environmental effects.

10 Liquids Processing

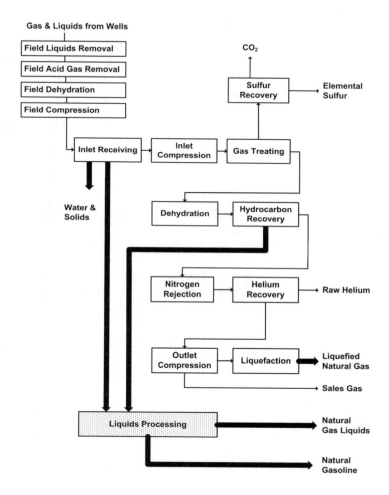

Gas & Liquids from Wells

- Field Liquids Removal
- Field Acid Gas Removal
- Field Dehydration
- Field Compression

Inlet Receiving → Inlet Compression → Gas Treating

CO₂

Sulfur Recovery → Elemental Sulfur

Water & Solids

Dehydration → Hydrocarbon Recovery

Nitrogen Rejection → Helium Recovery → Raw Helium

Outlet Compression → Liquefaction → Liquefied Natural Gas

→ Sales Gas

Liquids Processing → Natural Gas Liquids

→ Natural Gasoline

10.1 INTRODUCTION

The previous chapters focused on processing gas streams. This chapter discusses processing the hydrocarbon liquids. The two liquid streams processed in a gas plant are the condensate from inlet separators and demethanizer bottoms NGL stream. Condensate from field operations is brought by truck to the plant as well, but this condensate typically goes directly to storage and then to sale as natural gasoline or slop oil without further processing. This chapter discusses the processing required to produce marketable liquid products.

10.2 CONDENSATE PROCESSING

Figure 10.1 shows a schematic of one configuration for the condensate that leaves the inlet receivers of a plant. Processing the condensate involves two steps: water washing and condensate stabilization. Depending upon the associated water quality, the condensate may require a water wash to remove salts and additives. If they are not a problem, then incoming water is removed in the inlet receiver and a water wash is unnecessary. The water often contains high concentrations of methanol or ethylene glycol, added for hydrate prevention. Plants may recover the inhibitors onsite or ship the water offsite for inhibitor recovery before disposal. In most cases, waste water is injected into disposal wells (Kuchinski, 2005).

After removal of free water, the condensate goes to the stabilizer, where remaining lighter hydrocarbons are stripped and recombined with the gas that leaves the inlet receiver. The primary purpose of the stabilizer is to produce a bottoms product that has specifications to be sold as "natural gasoline," or "slop oil." Specifications on natural gasoline include:

- Volatility as measured by ASTM D323 (Standard Test Method for Vapor Pressure of Petroleum Products (Reid Method)) test. The values range from 10 to 34 psi (0.7 to 2.3 bar), but the common range is 9 to 12 psi (0.6 to 0.8 bar), when the liquid is trucked offsite (Kuchinski, 2005). The Engineering Data Book (2004a) lists the grades of natural gasoline on the basis of RVP and percentage of liquid evaporated at 140°F (60°C).

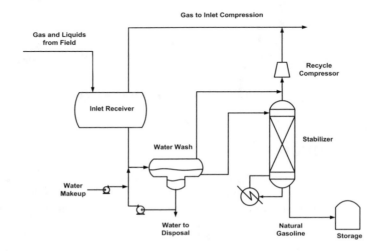

FIGURE 10.1 Inlet receiver condensate-processing schematic.

- Density set in the range of 65 to 85° API (650 to 720 kg/m³), with a price penalty for higher gravities (lower densities).
- Water content, defined as BS&W (bottoms sediment and water), that does not exceed 1 vol%, as determined by centrifuge. Higher values discount sales price of natural gasoline. Langston (2003) provides details on sampling and analysis for BS&W.
- Reactive sulfur, as determined by the copper strip test (see Chapter 5 for details).

The Engineering Data Book (2004a) lists additional specifications related to corrosion, distillation curve, color, and reactive sulfur.

The stripping column must operate at a temperature and pressure to ensure that the bottoms product meets the volatility and density specification. It usually operates in the 75 to 400 psig (5 to 27 barg) pressure range. Although lower pressure reduces the heat duty on the stabilizer reboiler, the operating pressure is usually dictated by the number of recycle compressor stages and available capacity (Kuchinski, 2005). It usually is more economical to provide additional reboiler duty than to add additional compression.

10.2.1 Sweetening

Ideally, the sulfur compounds go with the light gases stripped in the stabilizer, so that they can be removed in gas treating. Most of the H_2S and CO_2 will be removed along with COS, CS_2, and mercaptans. (Mercaptans are commonly used odorants in natural gas and propane.) To enhance acid gas removal, sweet natural gas can be used as a stripping gas. This process can reduce H_2S levels to the 10 ppmv range (Webber et al., 1984). However, the cost is increased recycle compression.

If the natural gasoline is sour, the plant can either treat it or take a price penalty and sell the liquid as sour crude. The price differential between sweet and sour crude drives the need for sweetening. In many cases, the liquid volumes are low and sweetening at the plant is not justified. Any of the sweetening processes for NGL, discussed in detail below in 10.3.1, can be used. However, some form of caustic wash is the most commonly used.

10.2.2 Dehydration

Dehydration of natural gasoline usually is unnecessary. When in storage, free water drops out and the remaining low water content should meet customer requirements, provided the water layer is not shipped, too. However, emulsions of water and natural gasoline occur, which require emulsion breakers to obtain a clean water–hydrocarbon separation.

10.3 NGL PROCESSING

Figure 10.2 shows a possible processing configuration for the demethanizer bottoms NGL stream, with extensive fractionation of NGL. The processing required depends upon the plant's feed quality and product slate and customer requirements.

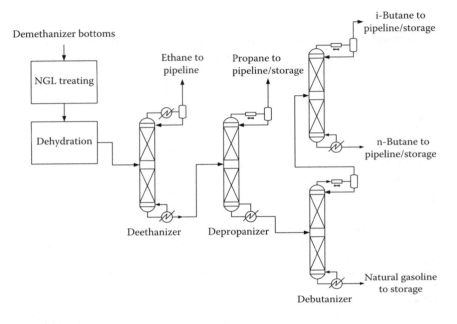

FIGURE 10.2 NGL fractionation train.

Processing demethanizer bottoms liquids ranges from no additional processing to sweetening, dehydration and, extensive fractionation. Figure 10.2 implies that all of the NGL is treated, which is not always the case. Feed quality and product slate dictate which streams require treatment.

10.3.1 SWEETENING

As noted above, most of the sulfur compounds are stripped into the gas phase, which, after any compression, typically goes to an amine treater. Ideally, all of the sulfur compounds would be removed in the gas treating step. This step eliminates additional sweetening and additional dehydration, if aqueous systems (i.e., amines or caustics) are used. As noted in Chapter 5, gas treating units are extremely effective at removing the acid gases, but most do a poor job of removing other sulfur compounds, especially mercaptans, sulfides, and disulfides. Also, the acid gases and sulfur compounds concentrate in the demethanizer bottoms. Therefore, extremely low sulfur levels are required in gas treating to meet the product specifications in the liquid. If inlet gas contains both CO_2 and H_2S, COS formed in regeneration of molecular sieve (see Chapter 6) may be present as well. Any COS present concentrates in the propane product stream (Mick, 1976). This section discusses commonly used processes to remove sulfur compounds and residual CO_2 from NGL.

10.3.1.1 Amine Treating

Amine treating of liquids is the same process used for treating gases. Most remove only CO_2 and H_2S. Two amines, DGA and DIPA (see Chapter 5), will remove COS, but none remove CS_2 and mercaptans. Amine treating is often used upstream of caustic treaters to minimize caustic consumption caused by irreversible reactions with CO_2.

Amine treating is usually conducted in a packed tower, although sieve trays are effective as well. To avoid vaporization, contactor operating pressure should be 100 psi (7 bar) above the liquid bubble point, and the entire unit should operate at least 50 psi (3.4 bar) above the bubble point (Nielsen, 1995). The Engineering Data Book (2004d) recommends a minimum contactor packing depth of 20 feet (6 m) and flow rates not exceeding 20 gpm per foot squared of cross-sectional area (0.014 m³/s per m²).

Sargent and Seagraves (2003) and Veroba and Stewart (2003) provide additional details on contactor design. Nielsen (1995) notes that the lean-amine viscosity should not exceed 2 cP to ensure good amine/hydrocarbon separation. Even so, the expected normal concentration of amine in the hydrocarbon stream that leaves the unit is about 100 ppmw. Water wash is recommended to both recover amine and protect downstream caustic treaters. Veroba and Stewart (2003) and Nielsen (1995) provide specifics on the water wash.

10.3.1.2 Adsorption

Adsorption is a commonly used method for sweetening NGL and LPG streams. It has the advantage of being able to remove sulfur compounds down to low levels when no water is present, which makes it attractive for treating demethanizer bottoms. In contrast to the amine and caustic wash, no potential downstream dehydration is required. Either promoted alumina, 13X or 5A molecular sieve will remove H_2S, COS, and mercaptans from LPG. The larger pore 13X is usually preferred because it has a higher capacity (Richman, 2005). Mick (1976) discusses the limitations of using molecular sieve for COS removal to the 2 ppm level when concentrations are in the hundreds of ppm.

Hydrocarbon flow can be up or down during adsorption. However, upflow is preferred, if possible, so that the regeneration can be down, to make recovery of all liquid from the bed easier (Richman, 2005). Liquid velocities are usually 3 to 5 ft/min (0.9 to 1.5 m/s), and bed height should be at least 5 feet (1.5 m), to allow good distribution (Engineering Data Book, 2004c). Liquid adsorption beds use the same graduated support packing used in gas adsorption.

The bed must be drained of liquid, and a gas used for regeneration. If the fluid being treated is butane or lighter, sweet vapor of the fluid being treated can be used, which eliminates purging after regeneration. Refilling should be done with dry sweet liquid. The filling rate must be slow to prevent bed movement. Also, the bed needs to be cool to avoid flashing and bed movement (Richman, 2005). Bed rearrangement is detrimental because it induces channeling and causes attrition; the generated dust leads to plugging in the bed or downstream filters.

The Engineering Data Book (2004c) indicates that new adsorbents capacity and performance data are usually presented on the basis of static measurements. However, under flow conditions, the capacity may be about 50% of the static values, depending on the total amount of adsorbent in the bed. This discrepancy is because of the mass transfer zone and bed aging (see Chapter 6 for details).

10.3.1.3 Caustic Treating

A number of caustic processes, both regenerative and nonregenerative, are used to remove sulfur compounds from NGL. The simplest process is the use of a nonregenerative solid KOH bed. It has the advantage of taking out H_2S and water as well. According to available literature, the nonregenerative solid KOH process is not commonly used. Sambrook et al. (1997) provide some discussion on the solid-bed KOH system. Mick (1976) found KOH effective for removal of H_2S but not other sulfur compounds. However, by injection of methanol into the bed with the propane, COS, mercaptans, and CS_2 can be removed. The process has the advantage of reducing sulfur content, while still producing a dry product.

Eckersley and Kane (2004) suggest that nonregenerative metal oxides, such as ZnO, are economically attractive if the sulfur rate into the bed is less than 400 lb/d (180 kg/d). The bed operates at 300 to 700°F (150 to 370°C).

One of the most common processes for treating natural gasoline, and NGL, is the regenerative caustic wash with sodium hydroxide. Frequently, it follows treating with an amine system. The amine pretreatment eliminates the acid gases and reduces the load on the caustic wash that removes mercaptans.

The caustic wash process is similar to the hot potassium carbonate process discussed in Chapter 5. It is extensively used in refineries for treating sour streams. For H_2S removal, the process is regenerable, but CO_2 forms a nonregenerative carbonate salt. This by-product, along with the consumption of caustic for H_2S removal, is why the amine process precedes the caustic wash.

Figure 10.3 shows the process for a single-stage treatment along with a water wash. The water wash eliminates carry over of the treating solution in the hydrocarbon. Sand towers can be used to disengage the two liquid phases as well.

Hydrocarbon enters at the bottom of the contactor through a distributor to ensure that it is the dispersed phase. The tower contains either packing or sieve trays. The top 8 to 10 feet (2.4 to 3 m) of the contactor contains no packing or trays to permit the phases to separate (Engineering Data Book 2004d).

Multiple stages of caustic wash are used if essentially complete removal is required. For a single-stage unit, the caustic concentration is 10 to 15 wt%; if multiple stages are used, the first stage concentration is 4 to 6 wt% and the others are 8 to 10 wt% (Engineering Data Book 2004d). The contactor operates near ambient temperature and at a sufficiently high pressure to avoid vaporization. The regenerator bottoms temperature is usually around 275°F (135°C) (Kuchinski, 2005). This system can be highly corrosive if not operated properly or if oxygen is present in the stream being treated (McCartney, 2005).

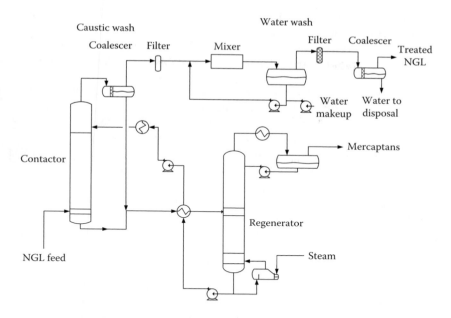

FIGURE 10.3 Regenerative caustic wash and water wash.

The chemistry involved in the wash is straightforward:

$$RSH + NaOH \leftrightarrow RSNa + H_2O \qquad (10.1)$$

$$H_2S + NaOH \leftrightarrow NaSH + H_2O \qquad (10.2)$$

$$NaSH + NaOH \leftrightarrow Na_2S + H_2O \qquad (10.3)$$

$$H_2S + 2NaOH \leftrightarrow Na_2S + 2H_2O \qquad (10.4)$$

$$CS_2 + 2NaOH \rightarrow 2NaSH + CO_2 \qquad (10.5)$$

$$CO_2 + 2NaOH \rightarrow Na_2CO_3 + H_2O \qquad (10.6)$$

The RSH in Equation 10.1 denotes a mercaptan, and their removal is a prime reason for use of a caustic wash. Unfortunately, as Equation 10.5 shows, carbon disulfide, which is usually at low concentrations compared with the other sulfur compounds, reacts with sodium hydroxide to form CO_2, which produces the nonregenerative carbonate salt in Equation 10.6.

Some variations on the caustic wash to extract mercaptans exist. One process, Merox® (Verachtert et al., 1990), forms the water-soluble mercaptan salt in

Equation 10.1. Air is added to the rich caustic, with a catalyst to oxidize the mercaptan to a disulfide:

$$2RSNa + 1/2 \ O_2 + H_2O \rightarrow RSSR + 2NaOH \qquad (10.7)$$

The disulfide is water insoluble and separated from the caustic before it returns to the wash. The paper describes process alternatives that reduce the required amount of caustic and treat different hydrocarbon streams. The separation of caustic and disulfide is somewhat difficult; although the phases are immiscible, the two liquids have almost identical densities (McCartney, 2005). Even when treating streams with no CO_2 present, the Merox™ caustic will eventually become spent from the CO_2 in the regeneration air stream.

The Engineering Data Book (2004d) describes a continuous process that oxidizes the mercaptan by reducing cupric chloride to cuprous chloride. Both the Merox™ and copper processes require the feed to be free of H_2S.

10.3.1.4 Other Processes

The focus of the above section was acid gas and mercaptan removal. However, none of the processes discussed address the problem of elemental sulfur removal. The Engineering Data Book (2004d) briefly describes a polysulfide solution prepared from elemental sulfur, water, sodium hydroxide, and sodium sulfide (Na_2S) for elemental sulfur removal.

As noted in Chapter 6, COS hydrolyzes with water to form CO_2 and H_2S. However, COS does not affect the copper strip sulfur test. A properly run copper strip test includes the addition of water in an attempt to hydrolyze the COS (Engineering Data Book, 2004a) but may be unreliable. Even when run properly, the lack of sample vessel conditioning can lead to erroneous results (Eckersley and Kane, 2004). Tests showed the rate of hydrolysis to be extremely slow in the copper strip test (McCartney, 2005). (McCartney [2005] notes that the Gas Processors Association is conducting a study of this problem.) Because of this problem, a shipment that contains water and COS can pass the test at departure but fail at delivery. Carbonyl sulfide unfortunately is fairly difficult to eliminate by conventional methods. Processes for its removal include treatment with either of the two amines diglycolamine (DGA) or diisopropanolamine (DIPA), or with adsorption by use of molecular sieve 3A, although the cheaper 4A will work if the regeneration gas contains no CO_2. Numerous nonregenerative metallic oxide processes are also available to remove COS from product streams. Some of these processes remove the COS directly and others require water to hydrolyze the COS to H_2S before it is reacted.

10.3.2 Dehydration

The acceptable water content in NGL streams varies from no free water present, for process streams, to product specifications of less than 10 ppmw, for commercial propane and HD-5 propane (Engineering Data Book, 2004a).

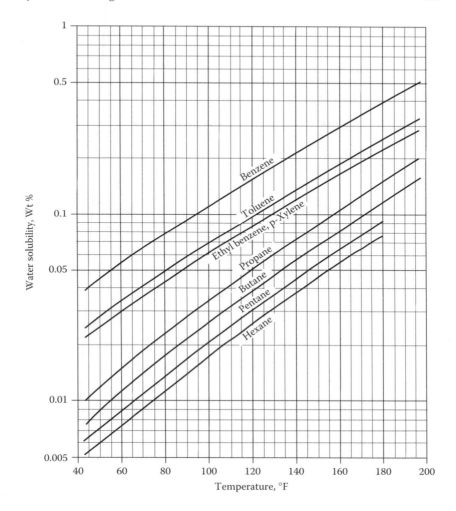

FIGURE 10.4 Solubility of water in hydrocarbons as a function of temperature. (Adapted from Engineering Data Book, 2004c.)

Figure 10.4, from the Engineering Data Book (2004c), gives the solubility of water in various hydrocarbon liquids as a function of temperature. Figure 10.5 shows the solubility of some hydrocarbons in water as a function of temperature. (Yaws et al. [1990] tabulates the solubility of C_4 to C_{20} hydrocarbons in water at 25°C.) However, customers sometimes require very dry liquid product, such as when it is used as feedstock to a process in which water is a catalyst poison.

Numerous processes are available for drying liquids, and many of the technologies discussed in Chapter 6 for dehydrating gases may be used for liquids.

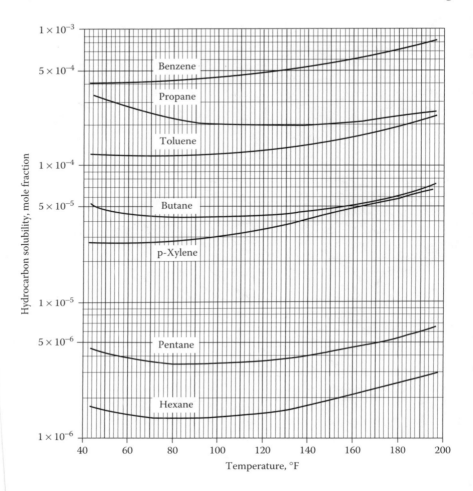

FIGURE 10.5 Solubility of liquid hydrocarbons in water. (Adapted from Engineering Data Book, 2004c.)

Therefore, this section presents only brief descriptions as needed and lists the pros and cons of each method as they apply to NGL.

10.3.2.1 Adsorption Processes

Section 10.3.1.1 described use of adsorbents in liquid service. When drying, to have the hydrocarbon flow up the bed to prevent refluxing during regeneration is particularly advantageous. Any of the adsorbents described in Chapter 6 provide adequate drying to meet the free water specification. If extremely low water contents are essential, then the 4A molecular sieve is required.

10.3.2.2 Desiccant Dehydration

A consumable desiccant such $CaCl_2$ is viable in remote situations, but disposal issues make it unattractive compared with the regenerative process in a plant situation where hydrocarbon liquids are dried. The Engineering Data Book (2004c) provides a brief description of a two-bed system that generates a solid waste instead of concentrated brine. The book notes that the process is less used now because of disposal problems.

10.3.2.3 Gas Stripping

Gas stripping is a simple process for liquid dehydration and is frequently used offshore to dry condensates before they go to the production platform (Engineering Data Book, 2004c). The liquid is usually stripped with dry natural gas in a tower with trays or packing. As shown by Equation 6.1, the process is feasible because the activity coefficient, γ, for water in liquid hydrocarbons is in tens of thousands, which makes the water volatile.

The process is low cost and space efficient. However, it requires a dry natural gas stream, which also strips some volatile hydrocarbons at the same time. This property makes it suitable for only condensates and natural gasoline. The stripping gas must be compressed and recycled to a dehydration unit or sent to fuel gas.

10.3.2.4 Distillation

Propane and butane mixtures can be dried by use of specially designed columns (Engineering Data Book 2004c). In these columns, the water goes overhead with hydrocarbon and is decanted from the overhead reflux drum. The condenser hydrocarbon returns as reflux. The bottoms product will be bone dry, and even a side draw three or four trays from the top will meet product specifications. The Engineering Data Book (2004c) notes that the side-draw liquid must be cooled, but the process can be cost effective. This factor should make it attractive for narrow boiling fractions.

10.3.2.5 Absorption

Dehydration of liquid product streams can also be done by use of glycol contacting. These systems are very similar to gas dehydration with glycol, which is covered in Chapter 6.

10.3.3 FRACTIONATION

Table 10.1 provides typical operating conditions for the various columns. The condensate stabilizer is included for completeness. As noted previously, the fractionation train depends upon the current economics and customer requirements. Frequently the first column takes an ethane–propane mixture (EP mix) overhead instead of pure ethane. Regardless of whether it is ethane or EP mix, this stream is pipelined to customers. All of the other products are shipped by tank car, truck, or pipeline.

TABLE 10.1
Typical Fractionator Operating Conditions

	Operating Pressure, psig (barg)	Number of Actual Trays	Reflux Ratio[a]	Reflux Ratio[b]	Tray Efficiency (%)
Deethanizer	375 – 450 (26 – 30)	25 – 35	0.9 – 2.0	0.6 – 1.0	60 – 80
Depropanizer	240 – 270 (16 – 19)	30 – 40	1.8 – 3.5	0.9 – 1.1	80 – 90
Debutanizer	70 – 100 (4.8 – 6.2)	25 – 35	1.2 – 1.5	0.8 – 0.9	85 – 95
Butane splitter	80 – 100 (5.5 – 6.9)	60 – 80	6.0 –14	3.0 – 3.5	90 – 100
Condensate stabilizer	100 – 400 (6.9 – 28)	16 – 24	Top feed	Top feed	50 –75

[a] Reflux ratio relative to overhead product, mol/mol.
[b] Reflux ratio relative to feed, gal/gal.

Source: Adapted from Engineering Data Book (2004b).

Distillation column design is beyond the scope of this book. However, the Engineering Data Book (2004b) provides extensive details on column design, column internals, and reboiler design.

10.4 SAFETY AND ENVIRONMENTAL CONSIDERATIONS

The major safety and environmental issues involved with treating the hydrocarbon liquids is the safe handling and disposal of the caustic compounds and solutions, if used. At present, sulfur specifications are set on the basis of corrosion. Gasoline and diesel will be required to meet increasingly stringent levels of total sulfur in fuels. By 2007, the maximum sulfur level in gasoline will be 30 ppm average (Environmental Protection Agency, 1999). The primary motivation for sulfur reduction is to reduce NO_X. NO_X reduction is accomplished by increasing the performance of the automobile's catalyst to destroy NO_X, which is poisoned by sulfur.

The push for low sulfur levels in fuel affects the natural gasoline and propane markets. Natural gasoline may not be accepted as a gasoline blending stock unless it contains 30 ppm or less of sulfur (McCartney, 2005). The other major market for the liquid is as a petrochemical feedstock, which often brings a lower sales price than as a gasoline blending stock. The commonly used propane fuel,

HD-5 propane, could face competition from other low-sulfur fuels. How signif-icantly this competition will affect the propane market and how gas processors will respond is unclear at this time.

REFERENCES

Eckersley, N. and Kane, J.A., Designing Customized Desulfurization Systems for the Treatment of NGL Streams, Proceedings of the Laurance Reid Gas Conditioning Conference, Norman, OK, 2004, 313.

Engineering Data Book, 12th ed., Sec. 2, Product Specifications, Gas Processors Supply Association, Tulsa OK, 2004a.

Engineering Data Book, 12th ed., Sec. 19, Fractionation and Absorption, Gas Processors Supply Association, Tulsa, OK, 2004b.

Engineering Data Book, 12th ed., Sec. 20, Dehydration, Gas Processors Supply Associ-ation, Tulsa, OK, 2004c.

Engineering Data Book, 12th ed., Sec. 21, Hydrocarbon Treating, Gas Processors Supply Association, Tulsa, OK, 2004d.

Environmental Protection Agency, EPA's Program for Cleaner Vehicles and Cleaner Gasoline, EPA420-F-99-051, 1999, www.epa.gov/tier2/frm/f99051.pdf, Retrieved June 2005.

Kuchinski, J. private communication, 2005.

Langston, L.V., *The Lease Pumper's Handbook,* 1st ed., The Oklahoma Commission on Marginally Producing Oil and Gas Wells, Norman, OK, 2003, http://www.mar-ginalwells.com/MWC/MWC/Searchable%20Text2.pdf, Retrieved May 2005.

McCartney, D.G., private communication, 2005.

Mick, M.B., Treat propane for COS removal, *Hydrocarbon Process.*, 55 (7) 137, 1976.

Nielsen, R.B., Principles of LPG Amine Treater Design, Proceedings of the Laurance Reid Gas Conditioning Conference, Norman, OK, 1995, 229.

Richman, P., private communication, 2005.

Sargent, A. and Seagraves, J., LPG Contactor Design and Practical Troubleshooting Tech-niques, Proceedings of the Laurance Reid Gas Conditioning Conference, Norman, OK, 2003, 125.

Sambrook, R.M., Dillingham, B., and Prugh, J., A Liquid Propane Desulfurization Study, Proceedings of the Laurance Reid Gas Conditioning Conference, Norman, OK, 1997, 158.

Verachtert, T.A., Cassidy, R.T., and Holmes, E.S., The UOP Merox Process for NGL Treat-ing, Proceedings of the Laurance Reid Gas Conditioning Conference, Norman, OK, 1990, 267.

Veroba, R. and Stewart, E., Fundamentals of Gas Sweetening, Proceedings of the Laurance Reid Gas Conditioning Conference, Norman, OK, 2003, 1.

Webber, W.W., Petty, L.E., Ray, B.D., and Story, J.G., Whitney Canyon Project, Proceed-ings of the Sixty-Third Annual Convention if Gas Processors Association, Tulsa, OK, 1984, 105.

Yaws, C. L., Yang, H-C., Hopper, J.R., and Hansen, K.C., Hydrocarbons: Water solubility data, *Chemical Eng.*, 97 (4) 1990, 177.

11 Sulfur Recovery

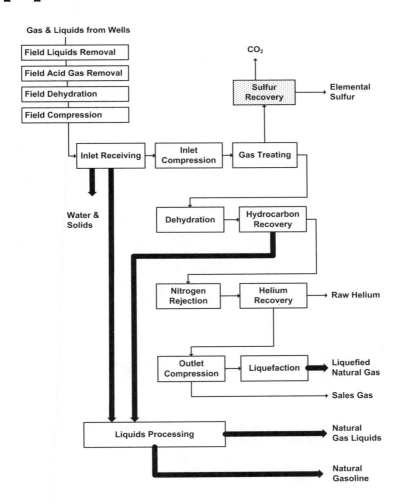

11.1 INTRODUCTION

Leppin (2001) points out that approximately 25% of the natural gas being brought into production from new sources requires H_2S removal and disposal. Consequently, sulfur removal processes discussed in Chapter 5 will play an increasingly larger role in future gas processing. Currently only two methods are available for dealing with large quantities of H_2S:

- Disposal of the gas by injection into underground formations
- Conversion of the H_2S into a usable product, elemental sulfur

Where environmental regulations permit and production of elemental sulfur is not economically attractive, injection wells provide a safe means for H_2S disposal. Jones et al. (2004) discuss design and operation of a facility that enriches an H_2S stream before injection into a disposal well. Oberding et al. (2004) provide additional details on the project.

However, more commonly, H_2S is converted into elemental sulfur, much of which goes into sulfuric acid production. As late as 1950, over half of the world's sulfur supply came from "voluntary producers," that is, companies whose principal purpose was to produce elemental sulfur. Now, these producers furnish less than 5% of the world's supply and "involuntary producers," primarily petroleum refineries and natural gas plants, are the major source of the element (Hyne, 2005).

The most common method of converting H_2S into elemental sulfur, is the Claus process or one of its modifications. We discuss two modified Claus configurations: straight through and split flow (Engineering Data Book, 2004).

The straight-through process is the preferred and simplest, and it can process feed streams that contain more than 55 mol% H_2S; with air or acid gas preheat, it can process 30 to 55 mol% H_2S in the feed. The split-flow configuration can process feeds that contain 5 to 30 mol% H_2S. Lower concentration feeds require other variations (Engineering Data Book, 2004) and are not discussed here. The straight-through process provides the highest sulfur-recovery efficiency.

Unfortunately, the exit stream from Claus plants usually cannot meet environmental emission requirements, and, consequently, a tail gas cleanup unit (TGCU) is often employed to eliminate the last of the sulfur compounds. According to available literature, the most commonly used processes are Shell Claus Offgas Treating (SCOT), SUPERCLAUS, and cold-bed adsorption (CBA). This discussion will be limited to these processes.

The most comprehensive information on sulfur recovery processes is available in the Conference Fundamentals Manual of the Laurance Reid Gas Conditioning Conference (2003), Maddox and Morgan (1998), Engineering Data Book (2004), Kohl and Nielsen (1997), Echterhoff (1991), and Leppin (2001). Strickland et al. (2001) provide a thorough discussion of TGCU processes, along with cost comparisons.

This chapter discusses the properties of elemental sulfur and then describes Claus and tail gas cleanup processes. It then briefly discusses sulfur storage.

11.2 PROPERTIES OF SULFUR

The thermophysical properties of sulfur are unusual. For a complete understanding of the Claus conversion process, a brief discussion of the relevant properties is necessary.

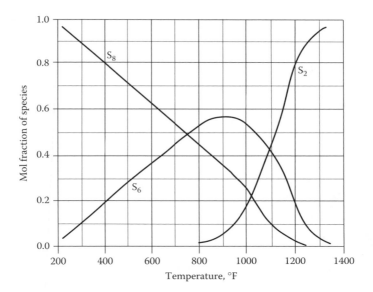

FIGURE 11.1 Sulfur-vapor species as a function of temperature. (Adapted from Engineering Data Book, 2004.)

Sulfur vapor exists as S_x, where x can have values from 1 through 8. Figure 11.1 shows the distribution of sulfur-vapor species as a function of temperature. Note that at lower temperatures, S_8 dominates, but as the temperature rises S_8 converts to S_6, and finally to S_2. The sequence is not unexpected, because increased temperature means increased energy for the molecules, which leads to the breakup of the clusters. The formation of sulfur clusters has a very pronounced effect on the physical properties that have a significant effect on processing operations, notably viscosity (fluid flow) and heat capacity (heat transfer). An excellent summary of liquid and vapor properties is available in the Engineering Data Book (2004).

11.3 SULFUR RECOVERY PROCESSES

11.3.1 CLAUS PROCESS

All Claus units involve an initial combustion step in a furnace. The combustion products then pass through a series of catalytic converters, each of which produces elemental sulfur. We briefly discuss the basic chemistry and then describe process details.

11.3.1.1 Basic Chemistry

The Claus process consists of the vapor-phase oxidation of hydrogen sulfide to form water and elemental sulfur, according to the overall reaction:

$$3 \ H_2S + 3/2 \ O_2 \rightarrow 3 \ H_2O + (3/x) \ S_x. \tag{11.1}$$

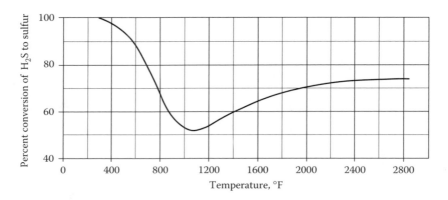

FIGURE 11.2 Equilibrium conversion of hydrogen sulfide to sulfur. (Adapted from Engineering Data Book, 2004.)

The above overall reaction does not represent the reaction mechanism or show intermediate steps. In practice, the reaction is carried out in two steps:

$$H_2S + 3/2 \, O_2 \leftrightarrow H_2O + SO_2 \qquad (11.2)$$

$$2 \, H_2S + SO_2 \leftrightarrow 2 \, H_2O + (3/x) \, S_x. \qquad (11.3)$$

The first reaction is a highly exothermic combustion reaction, whereas the second is a more weakly exothermic reaction promoted by a catalyst to reach equilibrium. Figure 11.2 shows the equilibrium conversion obtained for H_2S into elemental sulfur by the Claus reaction. Kohl and Nielsen (1997) state that the unusual shape of the equilibrium curve is caused by the existence of different sulfur species at different reaction temperatures. They point out that at a sulfur partial pressure of 0.7 psia (0.05 bar) and temperatures below 700°F (370°C), the vapor is mostly S_6 and S_8, but at the same partial pressure and temperatures over approximately 1,000°F (540°C), S_2 predominates. This shift in species causes the equilibrium constant in the reaction to shift from a downward slope to an upward slope, as shown in Figure 11.2. This behavior has a significant effect on the operation of the Claus process.

The melting point of amorphous sulfur is 248°F (120°C), and its normal boiling point is 832°F (445°C). Figure 11.2 shows that the maximum conversion to sulfur by reaction 11.1 is obtained at temperatures near the melting point of sulfur, but to maintain sulfur in the vapor state, relatively high temperatures are required. Consequently, if the catalytic converters are to operate under conditions in which the sulfur does not condense on the catalyst, they cannot operate at optimum equilibrium conversion. This is the reason for having a series of converters, with the sulfur product withdrawn from the reacting mixture between converters. Withdrawing the sulfur product causes the reaction 11.3 to shift to the right, which results in more sulfur product.

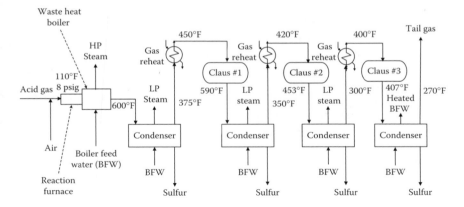

FIGURE 11.3 Straight-through Claus unit. (Adapted from Echterhoff, 1991.)

11.3.1.2 Process

Figure 11.3 and Figure 11.4 show simplified flow diagrams for the two common configurations, straight through and split flow, respectively. In the straight-through configuration, the first reaction takes place in a combustion furnace operating near ambient pressure (3 to 8 psig; 0.2 to 0.6 barg). The air flow rate is adjusted to react with one third of the H_2S, along with any other combustibles, such as hydrocarbons and mercaptans. The H_2S reaction is exothermic (637 Btu/scf, or 24,000 kJ/m^3, at 25°C at 1 atm) and is used to produce steam in a waste-heat boiler. Both reactions take place in the furnace–boiler combination, and the gases exit the waste-heat boiler in the range of 500 to 650°F (260 to 343°C) (Parnell, 1985), which is above the sulfur dew point, so no sulfur condenses in the boiler.

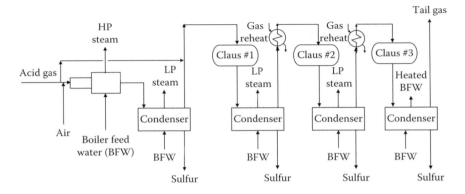

FIGURE 11.4 Split-flow Claus unit. Typically two thirds of feed bypasses the reaction furnace. (Adapted from Echterhoff, 1991.)

The combustion furnace–boiler is followed by several catalytic reactors in which only the second reaction takes place because all the O_2 has been consumed in the furnace. Each catalytic reactor is followed by a condenser to remove the sulfur formed. The gas is cooled to 300 to 400°F (149 to 204°C) in the condenser to remove elemental sulfur. The condenser generally achieves cooling by heat exchange with water to produce low-pressure steam. Vapor that leaves the condenser is at the sulfur dew point, so the gas is reheated before passing to the next converter to prevent sulfur deposition on the catalyst.

A combustion-furnace flame temperature of 1700°F (927°C) should be maintained (Kohl and Nielsen, 1997) because the flame is not stable below this value. The straight-through configuration cannot be used at H_2S concentrations below 55%, because the feed gas heating value is too low. Concentrations as low as 40% are acceptable (Parnell, 1985), if the air or acid gas is preheated.

For H_2S concentrations in the range 25 to 40%, the split flow configuration (Figure 11.4) can be utilized. In this scheme, the feed is split, and one third or more of the feed goes to the furnace and the remainder joins the furnace exit gas before entering the first catalytic converter. When two thirds of the feed is bypassed, the combustion air is adjusted to oxidize all the H_2S to SO_2, and, consequently, the necessary flame temperature can be maintained (Kohl and Nielsen, 1997). The split-flow process has two constraints (Kohl and Nielsen, 1997):

1. Sufficient gas must be bypassed so that the flame temperature is greater than approximately 1,700°F (927°C).
2. Maximum bypass is two thirds because one third of the H_2S must be reacted to form SO_2. (Note the stoichiometry in equations 11.2 and 11.3.)

If air preheating is used with the split-flow configuration, gases with as little as 7% H_2S can be processed (Parnell, 1985).

Generally, the sulfur recovery in the conventional plants discussed above varies from 90 to 96% for two catalytic converters. It increases to 95 to 98% for three catalytic converters (Lagas et al., 1989).

11.3.2 CLAUS TAIL GAS CLEANUP

Before sulfur emissions restrictions were imposed, the offgas from the Claus unit was flared to convert the remaining H_2S to SO_2. Now, environmental agencies demand higher sulfur recovery than can be achieved with a standard Claus unit, and additional treating of the Claus tail gas is needed. The tail gas cleanup entails either an add-on at the end of the Claus unit or a modification of the Claus unit itself. The processes for this final sulfur removal are generally divided into three categories (Kohl and Nielsen, 1997):

- Direct oxidation of H_2S to sulfur
- Sub-dew point Claus processes
- SO_2 reduction and recovery of H_2S

d'Haêne (2003) provides a recent overview of tail gas cleanup processes. This section provides an example process in each category.

11.3.2.1 Direct Oxidation of H$_2$S to Sulfur

SUPERCLAUS is an example of selective oxidation for final sulfur removal. The process as described by Lagas et al. (1988, 1989, 1994) involves a slightly modified two-stage Claus unit followed by a third-stage catalytic reactor to oxidize the remaining H$_2$S to elemental sulfur. Figure 11.5 shows the overall process.

Reactors 1 and 2 use the standard Claus catalyst, whereas the third reactor contains the selective oxidation catalyst. The Claus unit itself is operated with a deficiency of air so that the gas that exits the second reactor contains 0.8 to 3 vol% H$_2$S. Sufficient air is added to this exit gas to keep the oxygen level in the 0.5 to 2 vol% range. The mixture then goes to the third reactor, where the following catalytic reaction occurs:

$$2H_2S + O_2 \rightarrow 2S + 2H_2O \qquad (11.4)$$

The selective oxidation catalyst in the third reactor does not promote the reaction

$$2H_2S + 3O_2 \leftrightarrow 2SO_2 + 2H_2O \qquad (11.5)$$

or the reverse reaction of sulfur with H$_2$O

$$3S + 2H_2O \leftrightarrow 2H_2S + SO_2 \qquad (11.6)$$

and, consequently, a total recovery rate of 99% or higher can be reached. Lagas, et al. (1988) describes a modification of SUPERCLAUS that can reach a sulfur recovery rate of 99.5%.

11.3.2.2 Sub-Dew Point Claus Processes

This category has several processes, but the only one considered here is the cold-bed adsorption (CBA) process because it is the most widely used sub-dew point process (Ortloff, undated). Figure 11.6 shows one of its many possible variations (Goddin et al., 1974). The front end of the unit is a Claus reactor, and the sub–dew point process takes place in the final two catalytic converters, CBA #1 and CBA #2. One converter is in service while the other is being regenerated. After the gas leaves the final condenser, it is not reheated but is sent to the third converter (CBA #2), which is operated at a temperature well below the sulfur dew point. This flow results in a better equilibrium conversion but deposits sulfur on the catalyst and causes a gradual loss of activity. During the period that CBA #2 is in service, CBA #1 is being heated and regenerated with a slip stream from Claus #1. After regeneration is complete, the reactor can be cooled with a slip stream from the final condenser. When the catalyst in CBA #2 is exhausted, valves are switched, and CBA #1 becomes the adsorbing reactor while CBA #2 under-goes regeneration. Total sulfur recoveries greater than 99% can be obtained with CBA processes (Clark et al., 2002)

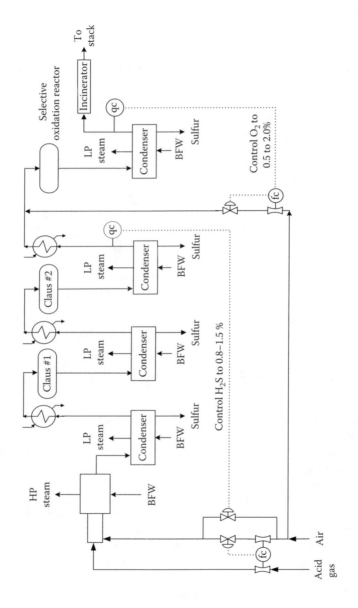

FIGURE 11.5 SUPERCLAUS unit. (Adapted from d'Haene, 2003.)

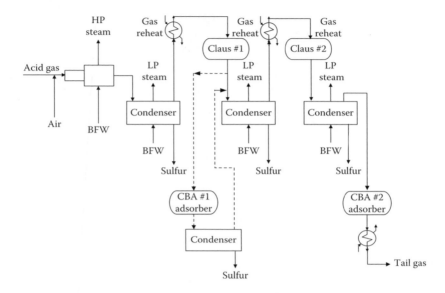

FIGURE 11.6 Cold-bed adsorption unit. Dashed line denotes regeneration stream. (Adapted from Goddin et al., 1974.)

11.3.2.3 Reduction of SO₂ and Recovery of H₂S

The SCOT process is an example of the process that reduces the SO_2 in the Claus plant offgas back to H_2S. It then uses amine treating to remove the H_2S, which is recycled back to the Claus plant for conversion to elemental sulfur. The process can produce an exit gas that contains 10 to 400 ppmv of total sulfur (Kohl and Nielsen, 1997), while increasing total sulfur recovery to 99.7% or higher. Figure 11.7 shows a simplified process flow diagram of the SCOT process.

The following description of the process has been adapted from Echterhoff (1991). The feed, offgas from the Claus unit, is heated to 575°F (302°C) in an inline

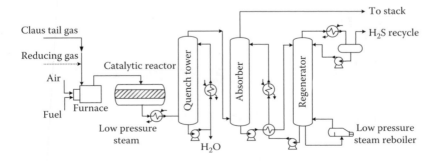

FIGURE 11.7 Shell Claus Offgas Treating (SCOT) unit. (Adapted from Echterhoff, 1991.)

burner, along with a reducing gas, H_2 or a CO and H_2 mixture. The reducing gas is supplied either from an outside source or generated by partial oxidation in an inline burner. The mixture then flows to the SCOT catalytic reactor (cobalt–molybdenum on alumina), where the sulfur compounds, including SO_2, CS_2, and COS are reduced to H_2S and water. The gas that leaves the reactor goes to a waste-heat exchanger, where it is cooled to about 320°F (160°C) and produces low-pressure steam.

The gas from the waste-heat exchanger then flows through a quench tower, where it is cooled to approximately 100°F (38°C) by externally cooled recycle water in countercurrent flow. The water from the tower is condensed, and the excess water is sent to a sour water stripper. Gas from the quench tower then contacts an aqueous amine solution in the absorption column. The amine is generally methyldiethanola-mine (MDEA) or diisopropylamine (DIPA) to absorb H_2S while slipping CO_2. The gas that exits the top of the absorber contains very little H_2S (10 to 400 ppm) and is sent to an incinerator. The rich amine that leaves the bottom of the absorber flows to the regenerator, where heat is applied to strip the H_2S from the amine solution. The overhead from the regenerator is cooled to condense the water, and the H_2S is recycled to the Claus unit. Lean amine is cooled and returned to the absorber.

11.4 SULFUR STORAGE

Sulfur from the Claus unit is withdrawn as a liquid and is generally stored and transported in the molten state. A number of potential problems are associated with sulfur storage, including release of H_2S dissolved in the molten sulfur and the possibility of sulfur fires, which will produce highly toxic SO_2. A brief summary, adapted from Johnson and Hatcher (2003a), is given below:

- Sulfur fires. Uncommon, but they can produce large amounts of SO_2.
- H_2S. Any H_2S dissolved in the molten sulfur from the condensers may be a significant hazard if appropriate degassing techniques are not used.
- Corrosion. A wet sulfidic atmosphere can lead to severe corrosion of carbon steel.
- SO_2. Highly toxic and it forms highly corrosive sulfurous acid in the presence of water.
- Static discharge. Because of the excellent insulating properties of mol-ten sulphur, static discharge may occur under certain conditions and lead to possible fires or explosions.

A comprehensive discussion of sulfur storage and degassing is beyond the scope of this book, but Johnson and Hatcher (2003a, 2003b) give an excellent presentation.

11.5 SAFETY AND ENVIRONMENTAL CONSIDERATIONS

Table 11.1 through Table 11.3, adapted from Johnson and Hatcher (2003a), sum-marize the more important physical properties of molten sulfur, hydrogen sulfide, and sulfur dioxide. The tables also address safety and environmental concerns.

TABLE 11.1
Important Properties, Safety, and Environmental Concerns of Molten Sulfur

Physical Property or Characteristic	Hazards
Melts at 240°F (116°C)[a]	Possible severe burns
Boils at 851°F (455°C)[a]	Exhibits measurable vapor pressure at molten sulfur handling process conditions
Flash point 334 – 369°F (168 – 187°C)	Not much higher that normal storage temperatures
Autoignites at 478 – 511°F (248 – 266°C)	Sulfur fires can easily occur in the presence of oxygen
Viscosity transition at 318°F (159°C)	Increasingly difficult to pump above this temperature Magnitude is dependent upon H_2S content of the molten sulphur
Highly flammable in air	Sulfur fires
Solid dust cloud autoignites at 375°F (191°C)	Important variable in solid-forming operations Deflagration possible
Excellent insulator	Static discharge at 11% or higher oxygen atmosphere
Toxicity	Highly toxic at low concentrations

[a] For S_8, Georgia Gulf Sulfur (2000) gives a melting point of 230 to 246°F (110 to119°C) and [a] Boiling point of 832°F (444°C).

Source: Johnson and Hatcher (2003a).

TABLE 11.2
Important Properties, Safety, and Environmental Concerns of H_2S.

Physical Property or Characteristic	Hazards
Highly flammable in air between approximately 3.4 and 45% at process conditions	Deflagration possible
Exposed carbon steel forms iron sulfide corrosion product under wet, reducing conditions	Excessive corrosion in carbon steel piping in some cases Iron sulfides are highly pyrophoric

Source: Johnson and Hatcher (2003a).

TABLE 11.3
Important Properties, Safety, and Environmental Concerns of SO_2

Toxicity	Highly toxic at low concentrations
Corrosive when wet	Forms sulfurous acid in presence of water at low temperatures

Source: Johnson and Hatcher (2003a).

REFERENCES

Clark, P., et al., Enhancing the Performance of the CBA Process by Optimizing Catalyst Macroporosity, Proceedings of the Laurance Reid Gas Conditioning Conference, Norman, OK, 2002, 79.

Conference Fundamentals Manual, Proceedings of the Laurance Reid Gas Conditioning Conference, Norman, OK, 2003.

d'Haêne, P.E., Tail Gas Treating, Proceedings of the Laurance Reid Gas Conditioning Conference, Norman, OK, 2003, 89.

Echterhoff, L.W., State of the Art of Natural Gas Processing Technologies, Gas Research Institute Report, GRI-91/0094, 1991.

Engineering Data Book, 12th ed., Sec. 22, Sulfur Recovery, Gas Processors Supply Association, Tulsa, OK, 2004.

Georgia Gulf Sulfur Corporation, www.georgiagulfsulfur.com/properties.htm, 2000, Retrieved September 2005.

Goddin, C.S., Hunt, E.B. and Palm, J.W., CBA process ups Claus recovery, *Hydrocarbon Process.*, 53 (10) 122, 1974.

Hyne, J.B., The sulfur bubble, *Hydrocarbon Eng.*, 10 (4) 23, 2005.

Johnson, J.E. and Hatcher, N.A., Hazards of Molten Sulfur Storage and Handling, Proceedings of the Laurance Reid Gas Conditioning Conference, Norman, OK, 2003a, 109.

Johnson, J.E. and Hatcher, N.A., A comparison of established sulfur degassing technologies, *Proceedings of the Laurance Reid Gas Conditioning Conference*, Norman, OK, 2003b, 131.

Jones, S.G., Rosa, D.R., and Johnson, J.E., Design, Cost and Operation of an Acid Gas Enrichment and Injection Facility, *Proceedings of the Laurance Reid Gas Conditioning Conference*, Norman, OK, 2004, 81.

Kohl, A. L. and Nielsen, R.B., *Gas Purification,* Gulf Publishing, Houston, TX, 1997.

Lagas, J.A., Borsboom, J., and Berben, P.H., The SUPERCLAUS Process, Proceedings of the Laurance Reid Gas Conditioning Conference, Norman, OK, 1988, 41.

Lagas, J.A., Borsboom, J., and Heijkoop, G., Claus process gets extra boost, *Hydrocarbon Process.*, 68 (4) 40, 1989.

Lagas, J.A., Borsboom, J., and Goar, B.G., SUPERCLAUS, Five Years of Operating Experience, Proceedings of the Laurance Reid Gas Conditioning Conference, Norman, OK, 1994, 192.

Leppin, D., Large-scale sulfur recovery, GasTIPS, 7, 26, 2001.

Maddox, R.N. and Morgan, D. J., *Gas Conditioning and Processing*, Vol. 4, Gas Treating and Sulfur Recovery, Campbell Petroleum Series, Norman, OK, 1998.

Oberding, W., et al., Lessons Learned and Technology Improvements at the Lost Cabin Gas Plant, Proceedings of the Laurance Reid Gas Conditioning Conference, Norman, OK, 2004, 273.

Ortloff Engineers, Ltd, www.Ortloff.com, Retrieved July 2005.

Parnell, D., Look at Claus unit design, *Hydrocarbon Process.*, 64 (9) 114, 1985.

Strickland, J.F., et al., Relative capabilities and costs of tail gas clean-up processes, GasTIPS, 7 9, 2001.

12 Transportation and Storage

12.1 INTRODUCTION

Whereas raw natural gas is usually located relatively close to gas plants, the processed natural gas and natural gas liquid products must be transported to the end-user, industrial or residential. Except for the less common situation in which the gas plant is dedicated to a single commercial customer, product storage is required. In the case of natural gas, transportation and storage problems are more difficult than with other common forms of energy, such as coal and oil, because the energy density of natural gas is so low at ambient temperatures and pressures. For an equal volume of these three fuels at ambient temperature and pressure, the energy content of gasoline is approximately 1,000 times greater than that of natural gas. The same volume of coal (anthracite, bulk density 50 to 58 lb/ft³ [800 to 940 kg/m³]), with a heating value 13,500 Btu/lb (31 MJ/kg), contains 700 times more energy as a fuel. This large difference in energy density compared with natural gas highlights two major problems with natural gas. First, a relatively high pressure is required to increase the gas density and raise the energy content per unit volume so that the gas can be transported economically by pipeline (common pressures are approximately 800 to 1,500 psig [60 to 100 barg]). Second, large quantities of natural gas cannot be stored in relatively simple and inexpensive aboveground facilities similar to those used for liquid-petroleum products. Note that for methane, the primary component of natural gas, the critical temperature is −118°F (−83°C), and, consequently, no amount of pressure can convert methane into a liquid at 60°F (15°C).

As discussed in Chapter 1, many natural gases contain significant quantities of hydrocarbons that can be removed and liquefied (NGL) to produce the dry natural gas distributed to industrial and residential customers. (Dry in this context means that the NGL content is sufficiently low so that liquid will not drop out at pipeline temperatures and pressures.) Although more volatile, NGL is transported and stored in much the same way as crude oil and refined petroleum products.

Natural gas can be liquefied at cryogenic temperatures as liquefied natural gas (LNG). The following chapter discusses the unique issues associated with transporting and storing LNG.

12.2 GAS

12.2.1 TRANSPORTATION

Depending on the gas plant's function, any of several possible gas streams are transported from the plant to customers. Producers ship less common gas products such as hydrogen sulfide, carbon dioxide, nitrogen, and raw helium to customers via dedicated pipelines. These lines tend to be short because the customer usually is located near the gas plant. Although some plants have a single dedicated customer for the sales gas, most gas goes to an extensive pipeline network dedicated to natural gas service. Because product quality is so uniform within each country, many different gas processors utilize the same trunk pipelines, and many gas plants are connected to multiple pipelines.

The United States in 2004 had over 212,000 miles of interstate gas pipelines (Energy Information Administration 2005a), serviced by more than 1,200 compressor stations, and capable of transporting over 32 Tcf (910 Bm^3) of natural gas per year (Smith et al., 2005).

The average increase in pipeline capacity (excluding gathering system lines) between 1998 and 2004 was close to 9 Bcfd (0.25 Bm^3/d), but this capacity is expected to increase even further over the next 3 years (Energy Information Administration, 2005a). This increase corresponds to an average of about 2,300 miles (3700 km) of pipelines added each year during the same period. Of these new pipelines, six major pipeline systems were put into operation in the Gulf of Mexico in 2004. This capacity accounted for 1.8 Bcfd (0.05 Bm^3/d), which was 23% of the total increase (Energy Information Administration, 2005a).

Figure 12.1 shows the distribution of line diameters scheduled for construction in 2006 in both the United States and the world; however, information on operating pressures for the lines was not available. The smaller diameter lines are normally used in the gathering systems, discussed in Chapter 3, which bring the gas from the wellhead to the gas processing plant. The larger lines are generally long distance transmission lines designed to bring sales gas to customers.

The larger and longer lines are normally built with 30-inch or more (75 cm) diameters. For example, one proposal for a gas pipeline from Alaska to the lower 48 states calls for a 48-inch (120-cm) line that operates at 2,500 psi

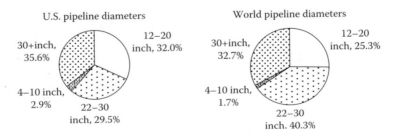

FIGURE 12.1 Distribution of line diameters in the United States and in the World Scheduled for Construction in 2006 (True, 2004).

(170 bar), and the West to East pipeline in China will have a mainline of 2,360 miles (3,800 km), with a diameter of 40 inches (100 cm), and 183 miles (294 km) of lateral lines, with diameters of 32 inches (813 mm), 20 inches (50 cm), and 16 inches (40 cm) (Gray et al., 2003). Major interstate pipelines are usually 24 to 36 inches (60 to 90 cm) in diameter, and the lateral lines that connect to the mainline are typically 6 to 16 inches (15 to 40 cm). More detailed information on pipeline transportation is available in the book by Manning and Thompson (1991).

Thirty gas pipeline compressor stations were built in the United States in 2003 (True and Stell, 2004), and 13 were built in 2004 (Smith et al., 2005). Capital costs averaged slightly over \$1,200/hp, but the average can be misleading because costs ranged from \$748/hp to \$3,436/hp. The large variation is expected, as the data do not consider important cost factors, such as location and type of compressor used. These cost figures include materials, labor, land, and miscellaneous items. Figure 12.2 shows the relationship between total cost and station size. A definite, expected relation exists between cost and horsepower. An approximate cost for new gas pipeline compressor stations can be determined from the following equation:

$$\text{Cost } (10^6 \text{ US \$}) = 2.6 + 1.1 \text{ (hp/1,000)}.$$

The cost for a compressor station on gathering systems may be as much as 50% lower (McCartney, 2005). Note that these costs are in 2004 dollars. Chapter 15 discusses how to correct for inflation.

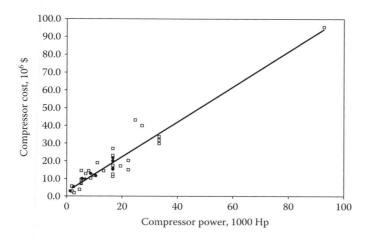

FIGURE 12.2 Total cost of gas pipeline compressor stations (millions of US dollars) as a function of compression horsepower (True and Stell, 2004 (open symbol); Smith et al., 2005 (closed symbol)). See text for details.

12.2.2 MARKET CENTERS

Market centers, or hubs, are usually located near the intersection of several major pipelines. They are an outgrowth of market restructuring in the 1980s and the requirement that interstate gas companies transport gas but not buy or sell the gas. They exist in the United States and in Canada. According to the Energy Information Administration (2003) "The defining characteristics of a natural gas market center are that it provides customers (shippers and gas marketers primarily) with receipt/delivery access to two or more pipelines systems, provides transportation between these points, and offers administrative services that facilitate that movement and/or transfer of gas ownership." In 2003, 37 operational hubs existed in the United States and Canada. Market center services vary widely between facilities, and no two hubs are identical, but Table 12.1 lists typical services provided. In 2003, all but seven hubs provided the services through the Internet; the others used fax-phone. The business is highly competitive, and like any business has had closings and mergers.

The best known (but not the largest) of the United States and Canadian market centers is undoubtedly the Henry Hub, located in Louisiana and serving an average of 180 large commercial customers each day through 14 interconnecting pipeline systems. The hub can handle over 1.8 Bscfd (0.05 Bm³) of gas, has two salt dome storage sites with a working capacity of 15 Bcf (0.42 Bm³), and also serves as the delivery point for New York Mercantile Exchange (NYMEX) natural gas futures contracts. Market centers are valuable in facilitating the management of transportation and storage of natural gas in the United States, and their numbers and importance will undoubtedly increase in the future. Additional information can be obtained from the Energy Information Administration (2003) and Beckman and Determeyer (1997).

12.2.3 STORAGE

12.2.3.1 Introduction

Before 1992, the natural gas market was regulated, and natural gas storage had two main purposes:

- To provide baseload storage to meet seasonal, weather-sensitive, demands above normal pipeline delivery capability
- To provide peak (or peak shaving) storage to smooth out the demand curve

After deregulation, storage facilities were open for additional uses, such as storing gas when prices were low and selling gas when prices were higher. The storage can be either baseload or peak shaving. Baseload storage provides a volume of working gas for a long-term steady withdrawal of the gas (typically 3 to 5 months), followed by a long-term injection period (typically 5 to 7 months). Peak shaving facilities, on the other hand, are designed to provide large quantities of gas over a relatively short time frame (hours or days). Figure 12.3 clearly

TABLE 12.1
Services Arranged and Coordinated in Market Hub Centers

Transportation/wheeling—Transfer of gas from one interconnected pipeline to another through a header (hub), by displacement (including exchanges), or by physical transfer over the transmission of a market-center pipeline.

Parking—A short-term transaction in which the market center holds the shippers gas for redelivery at a later date. Often uses storage facilities, but may also use displacement or variations in linepack.

Linepack—Refers to the gas volume contained in a pipeline segment. Linepack can be increased beyond the pipeline's certificated capacity temporarily and, within tolerances, by increased compression.

Loaning —A short-term advance of gas to a shipper by a market center that is repaid in kind by the shipper a short time later. Also referred to as advancing, drafting, reverse parking, and imbalance resolution.

Storage—Storage that is longer than parking, such as seasonal storage. Injection and withdrawal operations may be separately charged.

Peaking —Short-term (usually less than a day and perhaps hourly) sales of gas to meet unanticipated increases in demand or shortages of gas experienced by the buyer.

Balancing—A short-term interruptible arrangement to cover a temporary imbalance situation. The service is often provided in conjunction with parking and loaning.

Title transfer—A service in which changes in ownership of a specific gas package are recorded by the market center. Title may transfer several times for some gas before it leaves the center.

Electronic trading —Trading systems that either electronically match buyers with sellers or facilitate direct negotiation for legally binding transactions.

Administration—Assistance to shippers with the administrative aspects of gas transfers, such as nominations and confirmations.

Compression—Provision of compression as a separate service. If compression is bundled with transportation, it is not a separate service.

Risk Management—Services that relate to reducing the risk of price changes to gas buyers and sellers, for example, exchanges of futures for physicals.

Hub-to-hub transfers—Simultaneous receipt of a customer's gas into a connection associated with one center and an instantaneous delivery at a distant connection associated with another center.

Source: Energy Information Administration (2003).

illustrates the necessity for both types of storage by showing how the demand for natural gas in the United States cycled over a 3-year period. Considering the very large peaks and valleys, facilities capable of both meeting the long-term "average" demand and possessing storage to handle the peak seasonal demand are obviously necessary. Note that the spike in the summer months demand results from electric utilities that use gas turbines (the same as those used for gas compression) to meet peak power demands.

12.2.3.2 Facilities

Basically, two types of storage facilities exists for natural gas: (1) relatively small capacity (to 15 MMscf [400,000 Sm^3]) aboveground, floating-roof gas holders that

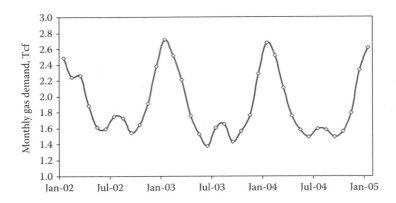

FIGURE 12.3 United States natural gas demand over 3-year period (Energy Information Administration, 2005b).

operate near ambient pressure and (2) much larger underground facilities (depleted oil and gas fields, salt caverns, and aquifers) that operate at elevated pressures. Overall, underground storage is more important and the discussion here is limited to this topic. The use of liquefied natural gas (LNG) storage for peak shaving is discussed in Chapter 13.

The United States Geological Survey first proposed use of underground storage for natural gas in 1909. The first North American facility was a depleted gas reservoir in Welland County, Ontario, Canada converted to storage use in 1915. This facility was followed by the Zoar field in Concord, New York in 1916 (Cates, 2001). In 1916, a German patent was issued on solution mining of salt caverns for storage of crude oil and distillates, and in 1950, this concept was applied to storage of NGL in the Keystone field, West Texas. The first use of a salt cavern for natural gas storage came in 1961, in Marysville, Michigan, and the first use of a leached salt dome for storage was in 1964, at Eminence, Mississippi (Beckman and Determeyer, 1997). The number of storage facilities has grown steadily in recent decades; Figure 12.4 shows the capacity and number of underground facilities in the United States in 2002. In the 2000 to 2003 period, the total underground storage capacity has remained constant at around 8.2 Tcf (220 Bm^3) (Energy Information Administration, 2005c).

Underground storage is prevalent throughout the world, with two exceptions, Japan and Korea. In these two countries, gas storage is primarily in the form of LNG. (International Energy Agency, 2002).

Aquifers are underground natural water reservoirs that can, under the right circumstances, be used for gas storage. However, aquifer storage is usually the most expensive and, thus, the least desirable underground storage method for six reasons (NaturalGas.org, 2004):

1. Geologic characteristics of a specific aquifer are generally not well known, which is usually not the case with a depleted gas or oil field,

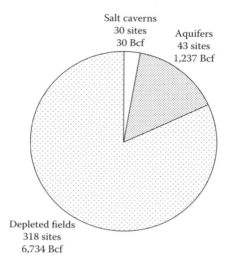

Salt caverns
30 sites
30 Bcf

Aquifers
43 sites
1,237 Bcf

Depleted fields
318 sites
6,734 Bcf

FIGURE 12.4 Capacity of underground storage facilities in the United States at the end of 2004 (Energy Information Administration, 2005c).

and, consequently, considerable resources must be expended to determine the suitability of the aquifer for gas storage.

2. Infrastructure (wells, pipelines, dehydration facilities, compression equipment, etc.) is unavailable at the aquifer site, whereas a depleted gas reservoir would have most of this infrastructure in place.

3. Considerable injection pressure may be required to displace the water with gas.

4. Withdrawn gas requires dehydration.

5. Aquifer formations generally require a much higher level of cushion or base gas (up to 80% of the total gas volume) than do depleted fields or salt caverns, and, thus, less of the reservoir volume is usable.

6. Environmental regulations govern the use of aquifers for gas storage.

All of the above factors increase both the capital cost and time necessary for development of aquifer storage. Consequently, depleted fields and salt caverns are normally preferred.

A common and relatively inexpensive technique for creating large storage facilities is solution mining of underground salt beds. After the salt bed has been located and the appropriate well or wells drilled, a coaxial pipe is inserted in the well bore. Water is then pumped down the annulus of the pipe, and the dissolved brine is withdrawn through the inner pipe (Figure 12.5). The cavern formed tends to be free from fractures that would permit gas leaks and is well suited for pressurized gas storage. Beckman and Determeyer (1997) present additional information on this subject.

In common with all engineering processes, storage gas reservoirs have their own terminology. Table 12.2 summarizes the most important terms.

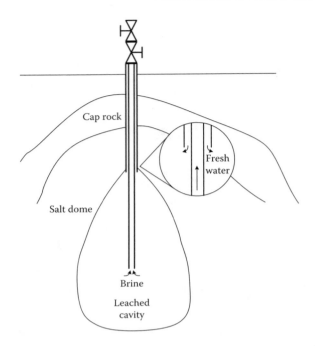

FIGURE 12.5 Generating underground-storage salt caverns by leaching with fresh water. (Adapted from Beckman and Determeyer, 1997.)

TABLE 12.2
Important Terminology for Underground Storage

Total gas storage capacity	Maximum volume of gas that can be stored in an underground storage facility
Total gas in storage	Volume of storage in the reservoir at a particular time
Cushion gas (or base gas)	Volume of gas intended as a permanent inventory in a storage reservoir to maintain adequate pressure and deliverability rates throughout the withdrawal season
Working gas capacity	Total gas storage capacity minus cushion gas
Deliverability	Most often expressed as measure of the amount of gas that can be delivered (withdrawn) from a storage facility on a daily basis; deliverability is variable and depends on such factors as the amount of gas in the reservoir at any time, the reservoir pressure, and compression capability available to the reservoir
Injection capacity (or rate)	Amount of gas that can be injected into a storage facility on a daily basis; injection rate is also variable; depends on the same factors as deliverability.

Source: Energy Information Administration (2004).

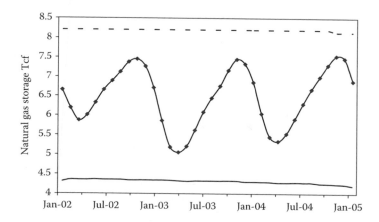

FIGURE 12.6 Underground gas storage volumes in the United States, 2002 through 2004. The lower solid line denotes the cushion gas volume and upper dashed line represents the total storage capacity (Energy Information Administration, undated).

The measures above are not absolute and are subject to change and interpretation. For example, in practice, a storage facility may be able to exceed certified total gas storage capacity by exceeding certain operational parameters. Also, the distinction between cushion and working gas is somewhat arbitrary, so facilities can withdraw some cushion gas for supply to markets in times of heavy demand.

Figure 12.6 shows the total gas in storage and the working gas for the 2002 to 2004 period. The pattern follows that for the demand shown in Figure 12.3. Figure 12.7 shows the total gas injected and withdrawn and the net gas withdrawn over the same time period. Note that even during the winter peak demand periods, some gas is being injected.

More detailed information on gas storage in depleted reservoirs and aquifers is available in Katz and Lee (1990).

12.3 LIQUIDS

12.3.1 TRANSPORTATION

In common with other petroleum products, liquids are transported by dedicated pipeline, by batching with other liquids in a petroleum pipeline, and by rail, truck, and marine transport. The most feasible and economical means of transport depends upon many factors, including:

- Product specifications, especially with respect to water and sulfur requirements
- Existing infrastructure
- Volatility
- Rate of product produced

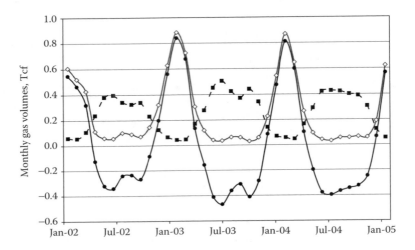

FIGURE 12.7 Volume of gas injected (■) and withdrawn (◇) and net gas withdrawn (●) from underground storage on monthly basis between January 2002 and December 2004 (Energy Information Administration, undated).

- Distance between plant and customer
- Geographic location

The first factor dictates the limitations of pipeline or storage container usage. Bulk containers and pipelines have different contamination problems. For containers, the possibility exists of a residual "heel" when one fluid is replaced with another. Even with a minimal heel, surface adsorption of water and sulfur compounds can contaminate the product being loaded. Many tank car loads are contaminated in these two ways.

In pipelining a fluid, contamination between different products transported in the same line is impossible to avoid. An interfacial zone exists between the two fluids, where mixing occurs because of convection. Pipeline operators offload the mixture, called transmix, into storage tanks at pipeline terminals. The transmix then is fractionated to regain the two products. To reduce the amount of transmix by up to 50%, pipeline operators use batching pigs (Webb, 1978).

Whereas pipeline operators know how much transmix to remove from the shipment, they have no idea of how much product can be contaminated by surface adsorption. Even worse, some fluids are shipped saturated with water. If the fluid temperature drops, water may condense and collect in line depressions. Large quantities of water-dry product can be contaminated this way. Therefore, pipeline operators try to monitor product quality when contamination can occur. When it happens, the pipeline goes out of service and "chasing fluids" are run through the line to remove contaminates. This process is costly to both the pipeline operator and those who use the pipeline.

Because of the diverse nature of the liquid products from a gas plant, we address transporting each of the products separately.

12.3.1.1 Natural Gas Liquids

Natural gas liquids (NGL) are the major liquid product from gas plants that do not further fractionate demethanizer bottoms. Usually, the quantity is large enough to dedicate a pipeline to the customers, who typically are operators of central NGL fractionation and storage hubs. Specifications exist on methane content, water content, and sulfur content, but the need for continuous flow is the major reason for the dedicated line.

12.3.1.2 Ethane-Propane Mixtures

The high vapor pressure of ethane–propane mixtures (E-P mix) and water-content specification make dedicated pipelines the common transport means for this liquid. Common carriers in Texas provide dedicated pipelines to transport E-P mixes to the major customers, refineries, and chemical plants. This also applies to high purity ethane.

12.3.1.3 Liquefied Petroleum Gas

Because it is a widely used fuel with strict specifications on water and sulfur, liquefied petroleum gas, or LPG, (more specifically HD-5 propane) must use dedicated transport. During the early years of the evolving LPG transportation system, the principal method was the railroad tank car, with pipelines playing a minor role, but by 1962, more than 40% of LPG was transported by pipeline and today it is the preferred method. A major reason for this shift was the construction of the Mid-America Pipeline in the early 1960s, a line added to serve the upper Midwest. This 2,200 mile (3,500 km) pipeline starts in eastern New Mexico and runs up to Wisconsin and Minnesota. Since then, additional LPG pipelines have been added to service most of the eastern United States.

12.3.1.4 Butanes

No strong preference exists for how butanes are shipped from gas plants, because the quantities produced usually do not justify the cost of a pipeline. If available, railroad tank car is desirable. Butane is usually stored on-site to permit batch loading into tank cars or trucks to a hub. For distribution of the product long distances from the hub, it can be put into pipelines that transport gasoline and diesel. Contamination is not an issue in this case, because the butanes are normally blended with gasoline to adjust the gasoline vapor pressure.

12.3.1.5 Natural Gasoline

Natural gasoline production is typically so small that it is stored and then loaded into a tank car or truck. There is little contamination issue because the customer is usually a refinery which will use it for gasoline blending.

FIGURE 12.8 Above ground storage vessels for LPG and NGL.

12.3.2 STORAGE

Light gas-plant producer liquids can be stored aboveground in pressurized vessels or below ground in large caverns. Figure 12.8 shows a typical row of cylindrical storage vessels, and Figure 12.9 is an example of a spherical vessel for butane storage. Figure 12.10 shows an underground cavern being prepared for NGL storage (Cannon, 1993). Additional information on cavern storage is available in the book by Katz and Lee (1990).

FIGURE 12.9 Spherical storage tanks. (Courtesy of Duke Energy Field Services.)

FIGURE 12.10 Underground storage cavern being prepared for NGL service (Cannon, 1993). (Courtesy of Gas Processors Association.)

REFERENCES

Beckman, K.L. and Determeyer, P.L, Natural gas storage: historical development and expected evolution, GasTIPS, 3, 13, 1997.

Cates, H.C., Storage functions evolve to match changes in U.S. natural gas industry, *Oil Gas J.*, 99.41, 48, 2001

Cannon, R.E., *The Gas Processing Industry, Origins and Evolution*, Gas Processors Association, Tulsa, OK, 1993, 245.

Energy Information Administration, U.S. Department of Energy, Natural Gas Navigator, U.S. Natural Gas Gross Withdrawals and Production, undated http://tonto.eia. doe.gov/dnav/ng/ng_prod_sum_dcu_NUS_a.htm, Retrieved September 2005.

Energy Information Administration, U.S. Department of Energy, Office of Oil and Gas, Natural Gas Market Centers and Hubs: a 2003 update, 2003, www.eia.doe.gov/pub/oil_gas/natural_gas/feature_articles/2003/market_hubs/mkthubs03.pdf, Retrieved September 2005.

Energy Information Administration, U.S. Department of Energy, The Basics of Underground Natural Gas Storage, 2004, http://www.eia.doe.gov/pub/oil_gas/natural_gas/analysis_publications/storagebasics/storagebasics.html, Retrieved October 2005.

Energy Information Administration, U.S. Department of Energy, Office of Oil and Gas, Changes in U.S. Natural Gas Transportation Infrastructure in 2004, 2005a, www.eia.doe.gov/pub/oil_gas/natural_gas/feature_articles/2005/ngtrans/ngtrans.pdf, Retrieved September 2005.

Energy Information Administration, U.S. Department of Energy, Natural Gas Consumption by End Use, 2005b, http://tonto.eia.doe.gov/dnav/ng/xls/ng_cons_sum_dcu_nus_m.xls, Retrieved September 2005.

Energy Information Administration, U.S. Department of Energy, Underground Natural Gas Storage Capacity 2005c, http://tonto.eia.doe.gov/dnav/ng/xls/ng_ stor_ cap_ dcu_ nus_a.xls, Retrieved September 2005.

Gray, L.A., Hong, W., and Naxin, M. Construction of major Chinese gas pipeline enters final phase in 2003, *Oil Gas J.*, 101.7, 68, 2003.

International Energy Agency, Natural Gas Information 2002, Paris, 2002.

Katz, D.L. and Lee, R.L., *Natural Gas Engineering, Production and Storage*, McGraw-Hill, New York, 1990.

Manning, F.S. and Thompson, R.E., *Oilfield Processing of Petroleum*, Pennwell Publishing, Tulsa, Oklahoma. 1991.

NaturalGas.org, Storage of Natural Gas, 2004, WWW.naturalgas.org, Retrieved September 2005.

Smith, C.E., Surging U.S., steady Asia-Pacific lead construction plans, *Oil Gas J.*, 104.6, 57, 2006.

Smith,C.E., True, W.R., and Stell, J., U.S. Gas carriers see 2004 net jump: construction plans rebound, *Oil Gas J.* 103.34, 50, 2005.

True, W.R., Construction plans surge on prospects for gas use, *Oil Gas J.*, 102.3, 58, 2004.

True, W.R. and Stell, J., U.S. construction plans slide: pipeline companies experience flat 2003, continue mergers, *Oil Gas J.*, 102.32, 52, 2004.

Webb, B.C., Guidelines set out for pipeline plugging, *Oil Gas J.*, 76, 196, 1978.

13 Liquefied Natural Gas

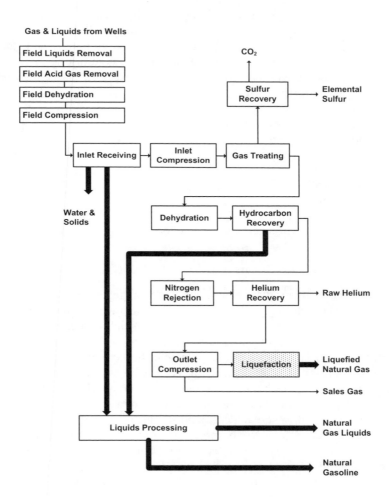

Gas & Liquids from Wells

Field Liquids Removal

Field Acid Gas Removal

Field Dehydration

Field Compression

Inlet Receiving → Inlet Compression → Gas Treating

CO₂

Sulfur Recovery → Elemental Sulfur

Water & Solids

Dehydration → Hydrocarbon Recovery

Nitrogen Rejection → Helium Recovery → Raw Helium

Outlet Compression → Liquefaction → Liquefied Natural Gas

→ Sales Gas

Liquids Processing → Natural Gas Liquids

→ Natural Gasoline

13.1 INTRODUCTION

The rationale for liquefying natural gas is simple: at atmospheric pressure, the liquid density at the normal boiling point of methane is approximately 610 times greater than that of the gas at ambient temperature and pressure. Consequently, a given volume of liquid contains over 600 times the heating value as the same volume of ambient gas. This density increase at ambient pressure makes it attractive to liquefy, transport, and store natural gas in large quantities and makes technically feasible the transport of the equivalent of several Bcf of gas per ship load from "stranded" gas fields to markets. Liquefaction and transport becomes economically feasible when the size of the reserves justify the capital investment of a liquefied natural gas (LNG) plant. Storage applications include storage at LNG terminals and, just as important, storage for peak shaving operations of gas utilities.

After a general discussion of peak shaving and baseload plants, this chapter focuses on eight topics:

1. Gas treating before liquefaction
2. Liquefaction cycles
3. Storage
4. Transport
5. Regasification and cold utilization
6. Economics
7. Plant efficiency
8. Safety and environmental considerations

13.1.1 Peak Shaving Plants and Satellite Facilities

In the previous chapter on Transportation and Storage, Figures 12.3, 12.6, and 12.7 clearly show the large seasonal shifts in gas demand that result in the need for gas storage facilities. Because natural gas fields are generally located far from residential and industrial consumers, storing large quantities of gas near the point of consumption to supplement the normal supply of pipeline gas during periods of peak demand (peak shaving) is essential. Chapter 12 discussed gas storage techniques, the most common of which is underground storage in depleted oil or gas fields, salt caverns, or abandoned mines that can be effectively sealed. Where underground storage is unavailable, aboveground storage of natural gas as LNG becomes attractive, and utilities use relatively small liquefaction, storage, and regasification plants to meet the demand. Plants that combine all three of these tasks are referred to as peak shaving plants. "Stranded" utilities, those not connected to the national pipeline grid, rely upon LNG received by truck to support their customers. Any LNG facilities like these, which contain only storage and regasification units, are called "satellite facilities."

Figure 13.1 shows a block diagram of the common steps involved in a peak shaving facility. Gas treating and compression are discussed in earlier chapters. This chapter discusses liquefaction, liquid storage, and regasification. Odorant injection may or may not be required at the peak shaving plant.

The first peak shaving plant built in the United States was in Cleveland, Ohio, in 1941 (Miller and Clark, 1941). Although the plant performed successfully for several years, in October 1943, a metallurgical failure in a storage tank resulted in a

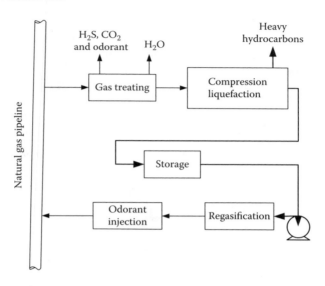

FIGURE 13.1 Schematic of peak-shaving facility.

fire and explosion (GAO, 1978) that destroyed the plant, with a heavy loss of life.*
Although this disaster was a major setback to the industry, in 2004 the United States
had 59 peak shaving plants, 39 satellite facilities, four LNG marine-import terminals,
and one LNG marine-export terminal (Energy Information Administration, 2004).

13.1.2 BASELOAD PLANTS AND STRANDED RESERVES

Baseload plants exist to provide the industrial world with gas from stranded reserves
in remote places. Stranded gas reserves are located where no economic use for the
natural gas exists at the point of origin and where transportation of the gas by pipeline
to a point of end use is not feasible. Romanow (2001) estimates that approximately
60% of the world's gas reserves are considered stranded. When compressed gas
pipelines are impractical or impossible, a limited number of conventional options are
open (Taylor et al., 2001), such as compression and transport of the gas in specially
built ships (Wagner, 2002), conversion of the natural gas into a liquid through gas-
to-liquid (GTL) technology, and liquefaction and shipment of the gas in specially built
LNG vessels. Leibon et al. (1986) as well as Taylor et al. (2001) evaluate the status
of several of the technologies. Hidayati et al. (1998) compare the cost of a compressed
gas pipeline to LNG carriers for a large Indonesian project. Some unconventional
methods that have been considered include conversion of the natural gas to hydrates
for shipping (Gudmundsson and Mork, 2001) and even use of a train of airships that
contain natural gas. Presently, LNG is the most viable option in almost all situations

---

* The Energy Information Agency (2004) states that "Modern LNG plants are designed and constructed
in accordance with strict codes and standards that would not have been met by the Cleveland plant.
LNG safety standards have been issued by the National Fire Protection Agency in NFPA 59A."

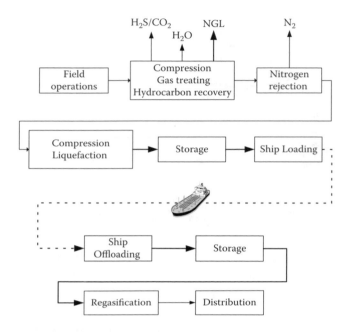

FIGURE 13.2 Schematic of a baseload plant combined with transporting, receiving, and regasification facilities.

involving stranded reserves, if the gas can be pipelined to a seaport. However, to economically justify a traditional baseload LNG plant requires reserves of approximately 3 Tcf (80 Bm^3). Newer designs have reduced the reserve volumes down to around 1 Tcf (30 Bm^3) (Price et al., 2000).

As Figure 13.2 shows, bringing the gas from the field to the customer involves four steps (Energy Information Administration, 2003b):

1. Gas production, gathering, and processing
2. LNG production, including gas treating, liquefaction, NGL condensate removal, and LNG storage and loading
3. LNG shipping
4. LNG receiving facilities, which include unloading, storage, regasification, and distribution

Depending on the specific situation, not all plants will have all the processes shown, and some plants may have additional processes.

Figures 13.3 and 13.4 show the worldwide movements of LNG. In 2002, four countries, Indonesia, Malaysia, Qatar, and Algeria, accounted for 64% of the world's exports, and two countries, Japan and South Korea, accounted for 63% of the world's imports (Energy Information Administration, 2003b).

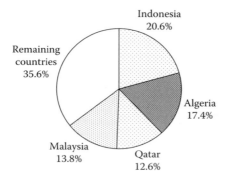

FIGURE 13.3 Worldwide LNG exports in 2002. Total exports were equivalent to 5.37 Tcf (152 Bm³) of natural gas (Energy Information Administration, 2003b).

Figure 13.5 shows the trend of the price for LNG imported into the United States compared with the Henry Hub* price. The LNG price is higher or equal to that of Henry Hub, with the exception of the 2000 price. This parity results in part from the fact that most of the LNG importation facilities are on the East Coast of the United States and do not have to incur pipeline transportation fees from Henry Hub to highly populated end markets.

The liquefaction and storage facilities for baseload plants are necessarily quite large, complex, and expensive. Table 13.1 lists the plants in operation in 2003.

Table 13.2 lists an approximate breakdown of costs (stranded wellhead gas price through liquefaction, marine transportation, and regasification) for three existing LNG operations. In all three situations, the price of the stranded gas is

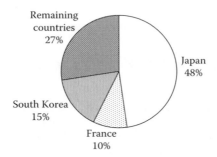

FIGURE 13.4 Worldwide LNG imports in 2002 (Energy Information Administration, 2003b).

---

* Henry Hub is one of a number of large marketing centers. See Chapter 12 for a discussion of these centers.

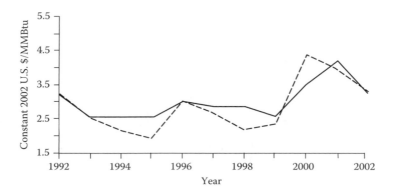

FIGURE 13.5 Price of imported LNG (–) compared with Henry Hub price (- -) (Energy Information Administration, 2003b).

less than 40% of the total cost of marketing the reserve. Cost distribution that includes offshore facilities, if they are developed, would be different.

The worldwide future of the LNG industry utilization of stranded reserves appears bright. Table 13.3 lists the projects that are under construction. Seven countries in different regions of the world accounted for an increase of over 60% in the capacity available in 1999.

Stranded gas exists offshore as well and can be either associated gas, with oil production, or nonassociated gas. Two papers, Finn, et al. (2000) and Wagner and Cone (2004), look at possible LNG liquefaction cycles for offshore applications. Finn et al. (2000) compare the various liquefaction cycles that might apply offshore. Wagner and Cone (2004) conclude that if the gas is nonassociated, reserve sizes comparable to those for onshore baseload plants will be needed. Associated gas might be more economical at lower reserve volumes because the value of the combined gas–oil project must be considered. They note the interest in having offshore liquid storage facilities alongside the LNG plant.

13.2   GAS TREATING BEFORE LIQUEFACTION

Production of LNG requires temperatures as low as −258°F (−161°C), the normal boiling point of methane, and, consequently, the allowable impurity levels in a gas to be liquefied are much lower than that of a pipeline-quality gas. For example, gas to be pipelined and sold to residential or industrial customers may contain a maximum of 3 to 4 mol% carbon dioxide, but gas for liquefaction should have a carbon dioxide content of less than 50 ppmv. Table 13.4 compares compositional specifications for the two cases.

Obviously, gas processed for LNG must have much more aggressive removal of water, nitrogen, and carbon dioxide than does gas destined for pipelines. The tight specifications on all the above components, except for nitrogen and mercury, are needed to avoid solids deposition that will plug the heat exchangers. Nitrogen

TABLE 13.1
Baseload LNG Plants

Plant	Country	Capacity Bcf gas /yr (BSm3/yr)	Number of Trains	Start-up Date
Arzew GL1Z	Algeria	385 (10.9)	6	1978
Arzew GL2Z	Algeria	404 (11.4)	6	1981
Arzew (Camel) GL4Z	Algeria	44 (1.2)	3	1964
SkikdaGL1K phase I and II[a]	Algeria	292 (8.27)	6	1972/1981
Marsa el Brega	Libya	29 0.82)	3	1970
Bonny Island	Nigeria	321 (9.09)	2	1999
Bonny Island Train 3	Nigeria	141 (3.99)	1	2002
Withnell Bay	Australia	365 (10.3)	3	1989
Lumut 1	Brunei	351 (9.93)	5	1972
Arun Phase I, II, and III	Indonesia	331 (9.37)	4	1978/1984/1986
Bontang A-H	Indonesia	1,101 (31.2)	8	1977
Bintulu MLNG 1	Malaysia	370 (10.5)	3	1983
Bintulu MLNG 2	Malaysia	380 (10.7)	3	1994
Bintulu MLNG 3	Malaysia	166 (4.70)	1	2003
Qalhat	Oman	356 (10.1)	2	2000
Ras Laffan	Qatar	404 (11.4)	3	1996
Ras Laffan	Qatar	321 (9.09)	2	1998
Das Island I, II	UAE	278 (7.87)	3	1977/1994
Point Fortin	Trinadad and Tobago	482 (13.6)	3	1999/2003
Kenai	United States	68 (1.92)	2	1969
World total		6,589 (186.6)	66	

[a] An explosion and fire occurred January, 2004 that killed 27, injured 70, and shut down a portion of the plant (for details, see California Energy Commission [2004]).

Source: Energy Information Administration (2003b).

is a volatile diluent which, at higher concentrations, can raise the potential for stratification and rollover (discussed in section 13.4). Elemental mercury presents serious problems in cryogenic operations. As noted in Chapter 9 trace quantities of mercury condense in the cryogenic heat exchangers and form an amalgam with aluminum that can lead to exchanger failure. Consequently, mercury must be removed to a level of 0.01 $\mu g/Nm^3$. Techniques for achieving the desired levels for all of the above mentioned impurities are discussed in the previous chapters.

TABLE 13.2
Cost Breakdown in $/MM Btu for LNG Plants

Project	Wellhead Gas	Liquefaction	Transportation	Regasification	Minimum CIF Cost[a]
Qatar	0.50–0.75	0.40–0.60	1.10–1.20	0.40–0.60	2.45
North West Shelf (Australia)	0.65–0.95	0.40–0.60	0.75–0.95	0.35–0.55	2.15
Bontang, Indonesia	0.60–0.80	0.45–0.65	0.55–0.75	0.30–0.60	1.90

[a] CIF = Carriage + insurance + freight (i.e., cost delivered to customer).

Source: Troner (2001).

Table 13.5 shows the range of compositions and properties for 17 LNG samples. These compositions are for the LNG produced and not the feed gas to the plant.

13.3 LIQUEFACTION CYCLES

The two most common methods that have been used in engineering practice to produce low temperatures are Joule-Thomson expansion and expansion in an engine doing external work. This section discusses each of these processes in detail and analyzes them thermodynamically.

13.3.1 JOULE-THOMSON CYCLES

The Joule-Thomson coefficient is the change in temperature that results when a gas is expanded adiabatically from one constant pressure to another in such a way that no external work is done and no net conversion of internal energy to kinetic energy of mass motion occurs. Thermodynamically, it is an irreversible process that wastes the potential for doing useful work with the pressure drop. However, it is as simple as a valve or orifice and finds wide use in refrigeration cycles (see Chapter 7).

The thermodynamic definition of the Joule-Thomson coefficient is

$$\mu = \left(\frac{\partial T}{\partial P} \right)_h . \tag{13.1}$$

One of the more important thermodynamic relations that involves the Joule-Thomson coefficient is

$$\mu = -\frac{1}{C_P} \left(V - T \left(\frac{\partial V}{\partial T} \right)_P \right) = -\frac{1}{C_P} \left(\frac{\partial H}{\partial P} \right)_T . \tag{13.2}$$

TABLE 13.3
LNG Liquefaction Facilities in Planning, Engineering, and Construction Phases

Country	Location	Capacity Million tpy[a]	Comments
Angola	Soyo	5.0	Planning stage
Algeria	Arzew	5.2	Planning stage
Australia	Barrow Island	5.0	Planning stage
Australia	Darwin	3.0	Under construction
Australia	Karratha	3.2	Engineering
Egypt	Idku	3.6	Under construction
Equatorial Guinea	Bioko Island	3.4	Under construction
Indonesia	Berau Bay, Papua	7.0	Engineering
Nigeria	Brass Terminal	10.0	Engineering
Nigeria	Bonny Island	4.0	Under construction
Nigeria	Bonny Island	3.0	Engineering
Nigeria	Bonny Island	8.8 (2 projects)	Planning stage
Norway	Melkoya Island	4.0	Under construction
Oman	Qalhat, sur	3.3	Under construction
Peru	Pampa Melchorita	4.0	Planning stage
Qatar	Ras Laffan	12.5 (2 projects)	Under construction
Qatar	Ras Laffin	7.8	Engineering
Qatar	Ras Laffin	7.8	Planning stage
Russia	Prigorodnoye Sakhalin	9.6	Under construction
Trinidad and Tobago	Point Fortin	4.6	Under construction
Trinidad and Tobago	Point Fortin	6.0 (2 projects)	Planning stage
Venezuela	Gran Marischal De Ayacucho Sucre	6.8 (2 projects)	Planning stage
Yeman	Bal Haf	6.2	Planning stage

[a] tpy denotes metric tons per year.
Source: Adapted from LNG Observer (2005).

Combination of the above relation with the ideal gas law (PV = RT) gives $\mu = 0$, and thus no temperature change occurs when an ideal gas undergoes a Joule-Thomson expansion. For a real gas, the Joule-Thomson coefficient may be positive (the gas cools upon expansion), negative (the gas warms upon expansion), or zero. The locus of all points on a pressure–temperature plot where the Joule-Thomson coefficient is zero is known as the inversion curve. Figure 13.6 shows that the Joule-Thomson inversion curve for methane expansions must take place below the curve to produce refrigeration.

The behavior of several gases upon expansion from 101 bar (1,470 psia) to 1 bar (14.5 psia) is shown in Table 13.6. Two items should be noted. First, for both methane and nitrogen, the cooling effect upon expansion when started at ambient temperature

TABLE 13.4

Compositional Specifications on Feed to LNG Plant and on Pipeline Gas

Impurity	Feed to LNG Plant[a]	Pipeline Gas[b]
Water	< 0.1 ppmv[c]	150 ppmv, (7.0 lb/MMscf, 110 kg/Sm³)
Hydrogen sulfide	< 4 ppmv	0.25 − 0.30 gr/100 scf (5.7 − 22.9 mg/Sm³)
Carbon dioxide	< 50 ppmv	3 to 4 mole%
Total sulfur	< 20 ppmv	5 − 20 gr/100 scf
(H₂S, COS, organic sulfur)		(115 − 459 mg/Sm³)
Nitrogen	<1 mol%	3 mol%
Mercury	< 0.01 µg/Nm3	
Butanes	2 mol% max	
Pentanes+	0.1 mol% max	
Aromatics	< 2 ppmv[c]	

[a] Foglietta (2002).
[b] Engineering Data Book (2005a).
[c] McCartney (2005).

(80° F, 27°C) is relatively small. Second, the cooling effect increases significantly as the initial temperature is lowered. For helium, the expansion results in heating the gas rather than cooling. The temperature increase remains constant because the Joule-Thomson coefficient remains nearly constant over the temperature range considered.

TABLE 13.5

Typical LNG Compositions

Component	Composition Range (mol%)
Nitrogen	0.00 − 1.00
Methane	84.55 − 96.38
Ethane	2.00 − 11.41
Propane	0.35 − 3.21
Isobutane	0.00 − 0.70
n-Butane	0.00 − 1.30
Isopentane	0.00 − 0.02
n-Pentane	0.00 − 0.04
HHV gas	1021 − 1157
Btu/scf (kJ/Sm³)	(38,000 − 43,090)
Wobbe number	1353 − 1432
GPM, on C₂+ basis	0.71 − 4.08
(m³/1,000m³)	(0.094 − 0.543)

Source: McCartney (2003).

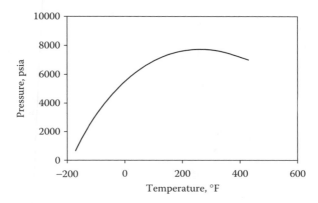

FIGURE 13.6 Joule-Thomson inversion curve for pure methane as a function of temperature and pressure (Goodwin, 1974).

Because methane, the principal constituent of natural gas, must be cooled to −258°F (−161°C) before it becomes a liquid at 1 atmosphere pressure, a liquefier that uses only a Joule-Thomson expansion requires more than a compressor and an expansion valve if it is to function at reasonable initial pressures. A counterflow heat exchanger needs to be added to make a complete system. A simple Joule-Thomson system suitable for natural gas liquefaction is shown in Figure 13.7.

The liquefaction cycle begins with natural gas being compressed and sent through the heat exchanger and expansion valve. Upon expansion, the gas cools (approximately 84°F [47°C] if the gas is principally methane and the expansion is from 1,500 to 14.7 psia [101 to 1 bar]), but none liquefies because a temperature

TABLE 13.6
Expansion from 1470 psia (101 bar) to 14.5 psia (1 bar)

	Initial Temperature	Final Temperature	$t_{final} - t_{initial}$
	°F (°C)	°F (°C)	°F (°C)
Methane	80 (27)	−4 (−20)	−44 (−47)
Nitrogen	80 (27)	46 (8)	−34 (−19)
Helium	80 (27)	91 (33)	11 (6)
Methane	−10 (−23)	−125 (−87)	−115 (−64)
Nitrogen	−10 (−23)	−60 (−51)	−50 (−28)
Helium	−10 (−23)	1 (−17)	11 (6)
Methane	−46 (−43)	−215 (−137)	−169 (−94)
Nitrogen	−46 (−43)	−107 (−77)	−61 (−34)
Helium	−46 (−43)	−35 (−37)	11 (6)

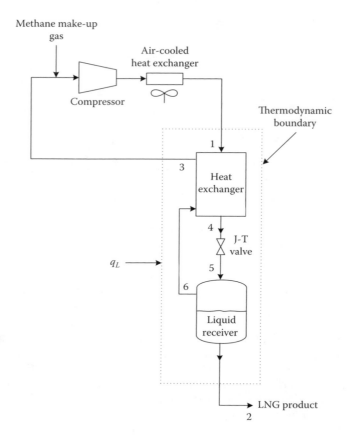

FIGURE 13.7 Simple Joule-Thomson liquefaction cycle.

drop of approximately 338°F (188°C) is required to convert the gas to a liquid. Thus, all of the chilled low-pressure gas is recycled through the heat exchanger for recompression. This cold low-pressure gas lowers the temperature of the high-pressure gas stream ahead of the expansion valve, which results in a lower temperature upon expansion.

As long as all of the gas being expanded is recycled through the counterflow heat exchanger to cool the high-pressure gas stream, temperatures will be progressively lower upon expansion. The process continues until liquid is formed during the expansion from high to low pressure. The liquid formed is separated from the low-pressure gas stream in the liquid receiver and is ultimately withdrawn as the product. The amount of low-pressure gas recycled to the compressor is now significantly reduced, which cuts back on the cooling effect in the heat exchanger. With the addition of makeup gas to the low-pressure side of the compressor to compensate for the liquid product being withdrawn, a steady-state is reached in the liquefaction system and no further cooling can be achieved.

The first law of thermodynamics* for a steady-state flow system is

$$0 = -\Delta[(h + KE + PE)\dot{m}] + \dot{m}q - \dot{m}w_s, \qquad (13.3)$$

where PE and KE are, respectively, the potential and kinetic energy per unit mass. The enthalpy, h, heat term, q, and work term, w_s, are on a mass basis, and $\dot{m}$ represents the mass flow rate. Application of the equation to the components inside the thermodynamic boundary of Figure 13.7 (heat exchanger, Joule-Thomson valve, and liquid receiver) gives the relation

$$\Delta h = q_L, \qquad (13.4)$$

where the overall enthalpy change of the gas, h, on a mass basis equals the heat leak, q_L, per unit mass of gas. On a per unit of mass flow of entering gas, and defining $f = \dot{m}_1 / \dot{m}_2$, the fraction of entering gas withdrawn as a liquid, the equation becomes:

$$fh_2 + (1 - f)h_3 - h_1 = q_L \qquad (13.5)$$

or

$$f = \frac{h_3 - h_1 - q_L}{h_3 - h_2} \qquad (13.6)$$

For a given system, h_2, h_3, and q_L are essentially fixed, so the only way to increase liquefaction is to decrease the inlet gas enthalpy, h_1, which is done by increasing the inlet pressure, assuming that the compressor outlet gas temperature remains constant. Thus, more compressor work should lead to more liquid production.

Example 13.1 Methane is to be liquefied in a Joule-Thomson cycle as shown in Figure 13.7. The methane enters the heat exchanger at 80°F and 1,500 psia and expands to 14.7 psia.

1. Calculate the fraction of methane entering the system that is liquefied.
2. Estimate the % decrease in production if a heat leak q_L of 15 Btu/lb of methane entering is present and if a temperature approach of 5°C is obtained at the warm end of the exchanger.
3. Calculate the fraction liquefied if the pressure is 2,000 psia.

---

* The first law equation for open, steady-flow systems is developed in most chemical and mechanical engineering textbooks. For example, see Smith et al. (2001).

Calculate the fraction liquefied—An ideal heat exchanger is assumed (no warm end ΔT and no pressure drop). From the methane pressure–enthalpy diagram and saturation table (Appendix B) the following values are obtained:

$$h_1 = 350 \text{ Btu/lb } (80°F, 1500 \text{ psia})$$

$$h_3 = 392 \text{ Btu/lb } (80°F, 14.7 \text{ psia})$$

$$h_2 = 0 \text{ Btu/lb } = (-259°F, 14.7 \text{ psia, liquid})$$

Then by use of Equation 13.6

$$f = \frac{h_3 - h_1 - q_L}{h_3 - h_2} = \frac{392 - 350 - 0}{392 - 0} = 0.107$$

Effect of heat leak on production—Use the same cycle but now have a 5°F temperature difference at the warm end of the heat exchanger (t_1, t_3) and a heat leak, $q_L = 15$ Btu/lb. This change lowers the recycle gas outlet temperature to 75°F and $h_3 = 390$ Btu/lb (75°F, 14.7 psia)

The liquid fraction generated now becomes

$$f = \frac{h_3 - h_1 - q_L}{h_3 - h_2} = \frac{392 - 350 - 15}{392 - 0} = 0.069$$

Effect of pressure on production—Determine how liquid production is affected by increasing the pressure on the inlet gas to 2,000 psia from 1,500 psia.

$$h_1 = 337 \text{ Btu/lb } = (80°F, 2,000 \text{ psia})$$

$$f = \frac{h_3 - h_1 - q_L}{h_3 - h_2} = \frac{392 - 337 - 0}{392 - 0} = 0.140$$

This example illustrates the effect of pressure and heat exchanger perfor-mance on liquid yield. For example, if the warm end ΔT for the heat exchanger is approximately 77°F (43°C), a very unlikely value, the liquid yield is reduced to zero, even if no external heat leaks are present.

This outcome raises the question of whether an optimum pressure exists. In Equation 13.6, f will be a maximum when $(h_3 - h_1 - q_L)$ is a maximum because the other terms are independent of inlet pressure. The enthalpy of the liquid, h_2, depends only on the liquid receiver pressure, which we hold constant at the lowest pressure (approximately 14.7 psia [1 bar]). Also q_L is independent of pressure and h_3 is fixed at the lowest pressure and the highest temperature (the inlet temperature for zero ΔT at the warm end of the heat exchanger). Thus we maximize f when h_1 is a minimum. The mathematical criterion is

$$\left(\frac{\partial h_1}{\partial P}\right)_T = 0 . \tag{13.7}$$

Because $\mu = -\frac{1}{C_p}\left(\frac{\partial H}{\partial P}\right)_T$, thermodynamic optimum pressure will occur when $\mu = 0$ or when the inlet conditions are on the inversion curve. However, many other factors must be considered in selecting the economically optimum inlet conditions.

Considerable improvement can be achieved in this simple Joule-Thomson cycle, but at the expense of added equipment and complexity of operation. The addition of an external source of refrigeration markedly improves efficiencies, as does the use of a double expansion of the high-pressure gas instead of a single expansion. Although both of these techniques are extensively used in air liquefaction plants, only the dual-expansion process has found favor in LNG processing.

Figure 13.8 shows the schematic of a commercial facility that used the Joule-Thomson cycle in a Richmond, B.C. plant, near Vancouver, B.C. This plant served a stranded utility, and its total production was transported over-land by truck. It was designed and built to allow easy movement to a new location.

Feed to the plant is obtained from a natural gas pipeline at 40°F (4°C) and pressures in excess of 300 psig (20 barg). The inlet gas is regulated to 300 psig (20 barg) and passed through a molecular sieve dryer to remove both water vapor and carbon dioxide. The gas then is compressed to 3,000 psig (210 barg) in an electrically-driven, two-stage reciprocating compressor. After passing through the three-stream heat exchanger, the gas undergoes a double Joule-Thomson expansion, first to 300 psig (21 barg), and then to 10 psig (0.7 barg) to liquefy the stream. The LNG is transferred to one of the two storage tanks at the facility, either a 21,000 gallon (Imperial) horizontal cylindrical tank that uses vacuum perlite insulation or a 35,000 gallon (Imperial) aluminum tank embedded in the ground. (See Section 13.4 for storage tank details.) The LNG was shipped to Squamish, British Columbia, by truck in a 5,000-gallon (Imperial) trailer. At Squamish, the LNG is transferred to a 21,000-gallon (Imperial) storage tank before regasification and distribution in the town's natural gas system.

Unfortunately, no specific details are available regarding the composition of the natural gas or the specific economics of the facility. Truck transport was apparently more economical than pipelining the gas between the two locations, a distance of about 40 miles. In a relatively small installation such as the Richmond facility, with a liquefaction capacity of 140 Mscfd (3.96 MSm³d), the low capital cost and simplicity of design in the Joule-Thomson cycle counterbalanced the inherent thermodynamic inefficiency, and thus made this cycle a logical choice.

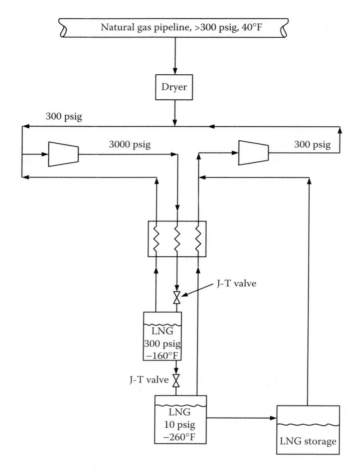

FIGURE 13.8 Joule-Thomson liquefaction plant (Blakely, 1968).

For larger installations, however, the increase in thermodynamic efficiency available in both the expander and cascade cycles (See section 13.3.3) becomes the determining factor in their selection.

13.3.2 EXPANDER CYCLES

The point was made during the discussion of the Joule-Thomson expansion that it was a thermodynamically irreversible process. Expansion of high-pressure gas to the lower pressure in a reversible or nearly reversible manner provides two distinct improvements over the Joule-Thomson expansion. First, in the reversible expansion, a large fraction of the work required to compress the gas can be recovered and used elsewhere in the cycle. This property provides an increase in cycle efficiency. Second, the reversible process will result in a much larger cooling effect. For example, reversibly and adiabatically expanding methane gas from 75 psia and 80°F (5.1 bar, 27°C) to 14.7 psia (1.01 bar) cools the gas to −94°F (−70°C), a temperature drop of 174°F (97°C). A Joule-Thomson expansion between the same pressure limits cools the gas approximately 4°F (2.2°C). As noted in Chapter 7, adiabatic reversible turboexpansion provides the most cooling possible over a given pressure range.

Several options are available for selection of expanders for LNG use, both in the type of expander and in the basic cycle itself. Expanders are basically compressors with the flow reversed and, as with compressors, positive displacement and dynamic expanders are available. In 1902, Georges Claude pioneered expander use in air liquefaction. Claude's expander was a reciprocating machine, as were most early machines used in cryogenic processes, such as those developed by Heylandt in 1912 and later by Collins (1947). Barron (1966) reports reciprocating-machine adiabatic efficiencies of 70 to 80%. He attributes reciprocating expander inefficiencies to four causes:

- Inlet and outlet valve losses
- Incomplete expansion
- Heat transfer
- Piston friction

Reciprocating machines are rarely used in LNG facilities.

Similar to dynamic compressors, dynamic expanders can be centripetal flow or axial flow. In centripetal machines (i.e., turboexpanders), the gas enters through nozzles around the periphery of the wheel, expands, and transmits work to the wheel, which causes it to rotate, and finally exhausts at low pressure at the axis of the machine. Chapter 7 provides more details. Axial-flow expanders have as their counterparts steam turbines. Centripetal machines have isentropic efficiencies on the order of 85 to 90%, whereas axial-flow expanders are about 80% efficient (Swearingen, 1968). Turboexpanders are high-speed machines, generally designed to operate from 10,000 to 100,000 rpm, depending on the throughput.

For design purposes, several techniques may be used to compute the expected enthalpy change, but the simplest and apparently satisfactory method is to use the ideal value from a P-H or T-S diagram, and correct this value with the anticipated turboexpander efficiency (Swearingen, 1968; Williams, 1970). The work generated in the expander must be removed from the system if the full thermodynamic efficiency of the cycle is to be realized. The general practice in large-scale operations is to couple the turboexpander to a gas compressor. Reciprocating expanders would naturally be coupled with reciprocating compressors, and turboexpanders coupled with centrifugal compressors. The available expander work can be very large. Swearingen (1968) states that a turboexpander handling 500 MMcfd (14 Sm³/d) at pipeline pressure would develop 10,000 hp (7,500 kW). Surprisingly, the turbine rotor would only be 18 inches in diameter. In small-scale operations, recovery of the expander work is often not economically feasible. In this case, the turboexpander is simply coupled to a braking device that dissipates the work. Expander–compressor combinations require considerable care in their selection and operation. Swearingen (1970) and the Engineering Data Book (2005b) discuss what must be considered in the selection, operation, and maintenance of turboexpanders.

As mentioned previously, several options are available in the type of expander cycle. All expander cycles fall into two groups: closed cycles and open cycles. Note that most expander cycles have J-T valves as well as turboexpanders.

13.3.2.1 Closed Cycles

In a closed expander cycle, the fluid being expanded is not the fluid to be liquefied; the expander simply acts as an external source of refrigeration, similar to the propane refrigeration discussed in Chapter 7. For example, in LNG production, nitrogen may be used in a closed expander system to liquefy natural gas. A very simple schematic of a closed cycle is shown in Figure 13.9. The compressed nitrogen is expanded, and the cold gas is then used to cool and liquefy the natural gas stream. Actual cycles for producing LNG are far more complex. Hathaway and Lofredo (1971) provide a complete process flow sheet for a plant that has four warm heat exchangers, one large nitrogen compressor, and three turboexpander/compressor combinations.

The closed cycle has several advantages over the open cycle, in which the natural gas itself is expanded. First, if nitrogen is used, safety is enhanced, because the closed cycle reduces the number of processing steps in which flammable natural gas is used. Second, the closed nitrogen cycle has been reported (Anonymous, 1970) to require simpler and less expensive shutdown procedures than its open cycle counterpart and appears to be the most economical process under many conditions. Finally, because the natural gas is not passing through the expander, the process purification system is not so critical. Gas passing through the high-speed expander must be free of condensed phases and any components that solidify at the expander exhaust temperature, because deposition on the rotor will destroy it.

13.3.2.2 Open Cycles

An open expander cycle uses the gas being liquefied as the expanding fluid and has the advantage over the closed cycle of being less complex. A basic expander cycle is shown in Figure 13.10. In this example, the cold exhaust stream from the expander is simply used as a source of refrigeration, and the high-pressure gas is liquefied as it expands through the Joule-Thomson valve. The first law of thermodynamics for a steady-state flow system applied to the two heat exchangers, the expander, and the liquid receiver gives the following equation:

$$f = \frac{h_3 - h_1 - q_L}{h_3 - h_2} + \frac{e(h_4 - h_6)}{h_3 - h_2} ,$$ (13.8)

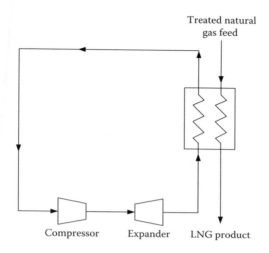

Treated natural
gas feed

Compressor   Expander   LNG product

FIGURE 13.9 Simple closed cycle liquefaction process.

where e is the fraction of the gas going to the expander and $(h_4 - h_6)$ is the work done by the expander. The quantities $\dot{m}$, $\dot{m}_f$, and $\dot{m}_e$ in Figure 13.10 represent the mass flow rate into the liquefier, the mass flow rate of liquefied product, and the mass flow rate to the expander, respectively. A more detailed analysis of the overall system is possible (Dodge, 1944). The cycle shown is only one of a number of possible process arrangements.

An industrial LNG facility that uses an open expander cycle (Figure 13.11) is the Chula Vista plant of the San Diego Gas and Electric Company (Hale, 1966). The plant receives 25 MMscfd (0.71 Mm³/d) from a natural gas pipeline at 300 psia (21 bar) and 90°F (32°C). The gas is first prepared for liquefaction by removal of CO_2, H_2S, and water by physical adsorption on a molecular sieve. The stream then splits, with about 21 MMscfd (0.59 Mm³/d) going to the expander to provide refrigeration. The refrigerant is initially cooled in the first heat exchanger before going to a separator. Liquid from the separator expands through a Joule-Thomson valve. It recombines with the vapor from the separator, which has been through a turboexpander where the pressure drops to 60 psia (4.1 bar) and the temperature drops to −175°F (−115°C), and through the second heat exchanger. This stream provides cooling to both the incoming refrigerant stream and the fraction to be liquefied in the first exchanger. The gas then is compressed to 82 psia (5.6 bar) before being odorized and sent to the local power plant. Compression comes from work done by the turboexpander.

The 4 MMscfd (113 MSm³/d) of gas in the liquefier stream passes through all three heat exchangers and a Joule-Thomson expansion valve. Liquid and vapor are then separated; the vapor stream passes through the heat exchangers and then goes to fuel for the power plant. Three-fourths of the gas that enters the liquefier becomes liquid.

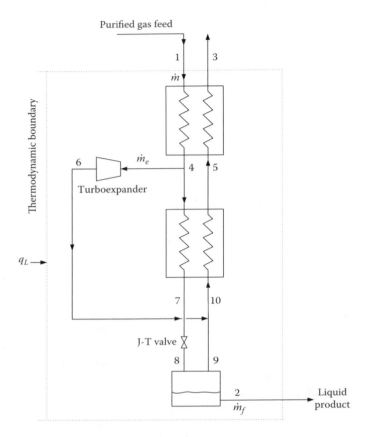

FIGURE 13.10 Open cycle expander plant.

When desired, the LNG is gasified by pumping the liquid to 460 psia (32 bar) and vaporizing it in a hot water heat exchanger. The gas, at 400 psia (27.5 bar) and 60°F (16°C), is then ready for distribution. The vaporizer capacity is 60 MMscfd (1.7 MMSm³/d).

The LNG is stored in a single 175,000-barrel (27,800 m³) aboveground storage tank but has 1 MMscfd (28 MSm³/d) of boil-off. The boil-off provides some refrigeration and is compressed and combined with the vapor from the separator before going to the power plant. With a net liquefaction rate of 2 MMscfd (57 MSm³/d), 315 days are required to fill the storage tank, but only 10.3 days are required to empty the tank if vaporization is at the maximum rate. This outcome matches the gas demand, as the company typically has surplus gas available about 300 days a year. During this period, the storage tank is filled. During the much shorter periods of peak demand, the LNG is vaporized and placed in the distribution system.

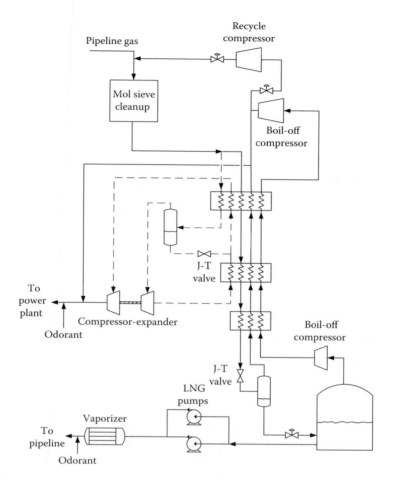

FIGURE 13.11 Peak-shaving plant with open cycle.

13.3.3 CASCADE CYCLES

One source of thermodynamic inefficiency is the finite temperature difference that must be present for heat transfer. The maximum thermodynamic efficiency in a liquefaction cycle is realized when the heating curve of the refrigerant corresponds to the cooling curve of the natural gas being liquefied. This relation means that both ΔT (and thus ΔS) are zero for the heat transfer process (Dodge, 1944). Figure 13.12 is a schematic of a cooling curve for a natural gas system and heating curves for a mixed refrigerant system and a three-fluid classical cascade system. The natural gas and mixed refrigerant curves show marked curvature because the fluids are mixtures. Thermodynamically, the mixed refrigerant

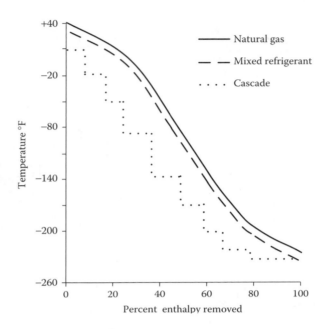

FIGURE 13.12 Cooling curve for natural gas (-) and the corresponding warming curves for the mixed refrigerant (- - -) and cascade ($\cdots$) cycles.

comes closest to a reversible process because it minimizes the temperature difference between the two fluids.

The classical cascade attempts to approximate the cooling curve by use of a series of refrigerants (usually three) in separate loops. Use of more than three refrigerants allows a closer approximation to the cooling curve but with the penalty of extra equipment and added cycle complexity. Both mixed refrigerant cascades and classical cascades will be discussed in more detail in subsequent sections.

The classical cascade was one of the earliest cycles developed and used in the liquefaction of the so-called permanent gases, helium and hydrogen, and was used both by Pictet in 1877 and by Kamerlingh-Onnes in 1892–94 for the liquefaction of air. The method was also used in the first LNG plant built in the United States at Cleveland, Ohio, in the 1940s. The classical cascade is attractive because it can be very efficient thermodynamically. Recently, the classical cascade has been modified with the introduction of the mixed refrigerant cascade. Because both types of cycles are in use for LNG production, we discuss the more important features of both processes.

13.3.3.1 Classical Cascade

The classical cascade process starts with a vapor that can be liquefied at ambient temperature by the application of pressure only. The liquid formed by pressurization is then expanded to a lower pressure, which results in a partial vaporization and cooling of the remaining liquid. This cold liquid bath is then used to cool a second gas so that it may also be liquefied by the application of moderate pressure and then expanded to a lower pressure. The temperature reached in the expansion of the second liquid will be substantially lower than that achieved by the expansion of the first liquid. In principle, any number of different fluids may be used and any desired temperature level can be reached by use of the appropriate number of expansion stages. In practice, however, three fluid and three levels of expansion are normal. In the liquefaction of natural gas, the custom now is to use a propane-ethylene (or ethane)-methane cascade, although the first LNG plant at Cleveland, Ohio, used an ammonia-ethylene-methane cascade (Miller and Clark, 1941).

A simplified process flow diagram for the first stage in a cascade cycle is shown in Figure 7.2, and the operation of the refrigeration system is discussed in Chapter 7. The cascade process differs from the previously described liquefaction techniques, in that the cooling is obtained principally from external circuits, with one final Joule-Thomson expansion for liquefaction. The Joule-Thomson expansion and many turboexpander processes use the gas to be processed as their refrigerating medium.

Figure 13.13 is a simplified schematic of a hypothetical two-fluid cascade designed to produce a liquid product at a low temperature. The process shown uses variations of a two-stage and three-stage refrigeration system (see Chapter 7), with two different working fluids. As noted above, plants that use this cycle normally use three fluids, but we discuss a two-fluid cycle to keep the schematic simple.

Following the flow from the compressor discharge (1) for the high-temperature working fluid, the pure fluid is cooled in heat exchanger E-1, expanded through a Joule-Thomson valve (2), and goes to liquid-vapor receiver R-1 (3). The liquid in the

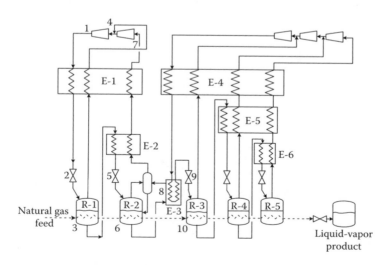

FIGURE 13.13 Schematic of a hypothetical two-fluid cascade cycle.

receiver cools the gas feed, and exiting vapor flows back through heat exchanger E-1 to the second-stage compressor suction (4). In addition to cooling the gas feed, part of the liquid from the receiver (3) flows through heat exchanger E-2, where it is cooled before expansion through a Joule-Thomson valve (5), which creates a liquid–vapor mixture in the receiver R-2 (6). Both the temperature and the pressure of the liquid–vapor mixture in receiver R-2 are lower than in receiver R-1. The liquid in receiver R-2 further cools the gas feed. The vapor passes through a liquid separator, two heat exchangers E-1 and E-2, and goes to the compressor suction (7). Liquid from receiver R-2 goes to heat exchanger E-3 (8) to cool the pure fluid in the second cascade before expansion through a Joule-Thomson valve (9) to create a cooler liquid–vapor mixture in receiver R-3 (10). Essentially the same process is repeated for the second working fluid, which goes through three stages of refrigeration. The feed gas progressively cools until it leaves receiver R-5. It then passes through the final Joule-Thomson valve, and becomes a liquid–vapor mixture at the final desired temperature and pressure. The liquid goes to storage, while the cold vapor is used for heat exchange before being recycled or used for fuel. For simplicity this loop is not shown.

Although the figure and its discussion are simplified, they do illustrate the fundamental features of a classical cascade. Industrial cascades for LNG production use three pure fluids with two or three stages of refrigeration for each fluid. Andress (1996) discusses an industrial application of the classical cascade. Note that for the classical cascade system to function, the vapor pressure curves of the pure fluids must overlap, as shown in Figure 13.14.

13.3.3.2   Mixed-Refrigerant Cascade

As mentioned earlier, the maximum thermodynamic efficiency in a liquefaction cycle is realized when the heating curve of the refrigerant corresponds to the cooling curve of the natural gas being liquefied. The classical cascade attempts

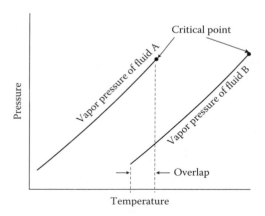

FIGURE 13.14 Qualitative plot of vapor pressure for two pure refrigerant fluids that shows the overlapping region needed for cascade cycles.

to approximate the cooling curve by use of three pure refrigerants in separate loops. However, replacement of several pure refrigerants in separate cycles with a single refrigerant composed of many components condensing at various temperatures in one cycle makes possible a more closely matched cooling curve of the natural gas. In addition, only a single compressor is required for the refrigerant. This feature is the principle for the mixed-refrigerant cascade, the MRC (Linnett and Smith, 1970), the autorefrigerated cascade, the ARC (Salama and Eyre, 1967), and the one-flow cascade OFC process (Forg and Etzbach, 1970).

The technical feasibility of mixed-refrigerant cycles was apparently first proved experimentally by Kleemenko (Forg and Etzbach, 1970) in 1959. Actual cycles that used mixed refrigerants appear in a variety of configurations, but they can be loosely grouped into two basic categories: closed cycles and open cycles.

Closed Cycles—Figure 13.15 shows a schematic of a simple closed cycle (single-flow mixed-refrigerant). The refrigerant would probably be a mixture of nitrogen, methane, ethane, propane, butane, and perhaps pentane; the exact composition depends upon the composition of the natural gas being liquefied. The refrigerant mixture is compressed and then partially condensed in a water-cooled exchanger. The refrigerant then undergoes a series of pressure reductions and liquid-vapor separations to provide the cold fluid needed in the heat exchangers to liquefy the natural gas. The temperatures attained in the various heat exchangers depend on the composition of the refrigerant and the pressure to which the gas is initially compressed. These operating parameters are selected to approximate the cooling curve of the natural gas being liquefied. In this cycle, the natural gas passes through all four heat exchangers in series and is then expanded into a separator, where the liquid and vapor fractions are separated. Figure 13.16 is an industrial example of this type system.

Figure 13.17 is a schematic of the popular closed cycle mixed refrigerant system, the propane precooled, mixed refrigerant system. It is the most commonly used cycle for LNG baseload plants. The process uses both external propane refrigeration and refrigeration from expanding the mixed refrigerant. A three-stage propane refrigeration cycle (see Chapter 7 and Engineering Data Book [2005c] for propane refrigeration systems) provides the initial cooling for both the natural gas and the mixed refrigerant in exchangers HX1, HX2, and HX3. Compressor C1 provides the work for the propane refrigeration. Further cooling is supplied by the expansion of the mixed refrigerant liquid and vapor leaving the cold separator S1. Both of these streams and the process gas are chilled by the expanded mixed refrigerant in HX4 and HX5. The low-pressure mixed refrigerant is recompressed by use of the two-stage mixed-refrigerant compressor C2. Perez et al. (1998) provide a complete description of a large propane precooled, mixed-refrigerant system.

Open Cycle—In the open cycle system shown in simplified form in Figure 13.18, the natural gas stream to be liquefied is physically mixed with the refrigeration cycle stream. This mixing can take place before, during, or after the compression process, depending upon the pressure at which the natural gas feed is available. After compression, the united gas streams are partially condensed in a water-cooled

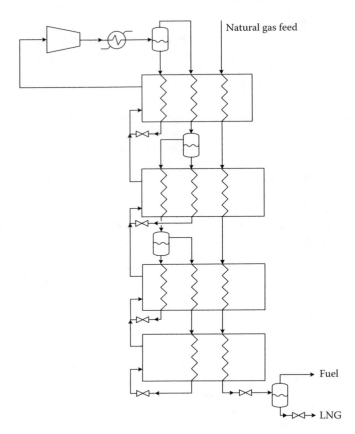

FIGURE 13.15 Schematic of simple closed-cycle, single-flow mixed refrigerant (Forg and Etzbach, 1970).

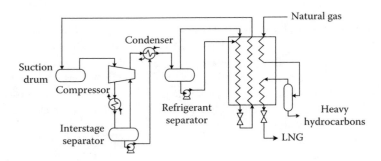

FIGURE 13.16 Schematic of commercial closed-cycle PRICO® system. (Adapted from Price et al. 2000.)

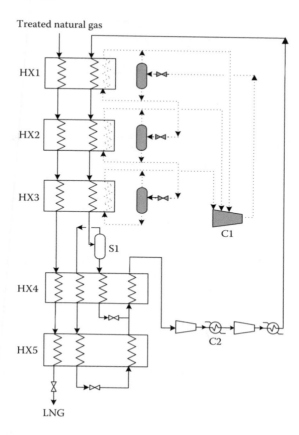

FIGURE 13.17 Schematic of a closed-cycle propane precooled mixed refrigerant system. Shaded areas and dotted lines represent propane cooling system. (Adapted from Foerg et al., 1998.)

or air-cooled heat exchanger, and then separated into liquid and vapor fractions in a separator. From this point, the process is similar to the closed cycle system. That is, the liquid fractions are expanded, which results in vaporization and cooling, and these cold streams are used as the coolant in the heat exchangers. The vapor from the last separator is condensed in the final heat exchanger, and then expanded and separated into an LNG product and a flash gas that would generally be used for plant fuel. To prevent heavy hydrocarbons from plugging in the low-temperature portion of the cycle a liquid slipstream may be withdrawn at a relatively high temperature.

Summary—The mixed refrigerant cycles discussed above possess several advantages over the classical cascade system. The principal advantage is the use of a single-compressor refrigerant system (excluding the propane precooling) in

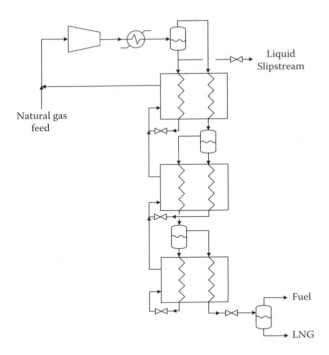

FIGURE 13.18 Open cycle schematic. Refrigerant and process stream are mixed (Forg and Etzbach, 1970).

place of the three refrigerant compressors and cycles of the standard cascade. This configuration provides simplification in instrumentation and piping and a better use of compression power. Further advantages include the ability to readily change the composition of the refrigerant for cycle optimization, should the composition of the feed gas change, and the ability to extract cycle refrigerants directly from the feed gas. The disadvantage when compared with the standard cascade is the necessity for having facilities to recover, store, and blend the components in the refrigerant cycle.

A number of studies compare the various cycles for a particular application. These studies include Vink and Nagelvoort (1998), Foerg et al. (1998), Linnett and Smith (1970), Finn et al. (2000), Foglietta (2002), and Kotzot (2003).

13.4 STORAGE OF LNG

Discussions of LNG storage facilities are normally divided into two major categories: aboveground and in ground. Each has three natural subcategories, and each is discussed below. Also, LNG storage involves a feature peculiar to the storage of cryogenic liquid mixtures, stratification. This section includes a discussion of this important phenomenon.

13.4.1 Cryogenic Aboveground Storage

Three basic types of aboveground storage vessels are in use:

- Steel
- Prestressed concrete
- Hybrid (combinations of steel and concrete)

Each type of construction will be discussed briefly.

13.4.1.1 Steel

Figures 13.19, 13.20, and 13.21 show three typical configurations for single-, double-, and full-containment steel tanks. Storage of LNG in aboveground metal tanks is a widely accepted method for both baseload and peak shaving uses. Nine percent–nickel steel is the most widely used metal for large LNG tanks. Because of its high cost, stainless steel is generally used only for small vessels, LNG plant piping, and heat exchangers. The conventional configuration is a double-walled, flat-bottomed tank, with the annular space between the walls filled with an insulating material. Long (1998) discusses two types of 9% nickel steel construction, the single-containment and the double-containment tank. The single-containment tank is obviously less expensive to build but has the obvious disadvantage of only one containment wall that is compatible with LNG. An economic study by Long (1998) indicated a 12 to 17% cost differential between the full- and double-containment

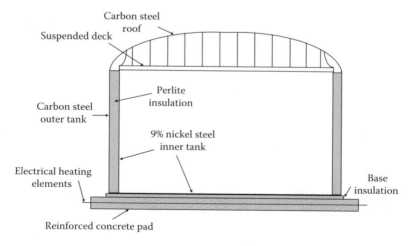

FIGURE 13.19 Single-containment tank. (Adapted from Long, 1998 and Kotzot, 2003.)

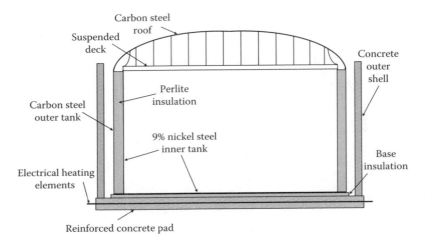

FIGURE 13.20 Double-containment tank. (Adapted from Long, 1998 and Kotzot, 2003.)

tanks, but his analysis did not include the additional cost of secondary impoundment or additional fire protection necessary for the single-containment design. Consequently, double-containment tanks may actually be the most economical choice. In a full-containment tank, the roof is also constructed of concrete.

13.4.1.2 Concrete

Hundreds of prestressed concrete tanks and reservoirs have been built for many uses, including the storage of liquid oxygen (Closner, 1968), which is both heavier and colder than LNG. These tanks have been in continuous use since 1968.

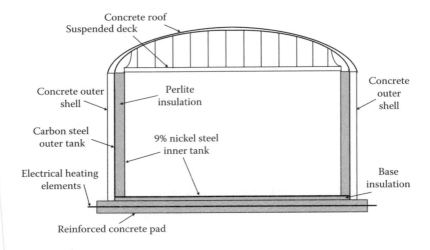

FIGURE 13.21 Full-containment tank. (Adapted from Long, 1998 and Kotzot, 2003.)

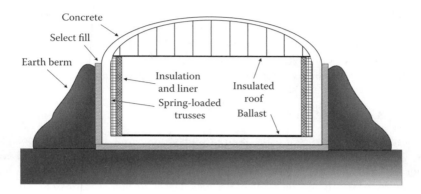

FIGURE 13.22 Prestressed concrete storage tank for LNG (Duffy et al., 1967).

Prestressed concrete tanks may be constructed at grade, below ground, or partially below ground, depending on site conditions or other factors.

Figure 13.22 shows typical installation of a prestressed concrete storage tank. The tank was constructed with its base at grade but was completely surrounded by an earthen berm. The tank had a capacity of approximately 2 billion cubic feet of gas (approximately 600,000 barrels [95,000 m³]) and was built for a peak shaving facility on Staten Island, New York. Prater (1970) provides details of the tank construction. The reinforced concrete walls were insulated on the inside of the tank with polyurethane. The LNG was isolated from the insulation with a thin laminate film that had additional protective insulation. The anticipated boil-off was 0.06% per day of the tank volume. Two centrifugal blowers, each capable of handling 2 MMcfd (60 MS m³/d) at atmospheric pressure and −200°F (−129°C) suction, were used to compress the boil-off vapors to 10 psig (0.7 barg), and the boil-off was subsequently reliquefied. The normal operating tank pressure was 4 inches (10 cm) of water, and a system of diaphragm-operated valves was used to add or release gas rapidly in the event of abnormal atmospheric conditions, such as those caused by a hurricane. The tank was destroyed in an unfortunate maintenance construction accident in 1973. During repairs to the inside of the storage tank, a fire started. The increase in pressure that resulted was rapid and lifted the concrete roof, which then collapsed and killed 37 workers inside the tank (Federal Energy Regulatory Commission, 2005).

Concrete tanks may be constructed where the concrete is protected from direct contact with the LNG by a membrane* or where the LNG directly contacts the concrete. The Portland Cement Association (Monfore and Lentz, 1962) showed that concrete cured at ordinary temperatures can be used to form tanks for holding cryogenic fluids. The mechanical properties of concrete are not significantly

---

* Membranes here denote sheets of impervious material, typically metal, that prevents the LNG and gas vapors from contacting the tank walls, not membranes for separation as discussed in Chapters 5, 6, and 7.

impaired at low temperatures. The compressive strength of concrete at −250 to −150°F (−156 to −101°C) is almost triple that of concrete at room temperature. Legatos et al. (1995) and Jackson and Powell (2001) provide details on the construction and economics of full-containment prestressed-concrete tanks. Jackson and Powell (2001) include data on the rate of LNG permeation of the concrete.

13.4.1.3 Hybrid Construction

Storage tanks constructed using both prestressed concrete and steel are popular. Nishizaki et al. (2001) discuss the construction of a 180,000-m^3 (1.2 MMBbl) LNG storage tank that is claimed to be the world's largest aboveground tank. The structure consists of an inner tank of 9% nickel steel, an outer prestressed-concrete tank, with a cryogenic insulation between the two tanks. Rapallini and Bertha (1998) discuss the conversion of two single-containment 9% nickel steel tanks into double-containment tanks by the addition of an outer concrete tank. Giribone and Claude (1995) describe a novel concept for an aboveground tank that features a 1.2-mm thick stainless-steel membrane combined with a prestressed concrete outer wall. In this configuration, the membrane acts to contain the LNG, but all the hydrostatic load is taken by the concrete.

13.4.2 CRYOGENIC IN GROUND STORAGE

Three basic types of in ground storage have been used:

- Conventional concrete or steel tanks in an underground configuration
- Tanks formed around a frozen-earth cavity
- Mined caverns

All three types will be discussed briefly.

13.4.2.1 Conventional Tankage

With aboveground tank storage discussed in the previous section, the walls must supply all of the mechanical strength. In ground tanks may use either the surrounding earth to provide mechanical support or an in-pit construction in which the tank is built as a separate unit and the pit provides containment in case of leakage or rupture.

As an example of the first type, two of the world's largest LNG tanks are 200,000-m^3 (1.3 MMBbl) in ground storage units installed at the Ohgishima LNG terminal of the Tokyo Gas Company (Umemura et al., 1998). The tanks are unique not only because of their size but also because of the fact that the entire tank, including the domed roof, is buried. Each of the cylindrical tanks is constructed entirely of reinforced concrete and has sidewalls of 2.2 m (7.2 ft) thickness, with a 9.8 m (32 ft) thick bottom slab, 78 m (255 ft) in diameter, designed to withstand ground water uplift. The domed roofs are also of reinforced concrete (1 m thickness at center and 2.5 m at circumference). Because the tanks are totally buried, each roof supports its own weight (15,000 metric tons) plus the weight of soil covering it (40,000 metric tons). The inside of the tanks, including the domes, are lined with cryogenic insulation and 2-mm stainless-steel membranes for liquid and gas tightness. Heating systems surround the sides and bottoms to control ground freezing.

Okamoto and Tamura (1995) discuss the first in-pit storage unit built in Japan, an 85,000-m³ (0.53 MMBbl) unit. In their design, they chose a storage tank identical to an aboveground, cylindrical and vertical, metallic double-shell, flat-bottom tank, and installed the tank inside an in ground reinforced-concrete pit. Dragatakis and Bomhard (1992) discuss in some detail the problem of tank construction in areas prone to severe earthquakes, and they discuss in-pit units at the Revythousa Island LNG facility, located 30 km from Athens, Greece. In their in-pit approach, the tank is freestanding in the pit and functions like an aboveground tank during a seismic event.

13.4.2.2 Frozen-Earth Cavities

Figure 13.23 shows a sketch of a frozen-earth storage container. The cavity is initially cooled by spraying LNG into the vapor space. The roof reaches its steady-state temperature rapidly. Because of the low thermal conductivity of the frozen earth, the surrounding soil may take several years to reach its steady-state temperature.

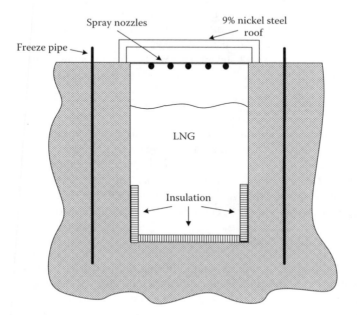

FIGURE 13.23 Sketch of frozen-earth LNG storage container.

The final effective thermal conductivity attained by the frozen earth depends strongly on the type of geologic formation and the moisture content of the earth.

Four early LNG plants incorporated frozen-earth cavities for storage. The first cavity was located in Arzew, Algeria; the next was connected with a peak shaving facility near Carlstadt, New Jersey. Two cavities were then constructed at the Tennessee Gas Pipeline Company's peak shaving facility near Hopkinton, Massachusetts. Finally, four frozen-ground storage units were built at the Canvey Island receiving facilities in England.

Of the four plants mentioned that incorporated frozen-earth cavities, in 1983, only the cavities at Arzew, Algeria were in service (Boulanger and Luyten, 1983). The two plants in the United States were abandoned in favor of other forms of storage because of the failure of the frozen-earth cavities to perform satisfactorily in peak shaving applications. The English facility is an offload site. Because of the lack of published operating data, assessment of differences in performance among these four sites and determination of the reasons for abandoning some of the cavities is not possible.

13.4.2.3 Mined Caverns

In this storage concept, a subterranean cavity is created to hold the LNG, with the cavity walls either in direct contact with the liquid or separated by an insulating wall. Boulanger and Luyten (1983) describe test results in a frozen-clay cavern at Schelle, Belgium. They showed that the horizontal cavern, 3 m (10 ft) in diameter and 30 m (100 ft) long, located 23 m (75 ft) beneath the surface in a layer of boom clay, held liquid nitrogen for at least 10 weeks.

Ahlin and Lindblom (1992) discuss a cavern model in which the LNG is stored in vertical rock caverns, with the LNG separated from the cavern by a layer of concrete cast against the wall and an inner wall of Invar steel on insulating panels.

Presently, neither of the above techniques has been applied commercially.

13.4.3 Rollover

In 1971, at the LNG terminal in La Spezia, Italy a sudden increase in pressure occurred, with subsequent substantial venting of an LNG tank that had received delivery some 18 hours earlier. Fortunately, no damage or injuries occurred. This incident was the first of several similar instances of large, sudden vapor releases that became known as "rollover" (Action and van Meerbeke, 1986). This section provides a short overview of the cause of the releases and how they are prevented.

13.4.3.1 Cause of Rollover

Storage of cryogenic liquids is unique in that the heat leak continually warms the liquid, whereas fluids stored at ambient temperature experience both heating and cooling. To dissipate the heat influx, the cryogenic liquid boils off. For pure liquids, such as liquid nitrogen, this process causes no problems because the liquid is pure and no composition change takes place.

As Table 13.5 shows, LNG contains many components with normal boiling points differing by hundreds of degrees. Of these liquids, only nitrogen and methane vaporize because they are the only components with a significant vapor pressure at LNG temperatures.

First, consider a tank that contains LNG that is well mixed, with no nitrogen present. As methane boils off, the LNG "weathers," which increases the concentration of the heavier components and, thus, increases the boiling point of the liquid. An increase in density occurs because of the loss of methane, which offsets the density decrease caused by temperature increase. Heat leak from the bottom and sides of the tank causes continual mixing through natural convection. Over time, this mixture slowly warms and increases in density.

Figure 13.24 shows a tank after the addition of a fresh batch of LNG on top of the existing liquid, which creates two stratified layers. Initially, the top layer is both lighter and cooler than the bottom layer. However, the top layer increases in density as methane vaporizes. At the same time, the bottom layer continues warming because of heat leak, but it cannot vaporize because of the hydrostatic head of the top layer. This pressure lowers the bottom-layer density. (Although mixing will occur between the layers at the interface, it is probably minimal, considering the typically large volumes of liquid in each layer.)

Continual weathering in the top phase, which increases its density and, concurrently, continual warming of bottom phase caused by heat leak, which lowers its density, can lead to the bottom phase being the lighter phase. If the density inversion becomes large enough to overcome the hydrostatic head, the potential exists for the phases to flip, or rollover, which brings the warm liquid to the top, with a sudden evolution of gas, because the hydrostatic head on the lighter phase is gone. Tank size and type (aboveground, in ground, concrete, or metal) apparently

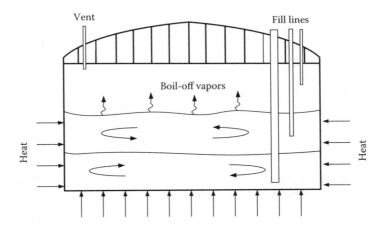

FIGURE 13.24 Schematic of tank that contains two stratified layers of LNG. Heat flux into tank promotes mixing in each layer by natural convection.

have no significant affect on the occurrence of rollover (Acton and van Meerbeke, 1986). The slight changes in liquid composition, along with density difference caused by temperature, are the important variables.

The heavy remnants (heel) in a tank are not a problem when lighter LNG is added, because the density of the heel is sufficiently high to preclude a density inversion. However, a thin, light layer on the top is the most prone to rollover, although the evolution of gas may not be as severe as a thick layer because the hydrostatic head is less.

Nitrogen at concentrations above 1 mol% in LNG mitigates the potential for rollover when it is in the top layer because the nitrogen losses lower the density instead of increasing it. However, if the nitrogen-rich layer is on the bottom, it enhances the potential for rollover.

13.4.3.2 Prevention of Rollover

Rollover prevention requires knowledge of the temperature and density pattern throughout the tank and the loading of new liquid accordingly. If the new charge is lighter than the existing liquid, it goes in at the bottom of the tank. This action forces immediate turnover of the liquids and provides the mixing required, avoiding stratification. (Most tanks are far too large to use pumping to turn over the liquid quickly, and pumping is kept to a minimum to reduce additional liquid heating.) Conversely, heavier feed is filled at the top. Modern tanks are instrumented with both density and temperature probes so that both can be monitored for potential signs of stratification and the potential for rollover.

To ensure that no danger is present, safety codes (e.g., NFPA-59A, 2006) require that storage tanks have vent sizes many times larger than the normal boil-off rate. A report by Groupe International des Importateurs de Gaz Natural Liquefie (G.I.I.G.N.L, undated) found that of the 41 rollovers, only a few exceeded 20 times the normal boil-off rate. European Standards require vent sizes to handle as much as 100 times the normal boil-off rate to ensure that tanks will not be overpressured.

The larger, offload storage facilities are less prone to rollover than peak shaving facilities, because peak shaving liquid storage times can be much longer than at the offload facility, allowing more time for "weathering and stratification."

13.5 TRANSPORTATION

As noted in section 13.1 for large gas reserves often the only alternative to pipelines is shipping by LNG. Three options are possible for transporting LNG:

- Truck transport
- LNG pipelines
- Marine carriers

Each method will be discussed, with the emphasis on marine carriers because they are the most important.

13.5.1 TRUCK TRANSPORT

Cryogenic liquids, including liquid helium, liquid hydrogen, liquid nitrogen, and liquid oxygen, are routinely moved by truck transport. Thus, over-the-road movement of LNG is a relatively simple, straight-forward process that requires no new technology. The major consumers of trucked LNG are vehicle fueling stations and "stranded local utilities," those who are not connected to the national network of natural gas pipelines (Energy Information Administration, 2003a). They consume a tiny faction of the LNG in the United States, 5 to 10 Bcf (140 MM to 280 MMSm^3) equivalent per year (Energy Information Administration, 2004).

13.5.2 PIPELINES

The concept of long-distance LNG pipelines was studied by Hoover (1970), Katz and Hashemi (1971), and Dimentberg (1971). Pumping liquid instead of compressing gas makes the concept seem attractive. Hoover's (1970) detailed study of LNG pipeline concluded that en route refrigeration, primarily for removing heat generated by friction losses, make pipelining economically feasible only in certain situations, at distances less than 200 miles (320 km).

13.5.3 MARINE TRANSPORT

The first ship, *Methane Pioneer*, sailed from Lake Charles, Louisiana, to the United Kingdom with 32,000 barrels (5,088 m^3) of LNG on January 28, 1959. Figure 13.25 shows that by the end of 2006 the LNG fleet will have 206 carriers

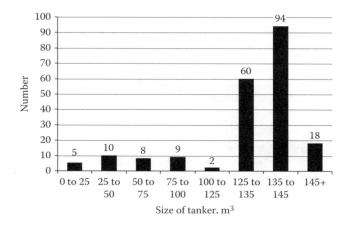

FIGURE 13.25 Size distribution of world's LNG carrier fleet in 2003 (Energy Information Administration, 2003b).

(Energy Information Administration, 2003b). Presently, the largest carriers in use or under construction hold 165,000 m³ (1 MMBbl), but ships that could hold 200,000 to 250,000 m³ (1.3 to 1.6 MMBbl) are under consideration. With little savings in economy of scale, reduced transportation costs are the prime motivation for the larger capacities.

We briefly discuss several of the more important design considerations, and then show details of some typical LNG carriers. Ffooks and Montagu (1967) present a detailed discussion of the evolution of LNG ship design, and they list several basic design criteria:

1. The low density of LNG and the requirement for separate water ballast containment require a large hull, with low draft and high freeboard*.
2. The low temperature of LNG requires the use of special and expensive alloys in tank construction. For free-standing tanks, only aluminum or 9% nickel steel are suitable, whereas for membrane tanks, stainless steel or Invar is used.
3. The large thermal cycling possible in the storage tanks demands special supporting arrangements for free standing tanks and membrane flexibility in membrane designs.
4. The hull of the vessel is carbon steel, so good thermal insulation is required between the tanks and the hull. In addition, for membrane tanks, the insulation must be capable of supporting the full weight of the cargo.
5. The cargo handling equipment must be carefully designed to account for thermal expansion and contraction.

Application of these principles in the design of LNG carriers resulted in a number of different LNG containment concepts, but today only three systems are in general use, and they may be grouped into two designs, independent tanks and membrane tanks, which use different membrane configurations (Marshall, 2002). Both designs are presently in use, and as reported by Harper (2002), approximately 130 ships are currently trading. More than half use independent tanks, and the remainder use membrane tanks. Approximately 60 ships are on order; 40 use the membrane configuration and 20 use the independent-tank design. Marshall (2002) has somewhat different figures: of 173 ships in service, 45% use independent tanks, 53% use membranes, and 2% use other configurations.

Presently, all LNG carriers are double-hulled. With two exceptions, they use steam-powered turbines fueled by boil-off natural gas. Movement has begun toward use of duel-fuel diesel engines, with efficiencies of 38 to 40%, compared with steam-powered turbines, with efficiencies of 28%. Diesel engines also have lower NO_X emissions (Anonymous, 2003).

---

* Draft is the depth of water necessary to float a ship, and freeboard is that part of the ship's side between the waterline and the main deck.

13.5.3.1 Independent Tanks

Independent tanks are self-supporting, do not form part of the ship's hull, and are not essential to the hull strength (Marshall, 2002). The principal system in use today is the Moss system, originally designed by Moss Rosenberg Verft (see discussions in Marshall [2002] and Harper [2002]). This system uses spherical aluminum tanks. Figures 13.26 and 13.27 show a tank ready to be placed in the ship's hull and a view of the vessel with three of the four tanks installed. Kato et al. (1995) describe a typical design for vessels and tanks used on two ships for the Indonesia–Japan route. The four aluminum-alloy tanks on each vessel are of the Moss-4 spherical design and can hold a total of 125,000 m^3 (0.8 MMBbl) of LNG, filled to 98.5% of capacity. The tanks are insulated on the external surface by an insulation composed of phenol resin and polyurethane foam and are designed for a boil-off rate 0.15% per day.

13.5.3.2 Membrane Tanks

Membrane tanks are non self-supporting and consist of a thin metal membrane, stainless steel or Invar (35% nickel steel), supported by the ship's hull through the thermal insulation. Presently, the Invar membrane (Gaz Transport) is more

FIGURE 13.26 Spherical LNG storage tank before installation on carrier hull.

FIGURE 13.27 Ship with three of four storage tanks installed.

popular than the stainless steel (Technigaz) membrane; 24 of the 40 membrane ships on order will use the Gaz Transport design (Harper, 2002). Invar owes its popularity to the fact that it has a very small coefficient of thermal expansion in the operating ranges of the tanks, which are approximately −260 to +180°F

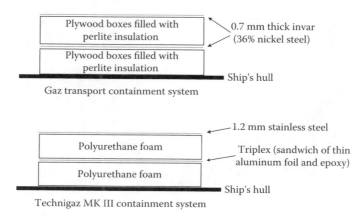

FIGURE 13.28 Gaz transport and Technigaz containment systems (adapted from Harper, 2002).

(−162 to +82°C). Figure 13.28 shows some details of both the Gaz Transport and the Technigaz containment systems (Harper, 2002).

13.6 REGASIFICATION AND COLD UTILIZATION OF LNG

13.6.1 REGASIFICATION

Regasification of the stored LNG is the final step in the operation of LNG peak shaving and off-load terminal storage facilities. The regasification or vaporization is accomplished by addition of heat from ambient air, ambient water, or integral-fired or remote-fired vaporizers. The cost of the regasification system generally represents only a small fraction of the cost of the storage plant; however, reliability of the system is most important because failure or breakdown would defeat the purpose of the facility. Assuming LNG to be pure methane, the energy required to gasify the liquid is almost 40% of the gross heating value.

The regasification section of a peak shaving plant is designed for only a few days of operation during the year to meet the extreme peak loads. To obtain adequate reliability, total send-out capacity may be divided into several independent parallel systems, each capable of handling all or a substantial fraction of the total demand. Designs for the regasification section at the offload terminal are similar to those for a peak shaving plant, but the capacities of these facilities are much higher, and many more spare items of equipment may be required to achieve adequate reliability.

Figure 13.29 is a simplified flow diagram of a typical LNG regasification system. Liquid is pumped from the storage container to the vaporizer. The pump discharge pressure must be high enough to provide the desired gas pressure for entry into the transmission or distribution system. Heat is added to vaporize the high-pressure LNG and to superheat the gas. Gas leaving the vaporizer must be odorized because the liquefaction process removes any odorant originally in the gas.

13.6.2 COLD UTILIZATION

A striking characteristic of the LNG industry is the waste of the vast amount of potential refrigeration available (as noted earlier, about 40% of the gross heating value). Accordingly, the problem of utilization of the refrigeration from LNG

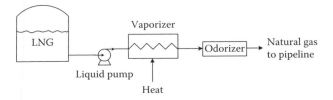

FIGURE 13.29 Simplified flow diagram of a typical LNG regasification system.

vaporization has been examined for years, but making economically viable use of the resource has been difficult. Two general approaches have been advanced for utilization of the cold in LNG vaporization:

- Extraction of work from a power cycle that uses the LNG cold as a heat sink
- Use of the LNG as a source of refrigeration

Only four of the more than 100 large LNG facilities in the world use the cold in LNG, and none are in the United States. These four commercial facilities are:

- An LNG receiving facility in Barcelona, Spain, which uses an LNG cryogenic process to recover heavy hydrocarbons from Libyan LNG (Rigola, 1970)
- An LNG receiving facility at La Spezia, Italy (Picutti, 1970)
- A Japanese facility that produces liquid oxygen and liquid nitrogen and provides warehouse refrigeration for frozen food as well (Kataoka et al., 1970); the plant received a 38% reduction in power requirement by use of the cold LNG
- An LNG-assisted air separation plant in France (Grenier, 1971)

With the expected increase in LNG production over the next decade, there will be more incentive to utilize the refrigeration available.

13.7 ECONOMICS

A report by the Energy Information Agency (2003b) provides the most recent information on global LNG costs and, unless otherwise noted, is the source of all data in this section. The four steps for bringing gas to the consumer are:

- Gas production. Transfer of the gas from the reservoir to the LNG plant (including any necessary processing) is 15 to 20% of the total cost.
- LNG plant. Treating, liquefaction, NGL recovery, storage, and LNG loading are 30 to 45% of the total cost.
- LNG shipping. Maritime shipping is 10 to 30% of the total cost.
- Receiving terminal. Unloading, storage, regasification, and distribution are 15 to 25% of the total cost.

We briefly look at some of the component costs of the last three items.

13.7.1 LIQUEFACTION COSTS

Liquefaction is the largest cost component of the LNG cost train and is typically high relative to other energy projects, for reasons such as:

- Remote locations
- Strict design and safety standards

TABLE 13.7
Representative LNG Shipping Rates in $/MMBtu ($/TJ) to LNG terminals the United States

Exporter	Everett	Cove Point	Elba Island	Lake Charles
Algeria	0.52 (0.49)	0.57 (0.54)	0.60 (0.56)	0.72 (0.68)
Nigeria	0.80 (0.75)	0.83 (0.78)	0.84 (0.79)	0.93 (0.88)
Norway	0.56 (0.53)	0.61 (0.57)	0.64 (0.60)	0.77 (0.72)
Venezuela	0.34 (0.32)	0.33 (0.31)	0.30 (0.28)	0.35 (0.33)
Trinadad and Tobago	0.35 (0.33)	0.35 (0.33)	0.32 (0.30)	0.38 (0.36)
Qatar	1.37 (1.29)	1.43 (1.35)	1.46 (1.38)	1.58 (1.49)
Australia	1.76 (1.66)	1.82 (1.72)	1.84 (1.74)	1.84 (1.74)

Prices are based on a 138,000 cubic-meter carrier at a charter rate of $65,000 per day

- Large amount of cryogenic material required
- Tendency to overdesign to ensure supply security

An estimate of generic liquefaction costs is $1.09 /MMBtu ($1.03/TJ) (based on fuel value) for a two-train, 8 million tpy* (tonne-per-year) Greenfield** project and $0.97 /MMBtu ($0.92/TJ) for an expansion train.

Major savings have been realized over the years by increasing the size of liquefaction trains, which requires fewer trains for a given output. In the early days of the industry, trains were 1.0 to 2.0 million tpy, but presently a 7.8-million tpy train is planned for Qatar.

13.7.2 SHIPPING COSTS

Costs are determined by the daily charter rate, which can vary widely from as low as $27,000 to as high as $150,000. The average rate for long-term charters is between $55,000 and $65,000 per day. Table 13.7 lists some representative shipping rates.

Prices are based on a 138,000 m³ (0.87 MMBbl) carrier at a charter rate of $65,000 per day. Purchase prices for LNG carriers vary significantly, but for a 138,000 m³ carrier in the mid -1980s, the price was approximately $280 million (nominal), whereas in November 2003, it had dropped to approximately $155 million. The main factor driving down the price is the increase in the number of shipyards that build the vessels. Figure 13.30 shows the construction price of LNG ships over the past 12 years and clearly documents the dramatic decrease in cost.

---

* The term tpy denotes metric ton per year. A ton is equivalent to 48,700 scf (1,380 Sm³).
** A Greenfield project is one built in an area without supporting infrastructure.

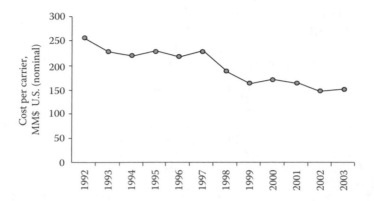

FIGURE 13.30 Construction price of LNG carriers. Price reflects a 125,000 m³ (0.78 MMBbls) carrier from 1992 to 2000 and 138,000 m³ (0.87 MMBbls) carrier from 2001 to 2003 (Energy Information Administration, 2003b).

13.7.3 REGASIFICATION TERMINAL COSTS

Terminal costs vary widely and naturally are site specific. A small terminal might cost as little as $100 million, whereas a state-of-the-art Japanese facility can be $2 billion or more. In the United States, prices are in the range of $300 million to $400 million for capacities of 183 to 365 Bcf (5.2 to 10.3 Bm³) (McCartney, 2005). The price of regasification is generally assumed to add about $0.30/MMBtu ($0.28/TJ) to the cost of imported LNG.

13.8 PLANT EFFICIENCY

Once a liquefaction plant is operational, a key economic number is plant thermal efficiency, η, defined as

$$\eta = \left[\frac{(\text{LNG delivered } + \text{ associated condensate}) \text{ in Btu}}{(\text{natural gas delivered}) \text{ in Btu}} \right] \times 100$$

Table 13.8 summarizes thermal efficiencies for several large baseload plants. Note that the thermal efficiencies of Table 13.8 are not the thermodynamic efficiencies.

TABLE 13.8
Thermal Efficiencies of Baseload Plants

Plant	Efficiency (%)	Reference
Badak (Indonesia)	87.6	Mahfud and Sutopo (1989)
Bintulu (Malaysia)	90.9	Peres and Punt (1989)
North West Shelf (Australia)	90.0	Bogani (1992)

13.9 SAFETY AND ENVIRONMENTAL CONSIDERATIONS

The major difference in safety of LNG facilities compared with that of normal gas plants is the larger equivalent amount of gas present on ships and in storage facilities. As noted earlier, current large LNG carriers hold the equivalent of 4.3 Bcf (120 MMSm^3). We first discuss two aspects relating to ships, and then touch on the controversy that surrounds the proposed new LNG terminals.

An aspect of LNG marine transport that has received much attention is safety. In addition to the usual problems in handling large quantities of flammable material, two safety problems are unique to LNG, and both are a function of the low temperature of the product.

The first problem involves materials of construction. Material for tanks and cargo-handling equipment must be chosen with care because of thermal expansion and contraction problems during the normal thermal cycling of the equipment. Many materials, notably carbon steel, become dangerously brittle at low temperatures and are unsatisfactory for normal use. This problem has been resolved as newer and bigger LNG carriers have been brought into service. According to the Federal Energy Regulatory Commission (FERC 2005) more than 33,000 LNG carrier voyages have taken place in the past 40 years, with no significant accidents.

The second problem is the so-called LNG–water pressure shock wave that may occur when large quantities of LNG are rapidly dumped on the ocean, as would be the case in a maritime accident. This phenomenon received much attention (Burgess et al., 1970; Nakanishi and Reid, 1971) because of its unique character. Nakanishi and Reid (1971) give an excellent summary of the problem and suggest an explanation based on the rapid, often violent boiling caused by the extreme large temperature difference between the LNG and the water. In more recent work, Cleaver et al. (1998) summarize the results of a collaborative industrial research program on the rapid phase transitions of LNG, and Conrado and Vesovic (1998) present the results of a study on the effect of composition on the vaporization rate of LNG on water.

Proposals have been made to establish at least 55 new LNG receiving facilities in North America (FERC, 2005). The most desirable locations are in areas of high population density (e.g., the northeast), so that pipeline costs are minimized. However, the combination of the perceived threat from terrorists and of officials not wanting any new facilities in local areas has slowed the implementation of new receiving facilities. An alternative is to locate the receiving facilities offshore. A project is underway to convert an existing LNG carrier into a floating storage and regasification unit (FRSU) (Anonymous, 2006). The facility, scheduled for completion in 2007, will be capable of delivering 2.75 BSm^3/yr ashore via subsea pipeline. Existing facilities plan to increase storage capacity from 18.8 to 35.2 Bcf (532 to 997 MMSm^3). Most of the expansion is on the Gulf Coast, far from the major population centers. This expansion might meet estimated demands up to 2008.

To meet the anticipated demands for energy and for LNG imports, people on all sides of the issues must work together to determine the true risks and decide if energy supplies are worth the risks.

REFERENCES

Acton, A., and Van Meerbeke, R.C., Rollover in LNG Storage—An Industry Perspective, Proceedings of the Eighth International Conference on Liquefied Natural Gas, Gas Technology Institute, Chicago, IL, 1986.

Ahlin, L. and Lindblom, U., A New Swedish Concept for Storing LNG Underground, Proceedings of the Tenth International Conference on Liquefied Natural Gas, Gas Technology Institute, Chicago, IL, 1992.

Andress, D.L., The Phillips Optimized Cascade LNG Process, A Quarter Century of Improvements, 1996, http://lnglicensing.conocophillips.com/NR/rdonlyres/FBB 538DA-256D-4B96-A844-5D147F4441CF/0/quartercentury.pdf, Retrieved October 2005.

Anonymous, LNG Liquefaction cycles, *Cryogen. Industrial Gases*, 5 (5) 33, 1970.

Anonymous, Future LNG Tanker Design Firming Up? *Nav. Architect*, Nov, 3, 2003, www.rina.org.uk/rfiles/navalarchitect/nov03editorial.pdf, Retrieved August 2005.

Anonymous, LNG carrier to become a storage, regas unit, Oil Gas J., 104.4, 10, 2006.

Barron, R., *Cryogenic Systems*, McGraw-Hill, New York, 1966.

Blakely, R., Remote areas of Canada can now be served by trucked LNG, *Oil Gas J.*, 66 (1) 60,1968.

Bogani, F., Initial Experience with the NWS LNG Plant, Proceedings of the Tenth International Conference on Liquefied Natural Gas, Gas Technology Institute, Chicago, IL, 1992.

Boulanger, A. and W. Luyten, W., Test Cryogenic Storage at Schelle, Results and Conclusions, Proceedings of the Seventh International Conference on Liquefied Natural Gas, Gas Technology Institute, Chicago, IL, 1983.

Burgess, D.S., Murphy, J.N., and Zebetakis, M.G, Hazards Associated with the Spillage of Liquefied Natural Gas on Water, U.S. Bureau of Mines Report of Investigations 7448, 1970.

California Energy Commission, Algerian LNG Plant Explosion Fact Sheet, 2004, www.energy.ca.gov/lng/news_items/2004-01_algeria_factsheet.html, Retrieved August 2005.

Cleaver, P., C., et al., Rapid Phase Transition of LNG, Proceedings of the Twelfth International Conference on Liquefied Natural Gas, Gas Technology Institute, Chicago, IL, 1998.

Closner, J.J., Prestressed Concrete Storage Tanks for LNG Distribution, Conference Proceedings, American Gas Association, May 8, 1968.

Collins, S. C., A helium cryostat, *Rev. Sci. Instr.*, 18, 157, 1947.

Conrado, C., and Vesovic, V., Vaporization of LNG on unconfined water surfaces, Proceedings of the Twelfth International Conference on Liquified Natural Gas, Gas Technology Institute, Chicago, IL, 1998.

Dimentberg, M., Development of LNG pipeline technology, *Cryogen. Industrial Gases*, 6 (5) 29, 1971.

Dodge, B. F., *Chemical Engineering Thermodynamics*, McGraw-Hill, New York, 1944.

Dragatakis, L., and Bomard, H., Revythousa Island LNG facility-LNG Storage Tanks for Seismically Highly Affected Areas, in Proceedings of the Tenth International Conference on Liquefied Natural Gas, Gas Technology Institute, Chicago, IL, 1992.

Duffy, A.R., et al., Heat transfer characteristics of belowground storage, *Chem. Eng. Progress*, 63, 55, 1967.

Energy Information Agency, U.S. Department of Energy, U.S. LNG Markets and Uses, 2003a, www.eia.doe.gov/pub/oil_gas/natural_gas/feature_ articles/20 03/lng/lng 2003.pdf, Retrieved October 2005.

Energy Information Administration, U.S. Department of Energy, The Global Liquefied Natural Gas Market: Status And Outlook, DOE/EIA-0637, 2003b, www.eia.doe. gov/oiaf/analysispaper/global/pdf/eia_0637.pdf, Retrieved October 2005.

Energy Information Administration, U.S. Department of Energy, U.S. LNG Markets and Uses: June 2004 Update, 2004, www.eia.doe.gov/pub/oil_gas/natural_gas/feature _articles/2004/lng/lng2004.pdf, Retrieved June 2005.

Engineering Data Book, 12th ed., Sec. 2, Product Specifications, Gas Processors Supply Association, Tulsa, OK, 2004a.

Engineering Data Book, 12th ed., Sec. 13, Compressors and Expanders, Gas Processors Supply Association, Tulsa, OK, 2004b.

Engineering Data Book, 12th ed., Sec. 14, Refrigeration, Gas Processors Supply Association, Tulsa, OK, 2004c.

Federal Energy Regulatory Commission, U.S. Department of Energy, www.ferc.gov/ industries/lng.asp, Retrieved October 2005.

Ffooks, R.C. and Montagu, H.E., LNG ocean transportation: experience and prospects, *Cryogenics*, 7, 324, 1967.

Finn, A.J., Johnson, G.L., and Tomlinson, T.R., LNG Technology for Offshore and Mid-Scale Plants, Proceedings of the Seventy-Ninth Annual Convention of the Gas Processors Association, Tulsa, OK, 2000.

Foglietta, J.H., Production of LNG Using Dual Independent Expander Refrigeration Cycles, Proceedings of the Eighty-First Annual Convention of the Gas Processors Association, Tulsa, OK, 2002.

Foerg, W., et al., A New LNG Baseload Process and the Manufacturing of the Main Heat Exchangers, Proceedings of the Twelfth International Conference on Liquefied Natural Gas, Institute of Gas Technology, 1998.

Forg, W. and Etzbach, V,. *Linde Rep. Sci. Technol.*, 15, 27, 1970.

General Accounting Office, Report to the Congress of the United States, Liquefied Energy Gases Safety, Chapter 10, 1978.

Giribone, R., and J. Claude, J., Comparative Safety Assessment of Large LNG Storage Tanks, Chapter 10 Proceedings of the Eleventh International Conference on Liquefied Natural Gas, Gas Technology Institute, Chicago, IL, 1995.

Goodwin, R.D., Thermophysical Properties of Methane, from 90 to 300 K at Pressures to 700 Bar, NBS Technical Note 1653, April, 1974.

Grenier, M.R., La Centrale d'Oxygene de Fos-sur-Mer, Proceedings of the XIII Congress of the International Institute of Refrigeration, Washington, D.C., 1971.

Groupe International des Importateurs de Gaz Natural Liquefie (G.I.I.G.N.L), www.giignl.org, undated, Retrieved August 2005.

Gudmundsson, J.S. and Mork, M., Stranded gas to hydrate for storage and transport, 2001 International Gas Research Conference, Amsterdam, November, 2001,http://www. ipt.ntnu.no/~ngh/library/paper11/Amsterdam2001Stranded.htm, Retrieved September 2005.

Hale, D., San Diego LNG plant puts "pressure energy" to work, *Am. Gas J*. 193 (1) 30, 1966.

Harper, I., Future Development Options for LNG Marine Transportation, Second Topical Conference on Natural Gas Utilization, Conference Proceedings, AIChE Spring National Meeting, March 10 – 14, 2002.

Hathaway, P. and Lofredo, A., Experience pays off for this peak shaving pioneer, *Cryogen. Industrial Gases*, 6 (5) 25, 1971.

Hidayati, L., Marlono, Y., and Mattteighianti, M., Comparison of LNG Transportation to JAWA Island: Tanker or Gas Pipeline, Proceedings of the Twelfth International Conference on Liquefied Natural Gas, Gas Technology Institute, Chicago, IL, 1998.

Hoover, T.E., Technical Feasibility and Cost of LNG Pipelines, Presented at the Second International Conference and Exhibition on LNG, Paris, 1970.

Jackson, G. and Powell, J., A Novel Concept for Offshore LNG Storage Based on Primary Containment in Concrete, Proceedings of the Thirteenth International Conference on Liquefied Natural Gas, Gas Technology Institute, Chicago, IL, 2001.

Kataoka, H., Fujisawa, S., and Inoue, A., Utilization of LNG Cold for the Production of Liquid Oxygen and Liquid Nitrogen, Proceedings of the Second International Conference and Exhibition on LNG, Paris, 1970.

Kato, M., et al., Flora and Vesta LNG Carrier Construction, Proceedings of the Eleventh International Conference on Liquefied Natural Gas, Gas Technology Institute, Chicago, IL, 1995.

Katz, D.L. and H.T. Hashemi, Pipeline LNG in the Dense Phase, *Oil Gas J.*, 69 (23) 55,1971.

Kotzot, H., Gas Treatment, Liquefaction and Storage: LNG Seminar, Proceedings of the Eighty-Second Annual Convention of Gas Processors Association, Tulsa, OK, 2003.

Legatos, N.A., et al., Very Large Prestressed Concrete LNG Tanks, Proceedings of the Eleventh International Conference on Liquefied Natural Gas, Gas Technology Institute, Chicago, IL, 1995.

Leibon, I., Davenport, S.T., and Muenzler, M.H., Comparison of Natural Gas Transportation Using LNG Pipelines, and Other Methods, Proceedings of the Eighth International Conference on Liquified Natural Gas, Gas Technology Institute, Chicago, IL, 1986.

Linnett, D. T. and Smith, K.C., Mixed refrigerant cascade liquefiers for natural gas, design and optimization, *Advances in Cryogenic Engineering*, Vol. 19, Timmerhaus, K.D. Ed., Plenum Press, New York, 1970, p. 18.

LNG Observer, *Oil Gas J.*, 103.37, 37, 2005.

Long, B., Bigger and Cheaper LNG Tanks? Overcoming the Obstacles Confronting Free-standing 9% Nickel Steel Tanks up to and Beyond 200,000 m³, Proceedings of the Twelfth International Conference on Liquefied Natural Gas, Gas Technology Institute, Chicago, IL, 1998.

Mahfud, H. and Sutopo, Availability and Efficiency Improvement of Badak LNG Plant, Proceedings of the Ninth International Conference on Liquefied Natural Gas, Gas Technology Institute, Chicago, IL, 1989.

Marshall, G., LNG Ships—An Important Link in the Chain, Second Topical Conference on Natural Gas Utilization, Conference Proceedings, AIChE Spring National Meeting, March 10–14, 2002.

McCartney, D.G., Gas Conditioning For Imported LNG, Proceedings of the Eighty-Second Annual Convention of the Gas Processors Association, Tulsa, OK, 2003.

McCartney, D.G., private communication, 2005.

Miller, R.W. and Clark, J.A., Liquefying natural gas for peak load supply, *Chem. Metallurgical Eng.* 48 (1) 74, 1941.

Monfore, G.E. and Lentz, A.E., Physical Properties of Concrete at Very Low Temperatures, Portland Cement Association Research Bulletin No.145, Chicago, IL, 1962.

NFPA 59A. Standard for the Production, Storage, and Handling of Liquefied Natural Gas (LNG) (2006), National Fire Protection Association, 2005.

Nakanishi, E. and Reid, R.C., Liquid natural gas-water reaction, *Chem. Eng. Progress.*, 67 (12) 36, 1971.

Nishizaki, T., et al., Largest Aboveground PC LNG Storage Tank in the World, Incorporating the Latest Technology, Proceedings of the Thirteenth International Conference on Liquefied Natural Gas, Gas Technology Institute, Chicago, IL, 2001.

Okamoto, H. and Tamura, K., Hatsukaichi LNG Receiving Terminal Construction-LNG In-Pit Storage Tank Construction, Proceedings of the Eleventh International Conference on Liquefied Natural Gas, Gas Technology Institute, Chicago, IL, 1995.

Peres, R. and Punt, A.R., Step by Step Optimization Of Bintulu LNG Plant, Proceedings of the Ninth International Conference on Liquefied Natural Gas, Gas Technology Institute, Chicago, IL, 1989.

Perez, V., et al., The 4.5 MMTPA LNG train—A Cost Effective Design, Proceedings of the Twelfth International Conference on Liquefied Natural Gas, Gas Technology Institute, Chicago, IL, 1998.

Picutti, E., The LNG Terminal at La Spezia, Proceedings of the Second International Conference and Exhibition on LNG, Paris, 1970.

Prater, P. B., Design considerations of an LNG peak shaving facility, *Cryogen.Industrial Gases*, 5 (3) 15, 1970.

Price, B.C., Winkler, R., and Hoffart, S., Developments in the Design of Compact LNG Facilities, Proceedings of the Seventy-Ninth Annual Convention of the Gas Processors Association, Tulsa, OK, 2000, 407.

Rapallini, R. and Bertha, G.J., Conversion of the LNG Storage Tanks at Panigaglia, Italy, to Double Containment Storage Systems, Proceedings of the Twelfth International Conference on Liquefied Natural Gas, Gas Technology Institute, Chicago, IL, 1998.

Rigola, M. Recovery of Cold in a Heavy LNG, Proceedings of the Second International Conference and Exhibition on LNG, Paris, 1970.

Romanow, S., Got gas? *Hydrocarbon Process.*, 80 (11) 11, 2001.

Salama, C. and Eyre, D.V., Multiple refrigerants in natural gas liquefaction, *Chem. Eng. Progr.*, 63 (6) 62, 1967.

Smith, J.M., Van Ness, H.C., and Abbott, M.M., *Introduction to Chemical Engineering Thermodynamics*, 6th ed.,, McGraw-Hill, New York, 2001.

Swearingen, J. S., Engineers Guide to Turboexpanders, Hydrocarbon Processing, April, 97, 1970.

Taylor, M., Alderson, T., and Mounsey, P., LNG Is King—But What Are the Alternatives? Proceedings of the Thirteenth International Conference on Liquefied Natural Gas, Gas Technology Institute, Chicago, IL, 2001.

Troner, A., Russian Far East natural gas searches for a home, *Oil Gas J.*, 99.10, 68, 2001.

Umemura, J., et al., Technological Challenges for the Construction of the Ohgishima LNG Terminal, Proceedings of the Twelfth International Conference on Liquefied Natural Gas, Gas Technology Institute, Chicago, IL, 1998

Vink, K.J. and Nagelvoort, R. K., Comparison of Baseload Liquefaction Processes, Proceedings of the Twelfth International Conference on Liquefied Natural Gas, Gas Technology Institute, Chicago, IL, 1998.

Wagner, J.V., Marine Transportation Of Compressed Natural Gas, Proceedings of the Second Topical Conference on Natural Gas Utilization, AIChE National Meeting, Spring 2002.

Wagner, J.V. and Cone, R.S., Floating LNG Concepts, Proceedings of the Eighty-Third Annual Convention of the Gas Processors Association, Tulsa, OK, 2004.

Williams, V.C., Cryogenics, *Chem. Eng.* 77, 92, 1970.

WEB SITES

The Center for LNG, www.lngfacts.org/: Homesite for organization that provides information on current issues relating to LNG. It has many links to other related sites. Operated by the American Petroleum Institute.

California Energy Commission, www.energy.ca.gov/lng/index.html: Home page for LNG at the state energy commission. Provides an overview of LNG beyond California and contains many links.

Federal Energy Regulatory Commision, U.S. Department of Energy, www.ferc.gov/industries/lng.asp.

LNG Journal home page, www.lngjournal.com/: Page contains recent news stories related to LNG on worldwide basis.

14 Capital Costs of Gas Processing Facilities

14.1 INTRODUCTION

Completely objective capital cost data for any kind of major plant facilities, for obvious reasons, are rarely available in the open literature. However, two papers by Tannehill (2000, 2003) provide estimated capital costs for grass-root facilities. These data are unique in that they represent consensus values based upon input from numerous engineering firms or actual plant cost data. With the author's permission, the data are presented here.

Numerous assumptions and qualifications to the costs are outlined below. These cost data should be used only as a quick guide to a first estimate of capital costs. An added contingency of 20 to 40% is realistic.

After a discussion on the basic premises that apply to all cost data, this chapter presents cost data along with the premises for the costs of the major components in a gas plant. The chapter concludes with some complete plant costs.

14.2 BASIC PREMISES FOR ALL PLANT COMPONENT COST DATA

Table 14.1 presents the basic premises that apply to all of the cost data presented in this chapter for plant components. These data apply only to new, grass-root facilities, and are not applicable to retrofits or used equipment. Tannehill (2000) estimates that costs excluded in Table 14.1, plus other contingencies, constitute an additional 25 to 40% to the total costs. All costs are based upon U.S. dollars in either 1999 or 2002.

14.3 AMINE TREATING

Figure 14.1 shows the capital cost for new facilities that remove various amounts of acid gases (CO_2 and H_2S) as function of plant capacity. The costs assume the following:

- Use of DEA (see Chapter 5) as solvent
- Operating pressure of 600 to 800 psig (41 to 55 barg)

TABLE 14.1
Premises and Assumptions for All Capital Cost Data

Costs Include	Costs Exclude
U.S. Gulf Coast location	Miscellaneous equipment associated with grass-roots plant
Those directly associated with process	Costs not directly associated with process
Two-month startup operating expense	Site and site preparation
Initial supplies and minimum spare parts	Owner home-office costs
Sales taxes	Interest on investment during construction
Contingency of 10%	Construction and bond costs
New facilities only	

Source: Tannehill (2000).

The costs exclude any dehydration of gas after treating. Also, no costs are allocated for BTEX containment. Operating and capital costs increase with acid gas volume to be removed because both absorber and regenerator capacities must increase to handle increased amine recirculation rates.

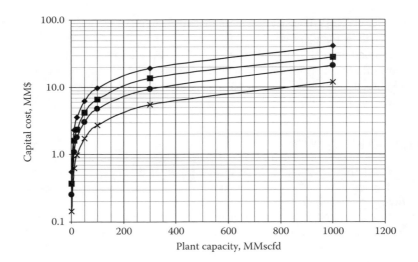

FIGURE 14.1 Capital cost of gas treating by use of DEA as a function of plant capacity in 1999. The lines denoted by x, ●,■,◆ denote 2, 5,10, and 20% acid gas removal, respectively. See Table 14.1 and text for premises (Tannehill, 2000).

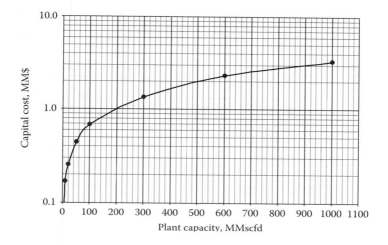

FIGURE 14.2 Capital cost of TEG dehydration in 1999. See Table 14.1 and text for premises (Tannehill, 2000).

14.4 GLYCOL DEHYDRATION

Figure 14.2 gives the capital cost of new glycol dehydration facilities (see Chapter 6). These costs are based upon the following:

- Use of TEG
- Having BTEX containment equipment
- Operating pressure of 600 to 800 psig (41 to 55 barg)

The cost is for a standard glycol unit and does not include enhancements to yield lean glycol concentrations of greater than about 98.6 wt%. Adding the capital costs for both amine and glycol processes gives the overall cost for gas treating and dehydration.

14.5 NGL RECOVERY WITH STRAIGHT REFRIGERATION (LOW ETHANE RECOVERY)

Figure 14.3 provides costs for a straight refrigeration process (see Chapter 7) to produce a single NGL product stream. The figure shows how an increase in the GPM (on C_3+ basis) affects the cost. Costs include limited storage and use of ethylene-glycol injection for hydrate inhibition. Glycol regeneration costs are included. However, the costs exclude any cost of upstream compression or upstream treating. Costs also exclude any fractionation of the liquid product but do include a stripping column to produce a truckable liquid product. It excludes any outlet-gas compression because the pressure drop is small through the unit.

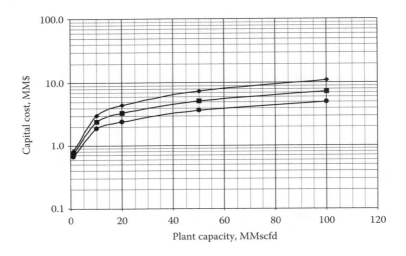

FIGURE 14.3 Capital cost of NGL recovery with straight refrigeration in 1999. See Table 14.1 and text for premises. The lines denoted by ●,■,◆ represent 1.5, 3, and 6 GPM on a C_3+ basis, respectively (Tannehill, 2000).

Going from a lean gas (1.5 GPM, 0.2 m³/Mm³) to a rich gas (6 GPM, 0.8 m³/Mm³) doubles the capital cost for large plants, primarily because of the increased refrigeration load required.

14.6 NGL RECOVERY WITH CRYOGENIC PROCESSING (HIGH ETHANE RECOVERY)

Figure 14.4 shows capital costs for recovering C_2+ by use of cryogenic processing, with and without additional cryogenic nitrogen rejection (see Chapter 8), based on 1999 costs. The figure also includes revised costs in 2002 dollars. Table 14.2 gives the premises used for the cost data.

The figure shows a roughly 10% increase in capital costs between 1999 and 2002. It also shows that including a nitrogen rejection unit (NRU) to a plant with a capacity of over 100 MMscfd (3 MMSm³/d) adds 60% to the cost of the combined unit.

14.7 SULFUR RECOVERY AND TAIL GAS CLEANUP

14.7.1 HIGH SULFUR RECOVERY RATES

For processing gas that produces more than about 50 long tons per day of elemental sulfur, modified Claus units, with or without tail gas cleanup, are the most economical. Figure 14.5 shows capital costs for a Claus sulfur recovery unit, with and

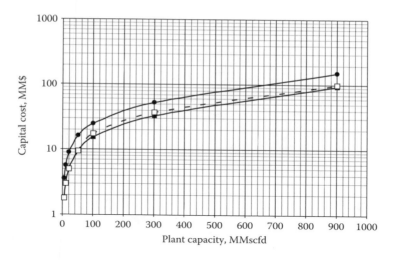

FIGURE 14.4 Capital cost of NGL recovery with cryogenic fractionation. Upper solid line with ● symbol includes cryogenic nitrogen recovery, and lower solid line omits nitrogen recovery. Both are based on 1999 costs (Tannehill, 2000). The dashed line with open symbols shows revised capital costs in 2002 dollars for recovery without nitrogen rejection (Tannehill, 2002). See Tables 14.1 and 14.2 for premises.

without tail gas cleanup (see Chapter 11), as a function of sulfur production. Table 14.3 gives the premises used for generating the costs. The data show that inclusion of a tail gas cleanup unit adds 50 to 90% to the cost of sulfur recovery.

14.7.2 LOW SULFUR RECOVERY RATES

For production of less than about 50 long tons per day of elemental sulfur, a number of processes that make a direct conversion of H_2S into elemental sulfur are economically viable. Amine treating may or may not be used. One commonly

---

TABLE 14.2
Premises for Cryogenic Processing Capital Costs

Costs Include	Costs Exclude
Molecular sieve dehydration	Feed gas treating
Limited storage	Compression of feed gas
Recompression of residue gas to plant inlet pressure	Additional fractionation of demethanizer bottoms product
	Recompression or further handling of nitrogen gas

Source: Tannehill (2000).

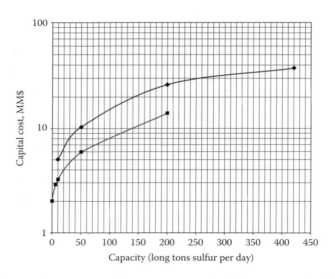

FIGURE 14.5 Capital cost of modified Claus sulfur recovery processes, with and without tail gas clean up, in 1999. See Tables 14.1 and 14.3 for premises (Tannehill, 2000).

used technology is liquid redox (see Chapter 11). Figure 14.6 shows the capital costs for one process, the autorecirculation Lo-Cat® process for sulfur production rates up to 20 long tons per day. The costs assume the feed is a concentrated H_2S stream coming from a conventional amine treater. Table 14.4 lists the premises

---

TABLE 14.3

Premises for Claus Unit Sulfur Recovery Process Capital Costs Along with Those for Tail Gas Clean up

Costs Include	Costs Exclude
For Claus unit	
Claus unit with 3 reactor beds and indirect heat	Treating natural gas to obtain H_2S
Feed gas H_2S content of > 80% with	Process for enriching feed gas
sulfur recovery rates of 94 to 97%	
Only boiler and condenser waste heat	
Incinerator and limited storage	
For tail gas clean up	
Amine-based clean up by use of selective solvent	Onsite disposal of purge water
(e.g., MDEA)	
Direct-fired heater	
Reducing gas generator	
Maximum use of air-cooled heat exchangers	
Heat recovery by steam generation	

Source: Tannehill (2000).

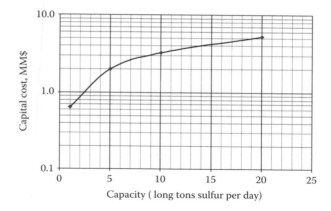

FIGURE 14.6 Capital cost of liquid-redox sulfur recovery process. See Tables 14.1 and 14.4 for premises (Tannehill, 2000).

for the costs. Incinerator costs could not be included because of the wide variation in the CO_2 and other inerts in the feed gas.

14.8 NGL EXTRACTION PLANT COSTS FOR LARGER FACILITIES

Tannehill (2003) provides cost data for grass-root NGL recovery facilities in the 75 to 300 MMscfd (2.1 to 8.5 MMSm³/d) size range as shown in Figure 14.7. The cost premises of Table 14.1 apply. Note the range in costs for the plants at a given size.

An important factor in all economic studies is estimated time for construction. It not only is important for planning but also affects economics in terms of when the plant will produce a revenue, tied up capital, and the interest cost. Figure 14.8 gives the time from when the construction contract was awarded until startup for

---

TABLE 14.4
Premises for Liquid-Redox Sulfur Recovery Process Capital Costs

Costs Include	Costs Exclude
Feed gas H_2S content of >80%	Preconditioning or filtering of feed gas
Sulfur melter	Treating natural gas to obtain H_2S
Initial supplies and minimal spare parts	Incinerator
Limited storage	

Source: Tannehill (2000).

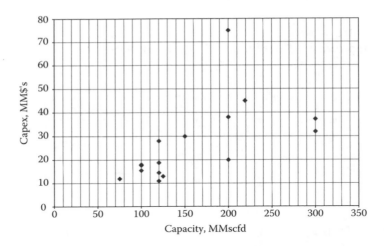

FIGURE 14.7 Capital costs of gas plants in the 75 to 300 MMscfd (2.1 to 8.5 MMSm³) range. See Table 14.1 for premises (Tannehill, 2003).

the plants in the study. The times also show wide variation. Plants smaller than 150 MMscfd (4.2 MMSm³/d) took 8 to 14 months to build, whereas the larger ones took 11 to 18 months. The one plant that took 24 months was a complex facility and involved large treating and sulfur recovery facilities, in addition to the NGL recovery facilities (Tannehill, 2003).

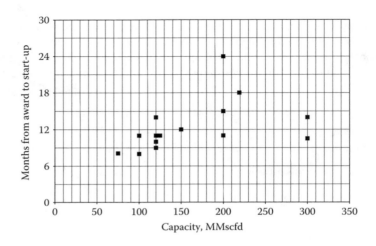

FIGURE 14.8 Time of construction for plants in Figure 14.7 (Tannehill, 2003).

TABLE 14.5
Approximate Distribution of Cost Components
for Gas Processing Facilities

	% of Total Cost	
Cost Component	NGL Recovery	Amine Treating, Dehydration, Sulfur Recovery
Pumps, compressors, etc.	20%	20%
Electrical machinery	5%	10%
Internal combustion engines	15%	0%
Instruments	5%	5%
Heat exchangers	10%	15%
Misc. equipment (average)	5%	5%
Materials	15%	20%
Labor	25%	25%

Source: Tannehill (2005).

14.9 CORRECTIONS TO COST DATA

A number of corrections are required for use of the above cost data for preparation of preliminary cost estimates. The two major factors are location and inflation. Providing these data is beyond the scope of this book. However, initial estimates of inflation can be made by use of cost-index data available in trade journals. These data normally include only costs for components such as pumps and labor. Table 14.5 list the approximate breakdown of costs for the various processes discussed above. Any cost index can be used if values for 1999 and 2002 are known.

REFERENCES

Tannehill, C.C., Budget Estimate Capital Cost Curves for Gas Conditioning and Process-ing, Proceedings of the Seventy-Ninth Annual Convention of the Gas Processors Association, Tulsa, OK, 2000, 141.

Tannehill, C.C., Update if Budget Estimate Capital Cost Curves for NGL Extraction with Cryogenic Expansion, Proceedings of the Eighty-Second Annual Convention of the Gas Processors Association, Tulsa, OK, 2003.

Tannehill, C.C., private communication, 2005.

15 Natural Gas Processing Plants

15.1 INTRODUCTION

The objective of this chapter is to show how the processes discussed in Chapters 3 through 11 are blended together to form a complete processing plant. This chapter discusses three modern plants with significantly different feed and product slates:

- A 500 MMscfd (14 MMSm³/d) plant with sweet gas feed and 98% ethane recovery
- A 279 MMscfd (7.9 MMSm³/d) plant with sour gas feed and both NGL and sulfur recovery
- 7.1 MMNm³/d (251 MMscfd) plant with sour gas feed, NGL recovery, and nitrogen rejection

Block schematics for each plant follow those shown in Figure 2.1 and at the beginning of most chapters. Details come from the open literature. Feed and product slates change with time, and plant modifications are common. Therefore, current plant configuration and operating conditions may differ from those discussed here.

15.2 PLANT WITH SWEET GAS FEED AND 98% ETHANE RECOVERY

Mallet (1988) provides an excellent description of the San Juan gas plant located in New Mexico. Figure 15.1 shows a block schematic of the plant, which is capable of processing 500 MMscfd (14 MMSm³/d) of low-sulfur gas, with the emphasis on obtaining extremely high ethane recovery. To accommodate limitations on equipment sizes and to minimize the loss of productivity during equipment downtime, gas is produced in two identical 250 MMscfd (7.1 MMSm³/d) streams. The streams split after the heat exchange section. Like many modern gas plants, it generates its own power by use of gas turbines and relies heavily upon centrifugal compressors.

15.2.1 OVERVIEW OF PLANT FEED AND PRODUCT SLATE

The San Juan plant feed is sweet, that is, it contains less than the maximum allowable limit of both sulfur compounds and carbon dioxide (refer to Chapter 1 for pipeline gas specifications). Table 1 lists the compositions of both the inlet gas and the products.

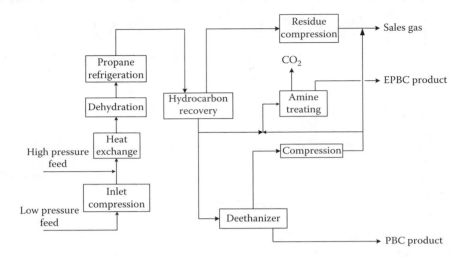

FIGURE 15.1 Block schematic of the San Juan gas plant. The light and heavy lines denote gas and liquid streams, respectively (Mallet, 1988).

While recovering 98% of the ethane, the plant also recovers 100% of the C_3+ for a total liquids production of 42,000 bpd (6700 m³/d). The plant is designed to produce both ethane-propane-butane-condensate (EPBC) liquid and propane-butane-condensate (PBC) liquid products. Relative gas and liquid prices dictate the optimum distribution of ethane between the three streams.

15.2.2 COMPRESSION

The plant receives both high-pressure (320 MMscfd [9 MMSm³/d], 850 psig [58 barg]) and low-pressure (120 MMscfd [3.4 MMSm³/d], 350 psig [24 barg]) inlet-gas feed. The 350 psig (24 barg) feed is pressurized to 850 psig (59 barg) by use of 1,100 hp (0.82 MW) of centrifugal compression and then enters the heat exchange block with the high-pressure gas.

15.2.3 HEAT EXCHANGE

The combined gas stream is cooled to 80°F (27°C) in a water-cooled heat exchanger to condense as much water as possible before the stream splits and enters the two parallel trains. (The gas that enters the plant contains water, but the water is not reported as a component in the inlet feed composition [Table 15.1]. Most gases that enter a gas plant contain water vapor, but standard practice is not to include water as a component when reporting the inlet feed composition.)

15.2.4 DEHYDRATION

Recovery of more than 60% of the ethane requires cryogenic processing, and, consequently, water must be removed to a dew point of −150°F (−101°C).

TABLE 15.1
Feed and Product Compositions for San Juan Plant

Component	Gas Compositions (mol%)	
	Feed Gas	Sales Gas
N_2	0.55	0.64
CO_2	1.13	0.54
C_1	85.57	98.67
C_2	7.60	0.15
C_3	3.03	0
iC_4	0.53	0
nC_4	0.78	0
C_5+	0.81	0
Gallons/Mscf (m^3 [liq]/MSm³[gas])	3.59 (0.481)	0.04 (0.005)
Btu/scf (kJ/MSm³)	1,154.9. (43,356)	978.9 (36,749)
Volume, MMscfd (MMSm³/d)	500 (14)	420 (12)

Component	Product Liquids	
	EPBC[a] (mol%)	PBC[b] (mol%)
N_2	0	0
CO_2	0	0
C_1	0.9	0
C_2	58.6	0.7
C_3	23.8	54.2
iC_4	4.2	11.2
nC_4	6.1	16.6
C_5+	6.4	17.3

[a] Liquid mixture of ethane, propane, butane, and condensate.
[b] Liquid mixture of propane, butane, and condensate.

Source: Mallet (1988).

Dehydration in this plant is accomplished by use of three molecular sieve adsorption beds operating in parallel. Two beds are always adsorbing on 16-hour cycles, while one bed is undergoing regeneration using 14 MMscfd (0.4 MMSm³/d) of dehydrated gas.

15.2.5 PROPANE REFRIGERATION

Because of the relatively high GPM of the feed, the turboexpander is incapable of supplying all of the necessary refrigeration. Propane refrigeration (propane chiller) supplies 64% of the plant requirements.

15.2.6 Hydrocarbon Recovery

This section of the plant is a proprietary design that uses cold reflux in the demethanizer. The turboexpander provides the remaining 36% of the required plant refrigeration. The expander drops the gas pressure from 850 psig (59 barg) to 350 psig (24 barg) to provide the −150°F (−101°C) feed to the top of the demethanizer. The turboexpander operates at 14,000 rpm and provides 4,800 hp (3.6 MW) of residue gas compression.

The demethanizer has 28 valve trays and operates at 350 psig (24 barg) and temperatures of −150°F (−101°C) at the top to 63°F (17°C) at the bottom. Four reboiler types, upper side, lower side, bottom, and trim, are used. (A "trim" reboiler is a second reboiler in series with the bottom reboiler.) By use of cold reflux in the top of the tower, the overhead residue gas product is 98.7% methane. The bottoms product is the EPBC stream that feeds the deethanizer and is also taken as a product stream. Although the CO_2 content of the inlet gas to the plant is only 1.13%, the CO_2 concentrates in the demethanizer bottoms, and the EPBC liquid stream contains 5% CO_2. Consequently, additional purification is required for the liquid stream.

15.2.7 Amine Treating

The EPBC stream from the demethanizer splits, with one branch feeding the deethanizer and the other flowing to the amine-treating unit. The amine treater operates at 655 psig (45 barg) and 85°F (29°C) and uses a 30 wt% diethanolamine solution to remove the CO_2. Activated alumina beds are used to remove water and amine in the liquid EPBC before it goes to storage. Rich amine passes through a hydrocarbon coalescer, flash vessel, and sock filter to remove hydrocarbons and impurities before being regenerated. This stream must have minimal hydrocarbons in it to avoid "stacking" or foaming in the regenerator.

15.2.8 Deethanizer

The tower contains 35 valve trays and operates at 400 psig (28 barg), with overhead and bottoms temperatures of 58°F (14°C) and 220°F (104°C), respectively. The PBC bottoms product goes to storage, whereas the overhead stream is partially condensed by use of propane refrigeration to provide reflux to the column. The vapor from a liquid–vapor separator is compressed and joins the sales gas stream or is sent to the amine treater.

15.2.9 Residue Compression

The demethanizer overhead is first compressed by the compressor coupled to the expander. The final sales gas pressure of 850 psig (59 barg) is obtained by using two 5,000 hp (3.7 MW) gas-driven centrifugal compressors operating in parallel.

15.3 PLANT WITH SOUR GAS FEED, NGL, AND SULFUR RECOVERY

The Whitney Canyon plant, located in Wyoming, is an excellent example of a large plant that processes a sour gas, extracts NGL, and recovers large quantities of sulfur. Figure 15.2 is a block diagram that shows only the major processes at Whitney Canyon. The discussion in this section focuses on those processes emphasized in this book and is consequently very limited. Webber et al. (1984) present a very detailed discussion of the entire project, and the material presented here is taken from their paper. This section discusses some of changes in processing required when relatively high H_2S concentrations are present.

15.3.1 OVERVIEW OF PLANT FEED AND PRODUCT SLATE

The plant was dedicated in late 1982 and lies near the producing field. The field contains 17 wells that produce from four different formations that range in depth from 9,000 ft (2,700 m) to 16,500 ft (5,000 m). The H_2S concentration in the gas varies from 1 to 19%, depending on the well and producing formation. Because of the large variability of the feed gas, no feed or product stream compositions were reported. The plant can handle 270 MMscfd (7.6 MMSm³/d) of feed and produce 200 MMscfd (5.6 MMSm³/d) of sales gas, 13 Mbbl/d (2 Mm³/d) of NGL, 6 Mbbl/d (1 Mm³/d) of condensate, and 1,200 long tons /d (1220 tonne/d) of elemental sulfur.

15.3.2 INLET RECEIVING

Both low-pressure and high-pressure gas from the gathering system enter inlet receiving, where water and condensate are separated from the gas. After removal

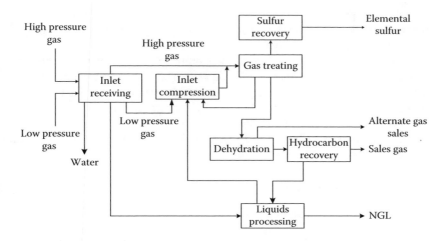

FIGURE 15.2 Block schematic of the Whitney Canyon plant (Webber, et al., 1984).

of dissolved H_2S, the water is sent to disposal wells. The condensate and gas enter the plant in separate lines that operate at approximately 1,000 psig (69 barg) and 250 psig (17 barg), respectively. An interesting feature of the plant is that all inlet-receiving facilities are located in a valley 800 feet (240 m) lower in elevation than the plant to avoid slug flow into the plant.

An important feature of the inlet facility is control of the temperature of gas and condensate to manage hydrocarbon dew point and prevent hydrate formation. The paper does not provide details of these processes or the stripping of H_2S from the water.

15.3.3 INLET COMPRESSION

The high-pressure gas from inlet receiving goes directly to gas treating, whereas the low-pressure gas is compressed by two 1,500 hp (1.1 MW) compressors before it joins the high-pressure stream at gas treating. Another pair of compressors process gas from condensate stabilization and from the amine treating flash tank, so that the gas can be blended with the high pressure gas. Both streams are discussed in following paragraphs.

15.3.4 GAS TREATING

Gas sweetening is accomplished in two identical parallel trains. Each consists of an amine unit that uses a 35 wt% solution of diethanolamine to remove the H_2S to less than one-quarter grain H_2S/100scf (6 mg/m³) and the CO_2 to a level sufficiently low to prevent freezing in the hydrocarbon recovery process. The amine circulation rate for each train is 2,700 gpm (170 l/s), with a rich-amine loading of 0.75 moles acid gas per mole of amine. This treating system differs from most in that both high-pressure and low-pressure contactors are used.

In the high-pressure contactor, sour gas contacts the amine at 95°F (35°C) and 1,000 psig (69 barg) and uses both lean and semi-lean amine from the regenerator. The lean amine enters at the top of the high-pressure contactor. The semi-lean amine is a partially regenerated stream from a side draw on the regenerator. It enters the contactor at a point lower than the lean amine. The high-pressure rich amine is expanded through a power-recovery turbine to 250 psig (17 barg) and goes to a low-pressure flash tank for removal of dissolved hydrocarbons.

However, the gas from the flash tank also contains H_2S. This gas goes to a low-pressure contactor (250 psig, 17 barg), where it contacts lean amine. The sweet gas then returns to inlet compression.

15.3.5 SULFUR RECOVERY

The acid gas from gas treating goes to sulfur recovery, where it is processed by use of a two-train, straight-through Claus process for primary (95%) sulfur recovery. The waste-heat boiler from the Claus process provides much of the plant steam. Each train has two Claus reactors and sulfur condensers.

Final recovery of the sulfur (tail gas cleanup) is accomplished by use of the cold-bed adsorption (CBA) process, with two CBA reactors per train. The overall recovery of sulfur is greater than 99%. The effluent from the CBA reactors is sent to the incinerator for conversion of all remaining sulfur compounds to SO_2, which is then vented into the atmosphere through a stack.

The molten sulfur is trucked 32 miles (51 km), in specially designed trailers, from the plant to a rail terminal. The sulfur then is shipped to Florida for manufacture of fertilizer. A rail line was not run to the plant because of concerns for wildlife.

15.3.6 Dehydration

After removal of acid gases in the gas treating section, the gas goes to dehydration. A glycol dehydrator and molecular sieve adsorption beds are used in series. The triethylene glycol dehydrator removes the bulk of the water, but it cannot achieve the necessary purity (less than 0.1 ppm water) for cryogenic processing, so the second dehydrator, a four-bed molecular sieve unit, is used for final purification. Three beds are always in service; two dehydrate while one undergoes regeneration, and the fourth bed acts as a spare. When NGL recovery is not required (no cryogenic units needed), the molecular sieve beds can be bypassed, and the gas can go to alternate gas sales.

15.3.7 Hydrocarbon Recovery

The plant recovers NGL and up to 88% of the ethane in the gas. The plant design allows operation in either an ethane recovery or ethane rejection mode, depending on the current economic situation. Refrigeration is obtained exclusively from gas expansion (the technique is not mentioned in the paper), and the cold fluids go to a demethanizer to separate methane.

15.3.8 Liquids Processing

This section of the plant has two separate units: condensate stabilization and NGL treating. Condensate stabilization consists of taking the liquids from inlet receiving, water washing them to remove salt, and then processing them in a stabilizer. The stabilizer removes the H_2S and light-hydrocarbon components to lower the vapor pressure of the condensate. A sweet stripping gas, combined with a direct-fired reboiler, is used to reduce the H_2S level to less than 10 ppm in the condensate. The H_2S stripping gas and light hydrocarbons go to inlet compression and then to gas treating.

The natural gas liquids are amine treated with a 30 wt% diethanolamine solution to remove trace contaminants, and then dehydrated, refrigerated, and transferred to storage. To perform satisfactorily, this amine must have lower impurity levels than required for gas treating.

15.4   PLANT WITH SOUR GAS FEED NGL RECOVERY, AND NITROGEN REJECTION

The Hannibal plant, located in Sfax, Tunisia, processes gas from the offshore Miskar field. In full operation since 1996, it provides 80% of the gas used in Tunisia. Jones et al. (1999) and Howard (1998) give complete discussions of the plant. Figure 15.3 is a block diagram of the major processing steps.

15.4.1   OVERVIEW OF PLANT FEED AND PRODUCT SLATE

The design feed gas composition is shown in Table 15.2. The plant processes 7.1 MMNm^3/d (265 MMscfd) of gas and produces 4.9 MMNm^3/d (180 MMscfd) of sales gas. The sales gas specifications include:

- No H$_2$S present (no concentration specification given)
- N$_2$ not to exceed 6.5 mol%
- H$_2$O less than 80 ppmw
- Hydrocarbon dew point less than −5°C (23°F)

Reduction of the N$_2$ to the specified level requires cryogenic processing in the nitrogen rejection unit (NRU). Consequently, the CO$_2$, H$_2$O, and BTEX must be removed to levels low enough to prevent freezing and plugging in the gas–gas exchangers.

15.4.2   INLET RECEIVING

The feed from the Miskar field that enters inlet receiving contains both the gas and the associated wet condensate. A slug catcher separates the liquid from the gas. The liquid goes to liquids processing, and the gas goes to gas treating. The slug catcher consists of six 48-inch (120 cm) diameter pipes that are 182 m (600 ft) long.

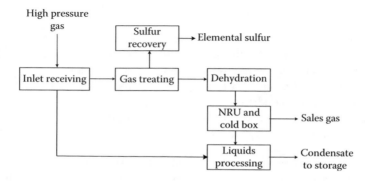

FIGURE 15.3 Block schematic of the Hannibal gas plant. (Jones; et al., 1999; Howard, 1998)

TABLE 15.2
Design Feed Gas Composition
for the Hannibal Plant.

Component	Mol%
Nitrogen	16.903
Carbon dioxide	13.588
Hydrogen sulfide	0.092
Methane	63.901
Ethane	3.349
Propane	0.960
i-Butane	0.258
n-Butane	0.286
i-Pentane	0.142
n-Pentane	0.147
Hexane	0.138
BTEX	0.121
C_7 fraction	0.057
C_8 fraction	0.019
C_9 fraction	0.005
$C_{10}+$	0.001

Source: Jones et al. (1999).

15.4.3 GAS TREATING

An amine unit that uses 50 wt% activated-MDEA solution removes both the H_2S and the CO_2. The H_2S is removed to sales gas specifications and the CO_2 to 200 ppmv. Some BTEX is also removed by the amine, however, not enough to prevent solid deposition in the cryogenic unit. Additional removal is required and is accomplished in the dehydration section.

15.4.4 SULFUR RECOVERY

The acid gases from the amine unit are treated for sulfur removal by use of the liquid-redox Lo-Cat® process. A Claus unit is not justified because the sulfur production rate is only approximately 22,000 lb/d (10 tonnes/d).

15.4.5 DEHYDRATION

Water removal is accomplished by use of two dehydrators, a triethylene glycol unit followed by molecular sieve adsorption beds. Although the glycol dehydrator reduces the load on the adsorption beds, it was designed for removing the bulk of the BTEX to prevent freezing in the cryogenic NRU. Thus, they turned what often is considered an environmental problem into an asset. The glycol regenerator operates at around 190°C (375°F). The overhead reflux drum runs at 0.2 barg (3 psig) and 28°C (82°F), where the water and BTEX are separated, and the BTEX goes to condensate storage.

15.4.6 NRU AND COLD BOX

The cold box contains a series of heat exchangers to condense the heavier hydrocarbons and two vapor–liquid separators that operate at different temperatures. After heat exchange, the liquid from the warm separator (−42°C, −44°F) goes to liquids processing, whereas the gas from this separator is chilled, partially condensed, and sent to the cold separator (−67°C, −89°F). The liquid from the cold separator ultimately combines with the liquid from the NRU to become part of the sales gas. The vapor from the cold separator is chilled, condensed, and expanded through a J-T valve into the NRU. In the NRU, cryogenic distillation is used to separate the N_2 from the hydrocarbons. However, the paper provides no details on the NRU. The N_2 is vented, and the sales-gas stream contains less than 6.5% N_2, which, thus, meets specifications.

15.4.7 LIQUIDS PROCESSING

In this plant, liquids processing consists of sending the condensate from inlet receiving and the liquids from the cold box to a distillation system, where the lighter components are removed to adjust the vapor pressure of the remaining liquid to a specific value (condensate stabilization). The stabilized condensate (Reid vapor pressure of 12 psig [0.8 barg]) is then sent to storage. The paper provides no details on the fate of the gas from the stripper or on the treating and disposition of the NGL stream.

REFERENCES

Howard, I., Hannibal's Experiences, Proceedings of the Laurance Reid Gas Conditioning Conference, 1998, 194.

Jones, S., Lee, S., Evans, M., and Chen, R., Simultaneous Removal of Water and BTEX from Feed Gas for a Cryogenic Plant, Proceedings of the Seventy-Eighth Annual Convention of the Gas Processors Association, 1999, 108.

Mallett, M.W., Conoco/Tenneco Gas Plant Meeting the Challenges, Proceedings of the Sixty-Seventh Annual Convention of the Gas Processors Association, Tulsa, OK, 1988, 150.

Webber, W.W., Petty, L.E., Ray, B.D., and Story, J.G., Whitney Canyon Project, Proceedings of the Sixty-Third Annual Convention of the Gas Processors Association, Tulsa, OK, 1984, 105.

Notation

A = area, ft^2 (m^2)

C = constant in equation, units depend upon equation; concentration, unit is use specific; percent clearance

C_P = heat capacity at constant pressure, Btu/lb mol-°F (kJ/kg-mol-°C)

C_{SS} = correction factor for adsorption of unsaturated gas (see Equation 6.6)

C_T = correction factor adsorption bed temperature (see Equation 6.7)

C_V = heat capacity at constant volume, Btu/mol-°F (kJ/kg-mol-°C)

D = diffusivity, ft^2/h (m^2/s)

D = outside diameter, ft (m)

d = inside diameter, ft (m)

D_P = particle or droplet diameter, ft(m)

F = factor in equations, units depend upon equation

g = acceleration due to gravity, 32.2 ft/s^2 (9.81 m/s^2)

f = fraction, dimensionless; fugacity, psia (bar)

h = enthalpy, Btu/lb (kJ/kg)

h_d = head, ft-lb$_f$/lb$_m$ (N-m/kg)

H = enthalpy, Btu/lb-mol (J/kg-mol)

H_{vi}^{id} = Btu/scf

J = molar flux, lb-mole/ft^2-h (kg-mole/m^2-s)

K = y/x, equilibrium constant, dimensionless

L = length, ft (m)

m = number of compressor stages, dimensionless; mass, lb (kg)

$\dot{m}$ = mass flow rate, lb/h (kg/s)

MW = molecular weight or molar mass, lb/lb-mol (g/g-mol)

n = number of moles, lb mols (g mols)

$\dot{n}$ = lb-mol/h (kg-mol/h)

N_{Re} = Reynolds number, dimensionless

P = pressure, psia (bara)

P = permeability, lb-mol/ft-psia-h (kg-mol/m-bar-s)

p_i = partial pressure of component i, psia (bar)

PD = piston displacement, ft^3/min (m^3/h)

PR = pressure ratio, also called compression ratio, dimensionless

q = heat, Btu/lb (kJ/kg)

q_L = heat leak into a system, Btu/lb

Q = volumetric flow rate, ft^3/min (m^3/h)

R = universal gas constant, 10.73 psia-ft^3/°R lb-mol (8.314 Pa-m^3/K kmol) (see Appendix B for values in other units)

R_K = relative ratio of K values, dimensionless (see Chapter 5)

s = entropy, Btu/lb-°F (J/kg-K)

S = solubility, lb-mol/psia-ft^3 (kg-mol/bara-m^3)

SpGr = specific gravity, dimensionless

T = absolute temperature, °R (K)

t = temperature, °F (°C)

V = volume ft^3 (m^3),

v = volume, ft^3/lb (m^3/kg) or ft^3/lb mol (m^3/g mol)

V_S = superficial velocity, ft/min (m/s)

V_T = terminal velocity ft/s (m/s)

W = water content of gas, lb /MMscf (mg/Sm^3)

w_S = Shaft work (positive if work done by system), Btu/lb (kJ/kg)

WB = Wobbe number, Btu/scf (kJ/m^3)

X = weight fraction, dimensionless

x = mol fraction in liquid phase, dimensionless

y = mol fraction in vapor phase, dimensionless

z = compressibility factor, PV/RT, dimensionless

GREEK LETTERS

α_{1-2} = selectivity, and is the ratio of permeabilities, P_1/P_2 in membranes, or K values K_1/K_2, in distillation, dimensionless

γ = ratio of heat capacities, C_P/C_V or liquid phase activity coefficient, dimensionless

η = efficiency, dimensionless

Θ = time, min (s)

κ = polytropic constant, dimensionless

λ = heat of vaporization, Btu/lb (kJ/kg)

μ = viscosity, cP (m-Pa-s); chemical potential, Btu/lb-mol (J/g-mol); Joule-Thomson coefficient, °F/atm (°C/ bar)

ρ = density, lb/ft^3 (kg/m^3)

φ = fugacity coefficient, dimensionless

SUBSCRIPTS

1 = entering

2 = exiting

A = actual

avg = average value

b = base value

c = critical property

g = gas phase

i = component i, initial value, inhibitor

in = inlet

out = outlet

IS = isentropic

m = mixture
MTZ = mass transfer zone
R = reduced value
L = liquid
P = polytropic
R = relative or reduced condition
rg = regeneration
si = sieve
st = steel
V = volumetric or vapor
W = water

SUPERSCRIPTS

L = liquid
Sat = saturation condition
V = Vapor

Appendix A

Glossary of Gas Process Terminology

Most terms in this glossary were obtained from GPA (1977) and the Engineering Data Book (2004a). Terms marked with an asterisk were added by the authors.

Absorber capacity: The maximum standard cubic feet per day of natural gas that can be processed through an absorber at specified absorption oil rate, temperature, and pressure without exceeding pressure drop or other operating limitations.

Absorption factor: A factor used in engineering calculations that expresses the propensity for a constituent in natural gas to be absorbed in a liquid solvent. This factor is generally found in the literature as $A = L/KV$, where L and V are the moles of liquid and vapor, respectively, from a tray or an average value for the section or total absorber. K is similarly the vapor–liquid equilibrium ratio for the particular component.

Acid gas: The hydrogen sulfide, carbon dioxide, or both contained in or extracted from gas or other streams.

Allowable: The maximum rate of production from an oil or gas well or group of wells that is allowed by a state or governing body. The rate is set by rules that vary among the different states or governing bodies.

Amine: Any of several compounds such as, but not limited to, monoethanolamine (MEA), $HOC_2H_4NH_2$, employed in treating natural gas. The amines are generally used in water solutions to remove hydrogen sulfide and carbon dioxide from gas and liquid streams.

*APC(Advanced process control): A "smart" control system trained to optimize plant operation. It supervises the basic digital control system (DCS).

Associated gas: Natural gas, commonly known as gas-cap gas, which overlies and is in contact with crude oil in the reservoir. Where reservoir conditions are such that the production of associated gas does not substantially affect the recovery of crude oil in the reservoir, such gas may be reclassified as nonassociated gas by a regulatory agency.

*BACT(Best available control technology): Usually required for emissions from a new or modified commercial emission source, such as a plant or combustion engine, in a clean area.

*Baseload plant: A liquefaction facility designed to convert gas from a stranded reserve to LNG for transit.

Barrel: A unit of liquid volume measurement that, in the petroleum industry, equals 42 U.S. liquid gallons for petroleum or natural gas products measured at 60°F and at equilibrium vapor pressure. Chemicals may be packaged in barrels or drums that have capacities of 55 U.S. gallons.

bcf: billion cubic feet

Blanket gas: A gas phase in a vessel above a liquid phase. The purpose may be to protect the liquid from air contamination, to reduce the hazard of detonation, or to pressurize the liquid. The source of the gas is external to the vessel.

Blow case: A small tank in which liquids are accumulated and drained by application of gas or air pressure above the liquid level. Such a vessel is usually located below a pipeline or other equipment at a location where an outside power source is not convenient for removing the drained liquids. Sometimes referred to as a drip.

Blowdown: The act of emptying or depressuring a vessel. This term may also refer to the discarded material, such as blowdown water from a boiler or cooling tower.

Bottom hole pressure (temperature): The pressure (temperature) measured in a well at a depth that is at the midpoint of the thickness of the producing zone.

B-P mix: A liquefied hydrocarbon product composed chiefly of butanes and propane. If it originates from a refinery, it may also contain butylenes and propylene. More specifically, it conforms to the GPA specifications for commercial B-P mixes as defined in GPA Publication 2140.

Breathing: The movement of vapor in or out of a storage tank because of the change of level of the storage liquid, a change in the temperature of the vapor space above the liquid, or of atmospheric pressure.

bs&w: Bottom sediment and water that collects in the bottom of storage tanks. It is also called tank bottoms.

*BTEX: benzene, toluene, ethyl benzene, and xylene

Butane, commercial: A liquefied hydrocarbon that consists predominately of butane, butylenes, or both and which conforms to the GPA specification for commercial butane defined in GPA Publication 2140.

Butane, field grade: A product that consists predominately of normal butane and isobutane that is produced at a gas processing plant. Also called "mixed butane."

Butane, normal: In commercial transactions, a product that meets the GPA specification for commercial butane and, in addition, contains a minimum of 95 liquid vol% normal butane.

Casinghead gas: The natural gas that is produced from oil wells, along with the crude oil. Also, it is an obsolete term for oil well gas.

Casinghead gasoline: An obsolete term for natural gasoline.

Charcoal test: A test standardized by the American Gas Association and the Gas Processors Association to determine the natural gasoline content of a given natural gas. The gasoline is absorbed from the gas on activated charcoal and then recovered by distillation. The test is described in Testing Code 101-43, a joint AGA and GPA publication.

Chromatography: A technique for separating a mixture into individual components by repeated adsorption and desorption on a confined solid bed. It is used for analysis of natural gas and NGL.

*CIF: Carriage, insurance, and freight (cost delivered to customer).

Claus process: A process to convert hydrogen sulfide into elemental sulfur by selective oxidation.

*CBM: Coal bed methane.

Color test: A visual test made against fixed standards to determine the color of petroleum or other product.

Compression ratio: The ratio of the absolute discharge pressure from a compressor to the absolute intake pressure. Also applies to one cylinder of a reciprocating compressor and one or more stages of a rotating compressor.

Compression tests: An AGA-GPA test to determine the natural gasoline product content of natural gas. Refer to AGA-GPA Testing Code 101-43.

Condensate: The liquid formed by the condensation of a liquid or gas; specifically, the hydrocarbon liquid separated from natural gas because of changes in temperature and pressure when the gas from the reservoir was delivered to the surface separators. Such condensate remains liquid at atmospheric temperature and pressure.

Condensate gas reservoir: A hydrocarbon reservoir with natural gas that will yield condensate.

Condensate well: A gas well producing from a condensate reservoir.

Connate water: Water that settled with the deposition of solid sediments and that has not existed as surface water at atmospheric pressure. Also, water in a particular formation that fills a portion of the pore space.

Convergence pressure: The pressure at a given temperature for a hydrocarbon system of fixed composition at which the vapor–liquid equilibria values of the various components in the system become or tend to become unity. The convergence pressure is used to adjust vapor–liquid values to the particular system under consideration.

Copper strip test: A test that uses a small strip of pure copper to determine qualitatively the corrosivity of a product.

Cryogenic plant: A gas processing plant that is capable of producing natural gas liquid products, including ethane, at very low operating temperatures.

Cushion gas: Refer to definition of "blanket gas."

Cycle gas: Gas that is compressed and returned to the gas reservoir to minimize the decline of reservoir pressure.

Cycling: The process in which effluent gas from a gas reservoir is passed through a gas processing plant or separation system, and the remaining residue gas is returned to the reservoir. The word "recycling" has also been used for this function, but it is not the preferred term.

*DDC (Direct digital controllers): See APC.

*DCS (Distributed control system): First-level control system used to manipulate control valves and other devices to maintain a desired setpoint.

DEA (Diethanolamine): Refer to the definition of "amine."

DEA unit: A treating system that uses DEA for reduction of hydrogen sulfide, carbon dioxide, carbonyl sulfide, and other acid gases from sour process streams.

Debutanizer (Deethanizer): A unit of equipment that separates butane (ethane), with or without lighter components, as debutanizer (dethanizer) overhead, from a mixture of hydrocarbons and leaves a bottoms product that is essentially butane (ethane) free.

Degree-day: A unit of temperature and time that shows the difference between a 65°F (18.3°C) base and a daily mean temperature when the later is less than 65°F. This temperature difference is the number of degree-days for a particular day.

Demethanizer: A unit of equipment that separates methane and more volatile components, as demethanizer overhead, from a mixture of hydrocarbons and leaves a bottoms product that is essentially methane free.

Depropanizer: A unit of equipment that separates propane, with or without lighter components, as depropanizer overhead, from a mixture of hydrocarbons and leaves a bottoms product that is essentially propane free.

Disulfides: Chemical compounds containing an —S—S— linkage. They are colorless liquids, completely miscible with hydrocarbons and insoluble in water and sweet to the doctor test. Mercaptans are converted to disulfides in treating processes that employ oxidation reactions.

Doctor test: A qualitative method to detect hydrogen sulfide and mercaptans in petroleum distillates. The test distinguishes between "sour" and "sweet" products.

Drip: Refer to definition of "blow case."

Drip gasoline: Hydrocarbon liquid that separates in a pipeline that transports gas from the well casing, lease separation, or other facilities and drains into equipment from which the liquid can be removed.

Ebullition: Boiling, as especially applied to remove heat from engine jacket water, wherein the water is permitted to boil and the evolved vapors are condensed in air-fin coolers.

*EPC: Engineering, procurement, and construction.

EP-mix (ethane-propane mix): A product that consists of a mixture of essentially ethane and propane.

Fast cycle unit (quick or short cycle unit): An adsorption plant that has adsorption cycles of relatively short duration.

Field condensate: Refer to definition of condensate.

Field processing unit: A unit through which a well stream passes before the gas reaches a processing plant or sales point. Field-processing units may be separator systems, LTX units, adsorption units, etc.

Field separator: A vessel in the oil or gas field that separates the gas, hydrocarbon liquid, and water from each other.

Formation gas: Gas initially produced from an underground reservoir.

*FPSO: Floating production, storage, and offloading vessel used in offshore operations.

Freeze valve: A specially constructed and calibrated valve designed and used solely to determine the water content in the propane product. See ASTM D 2713.

Full well stream: The total flow stream or effluent from a producing well that contains all the constituents of the reservoir fluids.

Gas cap: The natural gas zone located above an oil zone in a common reservoir.

Gas-cap gas: The gas produced from a gas cap. See "associated gas."

Gas condensate: Refer to definition of "condensate."

Gas condensate reservoir: Refer to definition of "condensate gas reservoir."

Gas drive: A manner of producing crude oil or other liquids from the reservoir wherein the required energy is provided by gas pressure.

Gas hydrate: Refer to the definition of "hydrate."

Gas injection: The injection of natural gas into a reservoir to maintain or increase the reservoir pressure or reduce the rate of decline of the reservoir pressure.

Gas lift: A method of bringing crude oil or water to the surface by injecting gas into the producing well bore.

Gas-oil ratio (GOR): The ratio of gas to liquid hydrocarbon produced from a well. This ratio may be expressed as standard cubic feet per barrel of stock tank liquid.

Gasoline: A product that by its composition is suitable for use as a fuel in internal-combustion engines.

Gasoline plant: A natural gas processing plant. This latter preferred term helps differentiate from a unit that makes gasoline within an oil refinery.

Gas reservoir: A geological formation that contains a single gaseous phase. When the gas is produced, the surface equipment may or may not contain condensed liquid, depending on the temperature, pressure, and composition of the single-reservoir phase.

Gas-well gas: The gas produced or separated at surface conditions from the full well stream produced from a gas reservoir.

Gas-well liquids: The liquid separated at surface conditions from the full well stream produced from a gas reservoir.

Glycol: A group of compounds used to dehydrate gaseous or liquid hydrocarbons or to inhibit the formation of hydrates. Commonly used glycols are ethylene glycol (EG), diethylene glycol (DEG), and triethylene glycol (TEG).

gpm: (1) gpm (gallons per minute); the term used to describe a pumping rate in gallons per minute for a liquid. (2) gpm; preferably Gal/MCF (gallons per thousand cubic feet): This term refers to the content in natural gas of components recoverable or that are recovered as liquid products. The preferred form of the term prevents confusion with "gallons per minute."

Gravity, API: An arbitrary scale that expresses the relative density of liquid petroleum products. The measuring scale is calibrated in degrees API, which is calculated by the following formula: $°API = [141.5/(Sp\ Gr.\ 60°F/60°F)] - 131.5$.

Gravity, Baume: An arbitrary scale that expresses the relative density of liquid products. For liquids lighter than water, the gravity Baume can be calculated from the following equation: degrees Baume = (140/Sp. Gr.) − 130. For liquids heavier than water, the formula is: degrees Baume = 145 − (145/Sp. Gr.).

*GTL: Gas to liquid; refers to plants designed to convert natural gas to hydrocarbon liquids.

Heating value (heat of combustion): The amount of heat developed by the complete combustion of a material. For natural gas, the heating value is usually expressed as the gross or higher heating value in Btu per cu. ft. of gas at designated conditions and either on the dry or water saturated basis. The gross, or higher, heating value (normally referred to in the United States) is that measured in a calorimeter, or computed from gas composition, when the heat of condensation of the water produced is included in the total measured heat. The net, or lower, heating value (normally referred to in Europe) is that obtained when the water produced by a combustion process is not condensed, as is the usual circumstance in equipment that burns fuel gas. The net value is the maximum portion of the heating value that can be utilized in usual equipment. The difference between the gross and net heating values is the heat that could be recovered if the water produced could be condensed.

High molecular weight absorption oil: Absorption oil that has a molecular weight in excess of 180. Such oil is ordinarily used in nonrefrigerated absorption plants that have distillation systems that employ stripping steam.

Hot carbonate process: A process for removing the bulk of the acid gases from a gas stream by contacting the stream with a water solution of potassium carbonate at a temperature in the range of 220°F to 240°F (104°C to 116°C).

Hydrate: A solid material that results from the combination of a hydrocarbon with water under pressure.

Hydrate off: Stoppage of the flow of fluid by formation of hydrate crystals.

Injection gas: Gas injected into the producing formation to maintain or increase the reservoir pressure or reduce the rate of decline of the reservoir pressure.

Injection well: The well through which the injection gas or other fluid flows into the underground formation.

Iron sponge process: The method for removing small concentrations of hydrogen sulfide from natural gas by passing the gas over a bed of wood shavings that have been impregnated with a form of iron oxide. The impregnated wood shavings are called "iron sponge." The hydrogen sulfide reacts with the iron oxide and forms iron sulfide and water. Regeneration, if desirable, may be accomplished by exposing the depleted sponge bed to the oxygen in the air.

Jumbo tank cars: Tank cars that have capacities of 30,000 gallons (114 cubic meters) or more. Another group of cars known as "small jumbo" have capacities that range from 18,000 to 22,000 gallons (68 to 83 cubic meters). Various other sizes are also used. "Standard" tank cars have a capacity of 10,000 to 11,000 gallons (38 to 42 cubic meters).

Knockout (liquid): Any liquid condensed from a stream by a scrubber after compression and cooling.

Lead-acetate test: A method for detecting the presence of hydrogen sulfide in a fluid by discoloration of paper that has been moistened with lead-acetate solution.

Lean amine: Amine solution that has been stripped of absorbed acid gases and gives a solution suitable for recirculation to the contactor.

*LEAR (Lowest Achievable Emission Rate): Required on major new or modified emission sources in nonattainment areas.

Lean gas: (1) The residue gas that remains after the recovery of natural gas liquids in a gas processing plant. (2) Unprocessed gas that contains little or no recoverable natural gas liquids.

Lean oil: Absorption oil purchased or made by the plant or oil from which the absorbed constituents have been removed.

Lift gas: Gas injected into a well to assist in raising liquid to the surface. See gas lift.

Low molecular weight absorption oil: Absorption oil that has a molecular weight below 155. Commonly used in refrigerated absorption plants that have dry distillation systems.

LNG (liquefied natural gas): The light hydrocarbon portion of natural gas, predominately methane, that has been liquefied.

*LPG (liquefied petroleum gas): Product streams that contain primarily propane and butane, with minor amounts of ethane.

LTX unit (low temperature extraction unit): A unit that uses the refrigerating effect of the adiabatic expansion of a gas for improved liquid recovery from streams that are produced from high pressure gas-condensate reservoirs. Sometimes called low temperature separators (LTS).

Mainline plant: A plant that processes the gas that is being transported through a cross-country transmission line. Also called pipeline, on-line, or straddle plant.

Make-up gas: (1) Gas that is taken in succeeding years that has been paid for previously under a "take or pay" clause in a gas purchase contract. The contract will normally specify the number of years after payment in which the purchaser can take delivery of make-up gas without paying a second time. (2) Gas injected into a reservoir to maintain a constant reservoir pressure and thereby prevent retrograde condensation. (3) In gas processing, a reduction in gas volume occurs because of plant loss (fuel and shrinkage). Some agreements between gas transmission companies and plant owners require plant losses to be made up or paid for. The same may apply to Btu reduction as well.

MCF (thousand cubic feet – 28.32 m³): A standard unit for measuring or expressing the volume of a thousand cubic feet of gas. The pressure and temperature conditions for the standard measurement must be defined.

MEA (monoethanolamine): An amine for treating gas. Refer to definition of "amine."

MER: Has two general meanings, the first of which is the more common. (1) Maximum efficient rate is the highest rate at which a well or reservoir may be produced without causing physical waste in the reservoir. (2) Most efficient rate is the highest rate at which a well or reservoir can be produced without either reservoir or surface physical waste. For example, a reservoir may be produced at the maximum efficient rate, but at such a rate, gas production will be in excess of the capacity of facilities in the field to handle the gas, so a lower, or most efficient, rate is set up for the reservoir to avoid surface waste of valuable hydrocarbons in the form of flared gas.

Natural gasoline: A mixture of hydrocarbons, mostly pentanes and heavier, extracted from natural gas, which meets vapor pressure, endpoint, and other specifications for natural gasoline as adopted by the GPA.

Natural gasoline plant: One of the terms, now obsolete, used to denote a natural gas processing plant. Refer to definition of "gas processing" and "gas processing plant."

NGL (natural gas liquids): Natural gas liquids are those hydrocarbons liquefied at the surface in field facilities or in gas processing plants. Natural gas liquids include propane, butanes, and natural gasoline. The raw product stream from a plant recovering C_2+ is known as "Y-grade" NGL.

Odorant: A highly odiferous fluid or gas, usually a light mercaptan, added to a gas or LP gas to impart to it a distinctive odor for safety precautions and to facilitate detection of leaks.

Oil-well gas: Gas that is produced from an oil well.

Operating factor: The percentage of time that a unit is performing its function; for example, if a unit runs 800 hours (on stream time), takes 100 hours for reconditioning and inspection, and takes 100 hours for starting up and shutting down, the operating factor is 80%. Refer to definition of "stream day."

Outage: The difference between the full interior volume of a storage vessel or sample container and the volume of liquid therein. For gasoline and lighter products, the regulatory bodies set a minimum limit for outage to provide space for expansion of the liquid.

Peak day requirements: Maximum requirement of gas for a 24-hour period. The quantity may be considerably greater than the daily average flow.

Peak shaving: The use of fuel and equipment to generate or manufacture gas to supplement the normal supply of pipeline gas during periods of extremely high demand.

*Peak shaving plant: A liquefaction facility designed to produce LNG for storage.

Pigging: A procedure of forcing a solid object through a pipeline for cleaning purposes.

Pipeline gas: Gas that meets a transmission company's minimum specifications.

*Pulsation dampeners: Internally baffled vessel attached to the suction and outlet of reciprocating compressors to dampen the pressure pulsations.

Propane, HD-5: A special grade of propane that consists predominately of propane and that conforms to the GPA specification for HD-5 propane.

*RACT (Reasonably Available Control Technology): Technology is usually required for emissions from a new or modified commercial emission source, such as a plant or combustion engine, in a clean area. RACT standards govern all major sources that have the potential to emit 100 tons per year (tpy) of VOCs and NO_x.

Raw gas: Unprocessed gas or the inlet gas to a plant.

Raw mix liquids: A mixture of natural gas liquids before fractionation. Also called "raw make."

Reclaimer: A system in which undesirable high-boiling contaminants of a stream are separated from the desired lighter materials; a purifying still.

Residue: The material that remains after a separation process. (1) Residue gas is the gas that remains after the recovery of liquid products. (2) Residue may also be the heaviest liquid or solid that remains after the laboratory distillation or some reclaiming process.

Retrograde condensation: Condensation that occurs in a pressure region when the pressure on the gas that contains methane and heavier hydrocarbons is reduced. This process is the reverse of usual behavior at lower pressures. Hence, the term "retrograde condensation" is used to describe this phenomenon.

Retrograde pressure region: The pressure region wherein the hydrocarbons exhibit increased volatility as the pressure increases.

Rich amine: The amine that leaves the bottom of the contactor. It is the lean amine plus the acid gases removed from the gas by the lean amine.

Rich gas: A gas that is suitable as feed to a gas processing plant and from which products can be extracted.

Rich oil: The oil that leaves the bottom of the absorber. It is the lean oil plus all of the absorbed products.

RVP (Reid vapor pressure): A vapor pressure specification for the heavier liquid products as determined by ASTM test procedure D-323. The vapor pressure is reported as pounds per sq. in. Reid. The pressure approximates, but is somewhat less than, the absolute vapor pressure of the liquid.

*SCADA (Supervisory Control and Data Acquisition): Computer system used to both control and monitor field and plant systems.

Shrinkage: (1) The reduction in volume, heating value, or both of a gas stream caused by the removal of some of its constituents. (2) Sometimes referred to as the unaccounted loss of products from storage tanks.

Slop, or slop oil: A term rather loosely used to denote a mixture of oil produced at various places in the plant that must be rerun or further processed to be suitable for use.

*Slug catcher: A separator that is designed to separate intermittent large volumes of liquids from a gas stream. It may be vessels or a manifolded pipe system.

*Snubber: See pulsation dampener.

Solution gas: Gas that originates from the liquid phase in the oil reservoir.

Sour: Liquids and gases are said to be "sour" if they contain hydrogen sulfide, mercaptans, or both over a specified level.

Sour gas: Gas that contains an appreciable quantity of carbon dioxide, hydrogen sulfide, or mercaptans.

Splitter: A name applied to fractionators, particularly those that separate isomers (i.e., butane splitter refers to a debutanizer).

***SPA:** Sales and purchase agreement

Sponge absorption unit: The unit wherein the vapors of the lighter absorption oils are recovered.

SRU: Sulfur recovery unit

Stabilized condensate: Condensate that has been stabilized to a definite vapor pressure in a fractionation system.

Stabilizer: A name for a fractionation system that stabilizes any liquid (i.e., reduces the vapor pressure so that the resulting liquid is less volatile).

Standard cubic feet (scf): This term refers to a gas volume measurement at a specified temperature and pressure. The temperature and pressure may be defined in the gas sales contract or by reference to other standards. The commonly used temperature and pressure is 60°F and 14.696 psia.

Straddle plant: Refer to definition of "mainline plant."

Stranded reserve: A gas reserve located where (1) no economic use exists for the gas at the point of origin, and (2) transportation of the gas by pipeline from the reserve to its point of end use is not feasible.

Strapping: A term applied to the process of calibrating liquid storage capacity of storage tanks by increments of depth.

Stream day: This terms refers to a basis for calculating plant production. A stream day is a day of full operation. This concept is different from a calendar day, which would be used to give average production for a full year.

Stripper: A column wherein absorbed constituents are stripped from the absorption oil. The term is applicable to columns using a stripping medium, such as steam or gas.

Stripping factor: An expression used to explain the degree of stripping. Mathematically, it is KV/L, the reciprocal of the absorption factor.

Substitute natural gas (SNG): Refer to definition of "synthetic gas."

Sweet: This term refers to the near or absolute absence of sulfur compounds in either gas or liquid as defined by a given specification standard.

Sweet gas: Gas that has no more than the maximum sulfur content defined by (1) the specifications for the sales gas from a plant or (2) the definition by a legal body such as the Railroad Commission of Texas.

Synthetic gas (SG): The preferred term to describe the salable gas product that results from the gasification of coal or gas liquids or heavier hydrocarbons.

Tail gas: The exit gas from a plant.

Tailgate: The point within the gas processing plant at which the residue gas is last metered. This point is usually at the plant residue sales meter or the allocation meter.

Take-or-pay clause: The clause that may be in a contract that guarantees pay to the seller for a gas even though the particular gas volume is not taken during a specified time period. Some contracts may contain a time period for the buyer to take later delivery of the gas without penalty.

TGCU (tail gas cleanup unit): A process unit that takes tail gas from an SRU and removes additional sulfur.

Therm: A unit of gross heating value equivalent to 100,000 Btu (1.055056×10^8 J).

Tonne: A unit of mass measurement, commonly used in international petroleum commerce; an expression for the metric ton, or 1,000 kilograms.

Ullage: See outage.

*****Vane pack:** An insert in gas-liquid separators, where gas goes through a torturous path to force mists to coalesce into droplets and drop out of the gas phase.

Vapor pressure gasoline: A descriptive phrase for natural gasoline that meets a specified vapor pressure.

Vapor pressure, GPA: Vapor pressure as specified by GPA procedures.

Weathering: The evaporation of liquid by exposure to the conditions of atmospheric temperature and pressure. Partial evaporation of liquid by use of heat may also be called weathering.

Weathering test: A GPA test for LP gas for the determination of heavy components in a sample by evaporation of the sample as specified.

Wet gas: (1) A gas that contains water, or a gas that has not been dehydrated. (2) A term synonymous with rich gas; that is, a gas from which the products have not been extracted. Refer to the definition of "rich gas."

White oil: A term for oil that has no color.

Wild gasoline: An obsolete term for natural gas liquids before fractionation and stabilization.

Wobbe number: A number proportional to the heat input to a burner at constant pressure. In British practice, it is the heating value of a gas divided by the square root of its gravity. Widely used in Europe, together with measured or calculated flame speed, to determine the interchangeability of gases.

Appendix B

Physical Constants and Physical Properties

CONTENTS OF TABLES AND FIGURES

B.1 UNIT CONVERSION FACTORS

Listed below are some of the more commonly used conversion factors.

Mass

1 pound (lb) = 7,000 grains = 16.0 ounces (oz)

= 453.5924 grams (g)

Length

1 foot (ft) = 12.0 inches (in)

= 30.480 centimeters (cm) = 0.30480 meters (m)

Temperature

$°C = (°F - 32)/1.8$

$K = °C + 273.15$

$K = °R/1.8$

$°F = 1.8(°C) + 32$

$°R = °F + 459.67$

$°R = 1.8(K)$

Volume

1 cubic foot (ft^3) = 7.48052 gallons (gal) = 1,728 cubic inches

= 0.1781076 barrels (42 U.S. gal) of oil (bbl)

= 28.31685 liters (L) = 0.02831685 cubic meters (m^3)

Density

1 lb$_m$/ft^3 = 0.1336806 lb$_m$/gal

= 0.01601846 g/cm^3 = 16.01846 kg/m^3

Force

1 pound (lb$_f$) = 4.448 × 10^5 dyne (dyn) = 4.448222 Newtons (N)

Pressure

1 atmosphere (atm) = 1.01325 bar = 14.696 lb$_f$/in^2 = 760 mm Hg (at 32°F)

= 1.013250 × 10^5 Pascal (Pa)

Energy

1 British Thermal Unit (Btu) (IT) = 252.1644 cal (tc) = 3.930148 × 10^{-4} hp

= 1.055056 × 10^3 joules (J) = 2.930711 × 10^{-4} kWh

(1 Btu [IT] = 1.00067 Btu [tc])

(Note: Customarily the Btu refers to the International Steam Table [IT] Btu, and the calorie refers to the thermochemical calorie [tc])

Flow

1 gal/min = 0.1336805 ft^3/min = 1.42857 bbl/h = 6.309020 × 10^{-5} m^3/s

Power

1 hp (US) = 2544.433 Btu (IT)/h = 550 ft lb$_f$/s = 745.6999 watts (W)

Specific Energy per Degree

1 Btu/lb$_m$-°F (IT) = 1.0 cal/g-°C (IT) = 4.186800 kJ/kg-K

B.2 GAS CONSTANTS AND STANDARD GAS CONDITIONS

B.2.1 UNIVERSAL GAS CONSTANTS

Metric/SI Units

8.31441 J/(mol-K)
1.98719 cal/(mol-K)
82.0568 cm³-atm/(mol-K)
62.3633 L-mm/(mol-K)
0.0820568 L-atm/(mol-K)
8.31441 m³-Pa/(mol-K)
8.31441×10^{-5} m³-bar/(mol-K)
0.0831441 L-bar/(mol-K)

Engineering and Mixed Units

1.98585 Btu/(lb-mol-°R)
0.730235 atm-ft³/(lb-mol-°R)
0.739911 bar-ft³/(lb-mol-°R)
1545.34 ft-lb$_f$/(lb-mol-°R)
10.7315 psi ft³/(lb-mol-°R)

B.2.2 STANDARD GAS CONDITIONS

Gas volumes given are for ideal gas only.

1. Normal: continental and scientific applications
    0°C, 1.01325 bar, gas volume = 22.4136 m³/kg-mol
    32°F, 14.696 psia, gas volume = 359.031 ft³/lb-mol
    (Nm³ based upon these conditions)
2. Standard (scf): U.S. engineering applications
    15.5556°C, 1.01325 bar, gas volume = 23.6900 m³/kg-mol
    60°F, 14.696 psia, gas volume = 379.49 ft³/lb-mol
    (scf = standard cubic foot based upon these conditions)
3. Standard: API 2564, SI 15°C, 101.325 kPa, gas volume = 23.6444 m³/kg-mol (Sm³ based upon these conditions)

B.3 THERMODYNAMIC AND PHYSICAL PROPERTY DATA

B.3.1 TABLE OF PHYSICAL CONSTANTS OF PURE FLUIDS

Table B.1 provides some of the most useful physical constants for fluids encountered in natural gas processing. For additional properties the reader should consult GPA Standard 2145-03 (2005). The standard includes a number of notes regarding the data and data sources.

TABLE B.1
Physical Properties of Compounds Typically Seen in Gas Processing

Component	Methane	Ethane	Propane	Isobutane	n-Butane	Isopentane	n-Pentane
Molar mass	16.042	30.069	44.096	58.122	58.122	72.149	72.149
Boiling Point, °F (K)	−258.67 (111.67)	−127.48 (184.55)	−43.72 (231.08)	11.08 (261.53)	31.09 (272.64)	82.11 (300.99)	96.98 (309.25)
Freezing Point, °F (K)	−296.45 (90.68)	−297.04 (90.35)	−305.73 (85.52)	−255.3 (113.54)	−217.05 (134.79)	−255.8 (113.26)	−201.5 (143.43)
Vapor Pressure at 100 °F (313.15K), psia (bar)	5000 (350)[a]	800 (55)[a]	188.69 (13.699)	72.484 (5.3012)	51.683 (3.7936)	20.456 (1.5136)	15.558 (1.1556)
Density of liquid							
Relative to water	0.3[a]	0.35643	0.50738	0.56295	0.58408	0.6246	0.63113
Density, lb/gal (kg/m³)	2.5 (300)[a]	2.9716 (358)	4.2301 (507.67)	4.6934 (563.07)	4.8696 (584.14)	5.2074 (624.54)	5.2618 (631.05)
Density of ideal gas							
Relative density, air = 1.0	0.55397	1.0383	1.5227	2.0071	2.0071	2.4914	2.4914
Density lb/1000 ft³ (kg/m³)	42.274 (0.67848)	79.237 (1.2717)	116.2 (1.8649)	153.16 (2.4581)	153.16 (2.4581)	190.12 (3.0514)	190.12 (3.0514)
Volume of liquid							
gal/lb-mol (cm³/mol)	6.417 (53.475)[a]	10.119 (83.992)[a]	10.424 (86.859)	12.384 (103.22)	11.936 (99.501)	13.855 (115.52)	13.712 (114.33)
Critical conditions							
Temperature, °F (K)	−116.66 (190.56)	89.924 (305.33)	205.92 (369.77)	274.41 (407.82)	305.546 (425.12)	368.98 (460.36)	385.75 (469.68)
Pressure, psia (bar)	667 (45.99)	706.6 (48.72)	615.5 (42.44)	527.9 (36.40)	550.9 (37.98)	490.4 (33.81)	488.8 (33.70)

	1	2	3	4	5	6	7
Gross heating value, ideal reaction							
Fuel as liquid Btu/lb, (MJ/kg)		22181 (51.594)	21490 (49.988)	21080 (49.033)	21136 (49.165)	20891 (48.594)	20923 (48.669)
Fuel as liquid, Btu/gal (MJ/m³)		65914 (18471)	90905 (25377)	98935 (27609)	102926 (28719)	108789 (30349)	110094 (30712)
fuel as ideal gas, Btu/lb (MJ/kg)	23892 (55.576)	22334 (51.952)	21654 (50.37)	21232 (49.389)	21300 (49.547)	21044 (48.95)	21085 (49.046)
Fuel as ideal gas Btu/ft³ (MJ/m³)	1010 (37.707)	1769.7 (66.067)	2516.2 (93.936)	3252 (121.4)	3262.4 (121.79)	4000.9 (149.36)	4008.7 (149.66)
Btu/gal, fuel as ideal gas	59730	66369	91599	99652	103724	109584	110946
Net heating value, ideal reaction, fuel as ideal gas							
Btu/ft³, (MJ/m³)	909 (33.949)	1619 (60.429)	2315 (86.419)	3000 (112.01)	3011 (112.4)	3699 (138.09)	3707 (138.38)
Heat of vaporization at 14.696 psia (313.15K, 1.01325 bar)							
Btu/lb, kJ/kg	219.8 (511.3)	210.3 (489.2)	183.4 (426.7)	157.2 (365.6)	166.3 (386.9)	147.7 (343.7)	154.4 (359.2)
Flammability limits at 100 °F, 14.696 psia (313.15K, 1.01325 bar) in air, vol%							
Lower limit	5	2.9	2	1.8	1.5	1.3	1.4
Upper limit	15	13	9.5	8.5	9	8	8.3

[a] hypothetical value because liquid does not exist at stated condition

(continued)

TABLE B.1 (Continued)

Component	n-Hexane	n-Heptane	n-Octane	n-Nonane	n-Decane	Carbon Dioxide
Molar mass	86.175	100.202	114.229	128.255	142.282	44.01
Boiling point, °F (K)	155.72 (341.88)	209.13 (371.56)	258.21 (398.82)	303.4 (423.93)	345.4 (447.26)	−109.12 (194.75)
Vapor pressure at 100°F (313.15K), psia (bar)	4.961 (37.3)	1.62 (12.336)	0.5366 (4.1403)	0.17 (1.3488)	0.0616 (0.4876)	
Density of liquid						
Relative to water	0.66405 (0.66448)	0.68819 (0.6886)	0.70698 (0.70737)	0.72186 (0.72224)	0.73406 (0.73442)	0.82203 (0.82195)
Density, lb/gal (kg/m³)	5.5363 (663.89)	5.7375 (687.98)	5.8942 (706.73)	6.0183 (721.59)	6.12 (733.76)	6.8534 (821.22)
Density of Ideal Gas						
Relative density, air = 1.0	2.9758	3.4601	3.9445	4.4289	4.9132	1.5197
Density lb/1,000 ft³ (kg/m³)	227.09 (3.6446)	264.05 (4.2378)	301.01 (4.831)	337.97 (5.4242)	374.93 (6.0174)	115.97 (1.8613)
Volume of liquid						
gal/lb-mol (cm³/mol)	15.566 (129.8)	17.464 (145.65)	19.38 (161.63)	21.311 (177.74)	23.249 (193.91)	6.4216 (53.59)
Critical conditions						
Temperature, °F (K)	453.83 (507.5)	512.87 (540.3)	564.22 (568.8)	610.8 (594.7)	652.2 (617.7)	87.8 (304.1)
Pressure, psia (bar)	436.9 (3012)	396.8 (2736)	360.7 (2487)	330.7 (2280)	304.6 (2100)	1070 (7377)

	Gross heating value, ideal reaction					
Fuel as liquid Btu/lb, (MJ/kg)	20783 (48.343)	20680 (48.104)	20601 (47.92)	20543 (47.785)	20494 (47.671)	
Fuel as liquid, Btu/gal (MJ/m³)	115060 (32094)	118654 (33095)	121428 (33866)	123634 (34481)	125424 (34979)	
fuel as ideal gas, Btu/lb (MJ/kg)	20944 (48.717)	20839 (48.474)	20760 (48.289)	20701 (48.153)	20652 (48.038)	
Fuel as ideal gas Btu/ft³ (MJ/m³)	4756 (177.55)	5502.5 (205.42)	6248.9 (233.29)	6996.4 (261.19)	7743 (289.06)	
Btu/gal, fuel as ideal gas	115951	119565	122363	124585	126388	
Net heating value, ideal reaction, fuel as ideal gas						
Btu/ft³, (MJ/m³)	4404 (164.4)	5100 (190.39)	5796 (216.37)	6493 (242.4)	7190 (268.39)	
Heat of vaporization at 14.696 psia (1.01325 bar)						
Btu/lbm, kJ/kg	144.1 (335.1)	136.7 (318.1)	130 (302.4)	124.4 (289.3)	119.3 (277.6)	246.5 (573.3)
Flammability limits at 100°F, 14.696 psia (313.15 K, 1.01325 bar) in air, vol%						
Lower limit	1.1	1	0.8	0.7	0.7	
Upper limit	7.7	7	6.5	5.6	5.4	

(continued)

TABLE B.1 (Continued)

Component	Hydrogen Sulfide	Nitrogen	Oxygen	Helium	Air	Water
Molar mass	34.082	28.013	31.999	4.0026	28.959	18.0153
Boiling Point, °F (K)	−76.52 (212.86)	−320.43 (77.355)	−297.33 (90.188)	−452.06 (4.23)	−317.64 (78.903)	211.95 (373.124)
Freezing Point , °F (K)	−121.81 (187.7)	−346 (63.151)	−361.82 (54.361)	−455.75 (2.177)		32 (273.15)
Vapor Pressure at 100 °F (313.15K), psia (bar)	395.55 (28.67)					0.95051 (0.073849)
Density of liquid						
Relative to water	0.80269 (0.80015)	0.80687 (0.8068)	1.1423 (1.1422)		0.87603 (0.87596)	1.0
Density, lb/gal (kg/m³)	6.6922 (799.4)	6.727 (806.1)	9.5236 (1141.2)		7.3036 (875.2)	8.3372 (999.103)
Density of gas						
Relative density, air = 1.0	1.1769	0.9673	1.105	0.1382	1.0	0.6221
Density lb/1000 ft³ (kg/m³)	89.811 (1.4414)	73.819 (1.1847)	84.322 (1.3533)	10.547 (0.1693)	76.311 (1.2247)	47.473 (0.76191)
Volume of liquid						
gal/lb-mol (cm³/mol)	5.0928 (42.63)	4.1643 (34.752)	3.36 (28.04)		3.965 (33.09)	2.1608 (18.031)

Critical conditions						
Temperature, °F (K)	212.8 (373.6)	−232.53 (126.19)	−181.43 (154.58)	−450.32 (5.195)	−220.97 (132.61)	705.1 (647.1)
Pressure, psia (bar)	1306.5 (90.08)	492.5 (33.96)	731.4 (50.43)	33.0 (2.275)	551.9 (38.05)	3200.1 (22.064)
Gross heating value, ideal reaction						
Fuel as liquid Btu/lb, (MJ/kg)	6897 (12825)					
Fuel as liquid, Btu/gal (MJ/m³)	46156 (16.042)					
Fuel as ideal gas, Btu/lb (MJ/kg)	7093.8 (16.501)					
Fuel as ideal gas Btu/ft³ (MJ/m³)	637.11 (23.785)					
Btu/gal, fuel as ideal gas	47473					
Net heating value, ideal reaction, fuel as ideal gas						
Btu/ft³, (MJ/m³)	586.8 (21.91)					
Heat of vaporization at 14.696 psia (1.01325 bar)						
Btu/lb, kJ/kg	234.4 (545.3)	85.63 (199.2)	91.6 (213.1)	8.922 (20.75)	86.81 (201.9)	970.12 (2256.5)
Flammability limits at 100°F, 14.696 psia (313.15K, 1.01325 bar) in air, vol%						
Lower limit	4.3					
Upper limit	45.5					

Unless stated otherwise, pressure-dependent properties are given at 14.696 psia (1.01325 bar); temperature-dependent properties are given at 60°F (15°C); temperature-dependent and pressure-dependent properties are given at 60°F (15°C) and 14.696 psia (1.01325 bar). *Source:* GPA (2005).

B.3.2 Temperature-Dependent Properties of Pure Fluids

Tables B.2 to B.11 provide constants for equations to calculate physical properties as a function of temperature. Pressure effect is not considered and liquid properties are given at the vapor pressure of the fluid (saturation boundary). The tables include the applicable temperature range and the estimated quality of the prediction. They also include equation numbers for which the coefficients apply. Table B.12 lists the equations. These data were made available by courtesy of the Design Institute for Physical Properties (DIPPR,2005), an Industry Technology Alliance of the American Institute of Chemical Engineers (AIChE), from the DIPPR 801 database, 2005 edition (www.aiche.org/DIPPR). The coefficients were regressed from critically reviewed data. The tables present 10 of the 15 temperature-dependent physical properties and 33 physical constants in the database. Nearly 1,800 compounds are included in the database.

TABLE B.2
Liquid Density (Computed value has units of $kmol/m^3$)

Compound	Equation Number	Temperature Range (K)	Quality	A	B	C	D	E
Methane	105	90.7 – 190.6	< 1%	2.9214E+00	2.8976E-01	1.9056E+02	2.8881E-01	
Ethylene	105	104 – 282.3	< 1%	2.0961E+00	2.7657E-01	2.8234E+02	2.9147E-01	
Ethane	105	90.4 – 305.3	< 1%	1.9122E+00	2.7937E-01	3.0532E+02	2.9187E-01	
Propane	105	85.5 – 369.8	< 1%	1.3757E+00	2.7453E-01	3.6983E+02	2.9359E-01	
Isobutane	105	113.5 – 407.8	< 1%	1.0631E+00	2.7506E-01	4.078E+02	2.758E-01	
n-Butane	105	134.9 – 425.1	< 1%	1.0677E+00	2.7188E-01	4.2512E+02	2.8688E-01	
Isopentane	105	113.2 – 460.4	< 1%	9.1991E-01	2.7815E-01	4.604E+02	2.8667E-01	
n-Pentane	105	143.4 – 469.7	< 1%	8.4947E-01	2.6726E-01	4.697E+02	2.7789E-01	
Benzene	105	278.7 – 562	< 3%	1.0259E+00	2.6666E-01	5.6205E+02	2.8394E-01	
n-Hexane	105	177.8 – 507.6	< 1%	7.0824E-01	2.6411E-01	5.076E+02	2.7537E-01	
n-Heptane	105	182.6 – 540.2	< 1%	6.1259E-01	2.6211E-01	5.402E+02	2.8141E-01	
Ammonia	105	195.4 – 405.6	< 1%	3.5383E+00	2.5443E-01	4.0565E+02	2.888E-01	
Water	116	273.2 – 647.1	< 1%	1.7863E+01	5.8606E+01	-9.5396E+01	2.1389E+02	-1.4126E+02
Oxygen	105	54.4 – 154.6	< 1%	3.9143E+00	2.8772E-01	1.5458E+02	2.924E-01	
Nitrogen	105	63.2 – 126.2	< 1%	3.2091E+00	2.861E-01	1.262E+02	2.966E-01	
Hydrogen	105	14 – 33.2	< 3%	5.414E+00	3.4893E-01	3.319E+01	2.706E-01	
Hydrogen sulfide	105	187.7 – 373.5	< 1%	2.7672E+00	2.7369E-01	3.7353E+02	2.9015E-01	
Carbon monoxide	105	68.2 – 132.9	< 1%	2.897E+00	2.7532E-01	1.3292E+02	2.813E-01	
Carbon dioxide	105	216.6 – 304.2	< 1%	2.768E+00	2.6212E-01	3.0421E+02	2.908E-01	
Helium	105	2.2 – 5.2	< 3%	7.2475E+00	4.1865E-01	5.2E+00	2.4096E-01	
Methanol	105	175.5 – 512.5	< 1%	2.3267E+00	2.7073E-01	5.125E+02	2.4713E-01	
Ethylene glycol	105	260.2 – 720	< 1%	1.315E+00	2.5125E-01	7.2E+02	2.1868E-01	
Diethylene glycol	105	262.7 – 744.6	< 3%	8.3692E-01	2.6112E-01	7.446E+02	2.422E-01	
Triethylene glycol	105	266 – 769.5	< 3%	5.9672E-01	2.6217E-01	7.695E+02	2.4631E-01	

TABLE B.3
Vapor Pressure (Computed value has units of Pa)

Compound	Equation Number	Temperature Range (K)	Quality	A	B	C	D	E
Methane	101	90.7 – 190.6	< 1%	3.9205E+01	-1.3244E+03	-3.4366E+00	3.1019E-05	2
Ethylene	101	104 – 282.3	< 1%	5.3963E+01	-2.443E+03	-5.5643E+00	1.9079E-05	2
Ethane	101	90.4 – 305.3	< 1%	5.1857E+01	-2.5987E+03	-5.1283E+00	1.4913E-05	2
Propane	101	85.5 – 369.8	< 3%	5.9078E+01	-3.4926E+03	-6.0669E+00	1.0919E-05	2
Isobutane	101	113.5 – 407.8	< 3%	1.0843E+02	-5.0399E+03	-1.5012E+01	2.2725E-02	1
n-Butane	101	134.9 – 425.1	< 3%	6.6343E+01	-4.3632E+03	-7.046E+00	9.4509E-06	2
Isopentane	101	113.2 – 460.4	< 3%	7.1308E+01	-4.976E+03	-7.7169E+00	8.7271E-06	2
n-Pentane	101	143.4 – 469.7	< 3%	7.8741E+01	-5.4203E+03	-8.8253E+00	9.6171E-06	2
Benzene	101	278.7 – 562	< 1%	8.3107E+01	-6.4862E+03	-9.2194E+00	6.9844E-06	2
n-Hexane	101	177.8 – 507.6	< 3%	1.0465E+02	-6.9955E+03	-1.2702E+01	1.2381E-05	2
n-Heptane	101	182.6 – 540.2	< 3%	8.7829E+01	-6.9964E+03	-9.8802E+00	7.2099E-06	2
Ammonia	101	195.4 – 405.6	< 1%	9.0483E+01	-4.6697E+03	-1.1607E+01	1.7194E-02	1
Water	101	273.2 – 647.1	< 0.2%	7.3649E+01	-7.2582E+03	-7.3037E+00	4.1653E-06	2
Oxygen	101	54.4 – 154.6	< 1%	5.1245E+01	-1.2002E+03	-6.4361E+00	2.8405E-02	1
Nitrogen	101	63.2 – 126.2	< 1%	5.8282E+01	-1.0841E+03	-8.3144E+00	4.4127E-02	1
Hydrogen	101	14 – 33.2	< 3%	1.269E+01	-9.4896E+01	1.1125E+00	3.2915E-04	2
Hydrogen sulfide	101	187.7 – 373.5	< 3%	8.5584E+01	-3.8399E+03	-1.1199E+01	1.8848E-02	1
Carbon monoxide	101	68.2 – 132.9	< 1%	4.5698E+01	-1.0766E+03	-4.8814E+00	7.5673E-05	2
Carbon dioxide	101	216.6 – 304.2	< 1%	1.4054E+02	-4.735E+03	-2.1268E+01	4.0909E-02	1
Helium	101	1.8 – 5.2	< 1%	1.1533E+01	-8.99E+00	6.724E-01	2.743E-01	1
Methanol	101	175.5 – 512.5	< 1%	8.2718E+01	-6.9045E+03	-8.8622E+00	7.4664E-06	2
Ethylene glycol	101	260.2 – 720	< 3%	8.409E+01	-1.0411E+04	-8.1976E+00	1.6536E-18	6
Diethylene glycol	101	262.7 – 744.6	< 10%	1.4245E+02	-1.505E+04	-1.6318E+01	5.9506E-18	6
Triethylene Glycol	101	266 – 769.5	< 10%	1.5248E+02	-1.6449E+04	-1.767E+01	6.4481E-18	6

Copyright 2005 by the Design Institute for Physical Properties (DIPPR), American Institute of Chemical Engineers (AIChE), and reproduced by permission of AIChE.

TABLE B.4
Heat of Vaporization (Computed value has units of J/kmol)

Compound	Equation Number	Temperature Range(K)	Quality	A	B	C	D
Methane	106	90.7 – 190.6	< 1%	1.0194E+07	2.6087E–01	–1.4694E–01	2.2154E–01
Ethylene	106	104 – 282.3	< 1%	1.8844E+07	3.6485E–01		
Ethane	106	90.4 – 305.3	< 1%	2.1091E+07	6.0646E–01	–5.5492E–01	3.2799E–01
Propane	106	85.5 – 369.8	< 3%	2.9209E+07	7.8237E–01	–7.7319E–01	3.9246E–01
Isobutane	106	113.5 – 407.8	< 3%	3.188E+07	3.9006E–01		
n-Butane	106	134.9 – 425.1	< 3%	3.6238E+07	8.337E–01	–8.2274E–01	3.9613E–01
Isopentane	106	113.2 – 460.4	< 3%	3.7593E+07	3.9173E–01		
n-Pentane	106	143.4 – 469.7	< 3%	3.9109E+07	3.8681E–01		
Benzene	106	278.7 – 562	< 1%	4.5346E+07	3.9053E–01		
n-Hexane	106	177.8 – 507.6	< 3%	4.4544E+07	3.9002E–01		
n-Heptane	106	182.6 – 540.2	< 3%	5.0014E+07	3.8795E–01		
Ammonia	106	195.4 – 405.6	< 1%	3.1523E+07	3.914E–01	–2.289E–01	2.309E–01
Water	106	273.2 – 647.1	< 1%	5.2053E+07	3.199E–01	–2.12E–01	2.5795E–01
Oxygen	106	54.4 – 154.6	< 1%	9.008E+06	4.542E–01	–4.096E–01	3.183E–01
Nitrogen	106	63.2 – 126.2	< 1%	7.4905E+06	4.0406E–01	–3.17E–01	2.7343E–01
Hydrogen	106	14 – 33.2	< 3%	1.0127E+06	6.98E–01	–1.817E+00	1.447E+00
Hydrogen sulfide	106	187.7 – 373.5	< 3%	2.5676E+07	3.7358E–01		
Carbon monoxide	106	68.1 – 132.5	< 1%	8.585E+06	4.921E–01	–3.26E–01	2.231E–01
Carbon dioxide	106	216.6 – 304.2	< 1%	2.173E+07	3.82E–01	–4.339E–01	4.2213E–01
Helium	106	2.2 – 5.2	< 10%	1.2504E+05	1.3038E+00	–2.6954E+00	1.7098E+00
Methanol	106	175.5 – 512.5	< 3%	5.0451E+07	3.3594E–01		
Ethylene glycol	106	260.2 – 720	< 3%	8.3518E+07	4.2625E–01		
Diethylene glycol	106	262.7 – 744.6	< 10%	1.0829E+08	5.4022E–01		
Triethylene Glycol	106	266 – 769.5	< 10%	1.2127E+08	5.8261E–01		

TABLE B.5
Liquid Heat Capacity (Computed value has units of J/kmol-K)

Compound	Equation Number	Temperature Range (K)	Quality	A	B	C	D	E
Methane	114	90.7 – 190	< 1%	6.5708E+01	3.8883E+04	-2.5795E+02	6.1407E+02	
Ethylene	100	104 – 252.7	< 3%	2.4739E+05	-4.428E+03	4.0936E+01	-1.697E-01	2.6816E-04
Ethane	114	92 – 290	< 1%	4.4009E+01	8.9718E+04	9.1877E+02	-1.886E+03	
Propane	114	85.5 – 360	< 1%	6.2983E+01	1.1363E+05	6.3321E+02	-8.7346E+02	
Isobutane	100	113.5 – 380	< 3%	1.7237E+05	-1.7839E+03	1.4759E+01	-4.7909E-02	5.805E-05
n-Butane	100	134.9 – 400	< 1%	1.9103E+05	-1.675E+03	1.25E+01	-3.874E-02	4.6121E-05
Isopentane	100	113.2 – 310	< 3%	1.083E+05	1.46E+02	-2.92E-01	1.51E-03	
n-Pentane	100	143.4 – 390	< 1%	1.5908E+05	-2.705E+02	9.9537E-01		
Benzene	100	278.7 – 353.2	< 3%	1.2944E+05	-1.695E+02	6.4781E-01		
n-Hexane	100	177.8 – 460	< 1%	1.7212E+05	-1.8378E+02	8.8734E-01		
n-Heptane	114	182.6 – 520	< 1%	6.126E+01	3.1441E+05	1.8246E+03	-2.5479E+03	9.3701E-06
Ammonia	114	203.2 – 401.2	< 3%	6.1289E+01	8.0925E+04	7.994E+02	-2.651E+03	
Water	100	273.2 – 533.2	< 1%	2.7637E+05	-2.0901E+03	8.125E+00	-1.4116E-02	
Oxygen	100	54.4 – 142	< 3%	1.7543E+05	-6.1523E+03	1.1392E+02	-9.2382E-01	2.7963E-03
Nitrogen	100	63.2 – 112	< 3%	2.8197E+05	-1.2281E+04	2.48E+02	-2.2182E+00	7.4902E-03
Hydrogen	114	14 – 32	< 5%	6.6653E+01	6.7659E+03	-1.2363E+02	4.7827E+02	
Hydrogen sulfide	114	187.7 – 370	< 3%	6.4666E+01	4.9354E+04	2.2493E+01	-1.623E+03	
Carbon monoxide	114	68.2 – 132	< 3%	6.5429E+01	2.8723E+04	-8.4739E+02	1.9596E+03	
Carbon dioxide	100	220 – 290	< 3%	-8.3043E+06	1.0437E+05	-4.3333E+02	6.0052E-01	
Helium	100	2.2 – 4.6	< 5%	3.8722E+05	-4.6557E+05	2.118E+05	-4.2494E+04	
Methanol	100	175.5 – 400	< 1%	1.058E+05	-3.6223E+02	9.379E-01		
Ethylene glycol	100	260.2 – 493.2	< 1%	3.554E+04	4.3678E+02	-1.8486E-01		3.2129E+03
Diethylene glycol	100	262.7 – 451.2	< 3%	1.2541E+05	4.0058E+02			
Triethylene Glycol	100	265.8 – 441	< 3%	1.538E+05	5.87E+02			

TABLE B.6
Ideal Gas Heat Capacity (Computed value has units of J/kmol-k)

Compound	Equation Number	Temperature Range (K)	Quality	A	B	C	D	E
Methane	107	50 – 1500	< 1%	3.3298E+04	7.9933E+04	2.0869E+03	4.1602E+04	9.9196E+02
Ethylene	107	60 – 1500	< 1%	3.338E+04	9.479E+04	1.596E+03	5.51E+04	7.408E+02
Ethane	107	200 – 1500	< 1%	4.0326E+04	1.3422E+05	1.6555E+03	7.3223E+04	7.5287E+02
Propane	107	200 – 1500	< 1%	5.192E+04	1.9245E+05	1.6265E+03	1.168E+05	7.236E+02
Isobutane	107	200 – 1500	< 1%	6.549E+04	2.4776E+05	1.587E+03	1.575E+05	7.0699E+02
n-Butane	107	200 – 1500	< 1%	7.134E+04	2.43E+05	1.63E+03	1.5033E+05	7.3042E+02
Isopentane	107	200 – 1500	< 1%	7.46E+04	3.265E+05	1.545E+03	1.923E+05	6.667E+02
n-Pentane	107	200 – 1500	< 1%	8.805E+04	3.011E+05	1.6502E+03	1.892E+05	7.476E+02
Benzene	107	200 – 1500	< 1%	4.4767E+04	2.3085E+05	1.4792E+03	1.6836E+05	6.7766E+02
n-Hexane	107	200 – 1500	< 1%	1.044E+05	3.523E+05	1.6946E+03	2.369E+05	7.616E+02
n-Heptane	107	200 – 1500	< 1%	1.2015E+05	4.001E+05	1.6766E+03	2.74E+05	7.564E+02
Ammonia	107	100 – 1500	< 1%	3.3427E+04	4.898E+04	2.036E+03	2.256E+04	8.82E+02
Water	107	100 – 2273.2	< 1%	3.3363E+04	2.679E+04	2.6105E+03	8.896E+03	1.169E+03
Oxygen	107	50 – 1500	< 1%	2.9103E+04	1.004E+04	2.5265E+03	9.356E+03	1.1538E+03
Nitrogen	107	50 – 1500	< 1%	2.9105E+04	8.6149E+03	1.7016E+03	1.0347E+02	9.0979E+02
Hydrogen	107	250 – 1500	< 1%	2.7617E+04	9.56E+03	2.466E+03	3.76E+03	5.676E+02
Hydrogen sulfide	107	100 – 1500	< 1%	3.3288E+04	2.6086E+04	9.134E+02	-1.7979E+04	9.494E+02
Carbon monoxide	107	60 – 1500	< 1%	2.9108E+04	8.773E+03	3.0851E+03	8.4553E+03	1.5382E+03
Carbon dioxide	107	50 – 5000	< 1%	2.937E+04	3.454E+04	1.428E+03	2.64E+04	5.88E+02
Helium	100	100 – 1500	< 1%	2.0786E+04				
Methanol	107	200 – 1500	< 1%	3.9252E+04	8.79E+04	1.9165E+03	5.3654E+04	8.967E+02
Ethylene glycol	107	300 – 1500	< 3%	6.3012E+04	1.4584E+05	1.673E+03	9.7296E+04	7.7365E+02
Diethylene glycol	107	200 – 1500	< 10%	8.79E+04	2.713E+05	1.3963E+03	1.7035E+05	6.2404E+02
Triethylene Glycol	107	300 – 1500	< 25%	9.04E+04	4.202E+05	1.2628E+03	2.7705E+05	5.311E+02

TABLE B.7
Liquid Viscosity (Computed value has units of Pa-s)

Compound	Equation Number	Temperature Range (K)	Quality	A	B	C	D	E
Methane	101	90.7 – 188	<3%	-6.1572E+00	1.7815E+02	-9.5239E-01	-9.0606E-24	10
Ethylene	101	104 – 250	<5%	1.8878E+00	7.8865E+01	-2.1554E+00		
Ethane	101	90.4 – 300	<3%	-7.0046E+00	2.7638E+02	-6.087E-01	-3.1108E-18	7
Propane	101	85.5 – 360	<5%	-1.7156E+01	6.4625E+02	1.1101E+00	-7.3439E-11	4
Isobutane	101	110 – 311	<5%	-1.3912E+01	7.9709E+02	4.5308E-01		
n-Butane	101	134.9 – 420	<3%	-7.2471E+00	5.3482E+02	-5.7469E-01	-4.6625E-27	10
Isopentane	101	150 – 310	<1%	-1.2596E+01	8.8911E+02	2.0469E-01		
n-Pentane	101	143.4 – 465.2	<3%	-5.3509E+01	1.8366E+03	7.1409E+00	-1.9627E-05	2
Benzene	101	278.7 – 545	<3%	7.5117E+00	2.9468E+02	-2.794E+00		
n-Hexane	101	174.6 – 406.1	<3%	-6.3276E+00	6.4E+02	-6.94E-01	5.6884E+21	-10
n-Heptane	101	180.2 – 432.2	<3%	-9.4622E+00	8.7707E+02	-2.3445E-01	1.4022E+22	-10
Ammonia	101	195.4 – 393.2	<5%	-6.743E+00	5.983E+02	-7.341E-01	-3.69E-27	10
Water	101	273.2 – 646.2	<3%	-5.2843E+01	3.7036E+03	5.866E+00	-5.879E-29	10
Oxygen	101	54.4 – 150	<25%	-4.1476E+00	9.404E+01	-1.207E+00		
Nitrogen	101	63.2 – 124	<5%	1.6004E+01	-1.8161E+02	-5.1551E+00		
Hydrogen	101	14 – 33	<5%	-1.1661E+01	2.47E+01	-2.61E-01	-4.1E-16	10
Hydrogen sulfide	101	187.7 – 350	<5%	-1.0905E+01	7.6211E+02	-1.1863E-01		
Carbon monoxide	101	68.2 – 131.4	<5%	-4.9735E+00	9.767E+01	-1.1088E+00		
Carbon dioxide	101	216.6 – 303.2	<10%	1.8775E+01	-4.0292E+02	-4.6854E+00	-6.9171E-26	10
Helium	101	2.2 – 5.1	<10%	-9.6312E+00	-3.841E+00	-1.458E+00	-1.065E-08	10
Methanol	101	175.5 – 337.8	<5%	-2.5317E+01	1.7892E+03	2.069E+00		
Ethylene glycol	101	260.2 – 576	<5%	-2.0515E+01	2.4685E+03	1.2435E+00	2.4998E+12	-5
Diethylene glycol	101	262.7 – 595.7	<10%	1.3011E+01	2.6481E+02	-3.4184E+00	4.843E+12	-5
Triethylene Glycol	101	266 – 615.6	<10%	-2.7963E+01	3.225E+03	2.2792E+00	1.8277E+17	-7

TABLE B.8
Vapor Viscosity at Low Pressure (Computed value has units of Pa-s)

Compound	Equation Number	Temperature Range (K)	Quality	A	B	C	D
Methane	102	90.7 – 1000	< 3%	5.2546E–07	5.9006E–01	1.0567E+02	
Ethylene	102	169.4 – 1000	< 5%	2.0789E–06	4.163E–01	3.527E+02	
Ethane	102	90.4 – 1000	< 5%	2.5906E–07	6.7988E–01	9.8902E+01	
Propane	102	85.5 – 1000	< 3%	4.9054E–08	9.0125E–01		
Isobutane	102	150 – 1000	< 5%	1.0871E–07	7.8135E–01	7.0639E+01	
n-Butane	102	134.9 – 1000	< 3%	3.4387E–08	9.4604E–01		
Isopentane	102	150 – 1000	< 5%	2.4344E–08	9.7376E–01	–9.1597E+01	1.872E+04
n-Pentane	102	143.4 – 1000	< 5%	6.3412E–08	8.4758E–01	4.1718E+01	
Benzene	102	278.7 – 1000	< 3%	3.134E–08	9.676E–01	7.9E+00	
n-Hexane	102	177.8 – 1000	< 5%	1.7514E–07	7.0737E–01	1.5714E+02	
n-Heptane	102	182.6 – 1000	< 5%	6.672E–08	8.2837E–01	8.5752E+01	
Ammonia	102	195.4 – 1000	< 10%	4.1855E–08	9.806E–01	3.08E+01	
Water	102	273.2 – 1073.2	< 1%	1.7096E–08	1.1146E+00		
Oxygen	102	54.4 – 1500	< 5%	1.101E–06	5.634E–01	9.63E+01	
Nitrogen	102	63.2 – 1970	< 3%	6.5592E–07	6.081E–01	5.4714E+01	
Hydrogen	102	14 – 3000	< 10%	1.797E–07	6.85E–01	–5.9E–01	1.4E+02
Hydrogen sulfide	102	250 – 480	< 3%	3.9314E–08	1.0134E+00		
Carbon-monoxide	102	68.2 – 1250	< 5%	1.1127E–06	5.338E–01	9.47E+01	
Carbon-dioxide	102	194.7 – 1500	< 5%	2.148E–06	4.6E–01	2.9E+02	
Helium	102	20 – 2000	< 3%	3.253E–07	7.162E–01	–9.6E+00	1.07E+02
Methanol	102	240 – 1000	< 10%	3.0663E–07	6.9655E–01	2.05E+02	
Ethylene-glycol	102	260.2 – 1000	< 10%	8.6706E–08	8.3923E–01	7.5512E+01	
Diethylene-glycol	102	262.7 – 1000	< 10%	6.7384E–08	8.489E–01	7.1139E+01	
Triethylene-Glycol	102	266 – 1000	< 10%	5.4291E–08	8.6024E–01	6.592E+01	

TABLE B.9
Liquid Thermal Conductivity (Computed value has units of W/m-K)

Compound	Equation Number	Temperature Range (K)	Quality	A	B	C	D	E
Methane	100	90.7 – 180	< 5%	4.1768E–01	–2.4528E–03	3.5588E–06		
Ethylene	100	104 – 280	< 5%	4.194E–01	–1.591E–03	1.306E–06		
Ethane	100	90.4 – 300	< 3%	3.5758E–01	–1.1458E–03	6.1866E–07		
Propane	100	85.5 – 350	< 3%	2.6755E–01	–6.6457E–04	2.774E–07		
Isobutane	100	113.5 – 400	< 10%	2.0455E–01	–3.6589E–04			
n-Butane	100	134.9 – 400	< 5%	2.7349E–01	–7.1267E–04	5.1555E–07		
Isopentane	100	113.2 – 368.1	< 10%	2.1246E–01	–3.3581E–04			
n-Pentane	100	143.4 – 445	< 5%	2.537E–01	–5.76E–04	3.44E–07		
Benzene	100	278.7 – 413.1	< 3%	2.3444E–01	–3.0572E–04			
n-Hexane	100	177.8 – 370	< 3%	2.2492E–01	–3.533E–04			
n-Heptane	100	182.6 – 371.6	< 3%	2.15E–01	–3.03E–04			
Ammonia	100	195.4 – 400	< 10%	1.169E+00	–2.314E–03			
Water	100	273.2 – 633.2	< 1%	–4.32E–01	5.7255E–03	–8.078E–06	1.861E–09	
Oxygen	100	60 – 150	< 5%	2.741E–01	–1.38E–03			
Nitrogen	100	63.2 – 124	< 3%	2.654E–01	–1.677E–03			
Hydrogen	100	14 – 31	< 10%	–9.17E–02	1.7678E–02	–3.82E–04	–3.3324E–06	1.0266E–07
Hydrogen sulfide	100	193.2 – 292.4	< 5%	4.842E–01	–1.184E–03			
Carbon monoxide	100	68.2 – 125	< 3%	2.855E–01	–1.784E–03			
Carbon dioxide	100	216.6 – 300	< 3%	4.406E–01	–1.2175E–03			
Helium	100	2.2 – 4.8	< 3%	–1.3833E–02	2.2913E–02	–5.4872E–03	4.585E–04	
Methanol	100	175.5 – 337.8	< 3%	2.837E–01	–2.81E–04			
Ethylene glycol	100	260.2 – 470.4	< 5%	8.8067E–02	9.4712E–04	–1.3114E–06		
Diethylene glycol	100	262.7 – 518	< 5%	6.4277E–02	7.8259E–04	–1.0562E–06		
Triethylene Glycol	100	266 – 561.5	< 5%	1.0753E–01	5.0392E–04	–7.2763E–07		

TABLE B.10
Vapor Thermal Conductivity at Low Pressure (Computed value has units of W/m-K)

Compound	Equation Number	Temperature Range (K)	Quality	A	B	C	D
Methane	102	111.6 – 600	< 5%	8.3983E–06	1.4268E+00	–4.9654E+01	
Ethylene	102	170 – 590.9	< 5%	8.6806E–06	1.4559E+00	2.9972E+02	–2.9403E+04
Ethane	102	184.6 – 1000	< 5%	7.3869E–05	1.1689E+00	5.0073E+02	
Propane	102	231.1 – 1000	< 5%	–1.12E+00	1.0972E–01	–9.8346E+03	–7.5358E+06
Isobutane	102	261.4 – 1000	< 10%	8.9772E–02	1.8501E–01	6.3923E+02	1.1147E+06
n-Butane	102	272.6 – 1000	< 5%	5.1094E–02	4.5253E–01	5.4555E+03	1.9798E+06
Isopentane	102	273.2 – 1000	< 5%	8.968E–04	7.742E–01	4.56E+02	2.3064E+05
n-Pentane	102	273.2 – 1000	< 5%	–6.844E+02	7.64E–01	–1.055E+09	
Benzene	102	339.2 – 1000	< 5%	1.652E–05	1.3117E+00	4.91E+02	
n-Hexane	102	339.1 – 1000	< 5%	–6.505E+02	8.053E–01	–1.4121E+09	
n-Heptane	102	339.2 – 1000	< 5%	–7.0028E–02	3.8068E–01	–7.0499E+03	–2.4005E+06
Ammonia	102	200 – 900	< 5%	9.6608E–06	1.3799E+00		
Water	102	273.2 – 1073.2	< 1%	6.2041E–06	1.3973E+00		
Oxygen	102	80 – 2000	< 10%	4.4994E–04	7.456E–01	5.6699E+01	
Nitrogen	102	63.2 – 2000	< 3%	3.3143E–04	7.722E–01	1.6323E+01	3.7372E+02
Hydrogen	102	22 – 1600	< 5%	2.653E–03	7.452E–01	1.2E+01	
Hydrogen sulfide	102	212.8 – 600	< 3%	1.381E–07	1.8379E+00	–3.5209E+02	4.6041E+04
Carbon monoxide	102	70 – 1500	< 5%	5.9882E–04	6.863E–01	5.713E+01	5.0192E+02
Carbon dioxide	102	194.7 – 1500	< 10%	3.69E+00	–3.838E–01	9.64E+02	1.86E+06
Helium	102	30 – 2000	< 5%	2.26E–03	7.305E–01	–1.863E+01	4.4E+02
Methanol	102	273 – 684.4	< 5%	5.7992E–07	1.7862E+00		
Ethylene glycol	102	470.4 – 1000	< 10%	–8.1458E–06	–3.0502E–01	1.8325E+09	–1.1842E+13
Diethylene glycol	102	518 – 1000	< 10%	2.0395E+03	9.0063E–01	1.2238E+10	
Triethylene Glycol	102	269.3 – 1000	< 25%	1.8738E–05	1.238E+00	4.7673E+02	

Copyright 2005 by the Design Institute for Physical Properties (DIPPR), American Institute of Chemical Engineers (AIChE), and reproduced by permission of AIChE.

TABLE B.11
Surface Tension (Computed value has units of N/m)

Compound	EquationNumber	Temperature Range (K)	Quality	A	B	C	D
Methane	106	90.7 – 190.6	< 3%	3.6557E–02	1.1466E+00		
Ethylene	106	104 – 282.3	< 3%	5.294E–02	1.278E+00		
Ethane	106	90.4 – 305.3	< 3%	4.8643E–02	1.1981E+00		
Propane	106	85.5 – 369.8	< 3%	5.092E–02	1.2197E+00		
Isobutane	106	113.5 – 407.8	< 3%	5.1359E–02	1.2532E+00		
n-Butane	106	134.9 – 425.1	< 3%	5.196E–02	1.2181E+00		
Isopentane	106	113.2 – 460.4	< 3%	5.0876E–02	1.2066E+00		
n-Pentane	106	143.4 – 469.7	< 3%	5.202E–02	1.2041E+00		
Benzene	106	278.7 – 562	< 3%	7.1815E–02	1.2362E+00		
n-Hexane	106	177.8 – 507.6	< 3%	5.5003E–02	1.2674E+00		
n-Heptane	106	182.6 – 540.2	< 3%	5.4143E–02	1.2512E+00		
Ammonia	106	195.4 – 405.6	< 3%	1.0162E–01	1.216E+00		
Water	106	273.2 – 647.1	< 1%	1.8548E–01	2.717E+00	–3.554E+00	2.047E+00
Oxygen	106	54.4 – 154.6	< 3%	3.8066E–02	1.2136E+00		
Nitrogen	106	63.2 – 126.2	< 3%	2.901E–02	1.2485E+00		
Hydrogen	106	14 – 33.2	< 1%	5.345E–03	1.0646E+00		
Hydrogen sulfide	106	187.7 – 373.5	< 5%	7.4256E–02	1.2997E+00		
Carbon monoxide	106	68.2 – 132.9	< 3%	2.7959E–02	1.133E+00		
Carbon dioxide	106	216.6 – 304.2	< 10%	8.071E–02	1.2662E+00		
Helium	106	2.2 – 5.2	< 5%	5.1136E–04	1.003E+00		
Methanol	100	273.1 – 503.2	< 1%	3.513E–02	–7.04E–06	–1.216E–07	
Ethylene glycol	100	260.2 – 470.4	< 1%	7.4516E–02	–8.9E–05	–3.9465E–17	
Diethylene glycol	100	293.2 – 518.2	< 1%	6.761E–02	–4.62E–05	–6.46E–08	
Triethylene Glycol	100	293.2 – 548.2	< 1%	6.7901E–02	–6.9536E–05	–2.4419E–08	

TABLE B.12
Equations Noted in Tables B.2 to B.11

Equation Number	Empirical Equation
100	$Y = A + BT + CT^2 + DT^3 + ET^4$
101	$Y = \text{Exp}[A + B/T + C\text{Ln } T + DT^E]$
102	$Y = \dfrac{AT^B}{1 + \dfrac{C}{T} + \dfrac{D}{T^2}}$
105	$Y = \dfrac{A}{B^{\left[1 + (1 - \frac{T}{C})^D\right]}}$
106	$Y = A[1 + T_r]^{\left(B + CT_r + DT_r^2 + ET_r^3\right)}$, where $T_r = T/T_c$
107	$Y = A + B\left[\dfrac{\dfrac{C}{T}}{\sinh\left(\dfrac{C}{T}\right)}\right]^2 + D\left[\dfrac{\dfrac{E}{T}}{\cosh\left(\dfrac{E}{T}\right)}\right]^2$
114	$Y = \dfrac{A^2}{t} + B - 2ACt - ADt^2 - \dfrac{C^2t^3}{3} - \dfrac{CDt^4}{2} - \dfrac{D^2t^5}{5}$, where $t = (1 - T_r)$
116	$Y = A + Bt^{0.35} + ct^{2/3} + dt + et^{4/3}$, where $t = (1 - T_r)$

TABLE B.13
Molar Ideal Gas Heat Capacities for Various Gases (Btu/(lb-mol-°R))

Gas	Formula	Mass	Temperature (°F)							
			0	50	60	100	150	200	250	300
Methane	CH_4	16.04	8.21	8.40	8.44	8.64	8.94	9.27	9.63	10.01
Ethylene	C_2H_4	28.05	9.32	9.91	10.04	10.56	11.23	11.90	12.57	13.22
Ethane	C_2H_6	30.07	11.31	12.07	12.23	12.90	13.77	14.64	15.50	16.35
Propane	C_3H_8	44.10	15.50	16.81	17.08	18.19	19.59	20.98	22.32	23.61
Isobutane	C_4H_{10}	58.12	20.19	22.02	22.40	23.94	25.87	27.75	29.56	31.28
n-Butane	C_4H_{10}	58.12	20.89	22.55	22.89	24.31	26.11	27.88	29.61	31.26
Isopentane	C_5H_{12}	72.15	24.57	27.00	27.50	29.47	31.89	34.21	36.41	38.50
n-Pentane	C_5H_{12}	72.15	25.47	27.46	27.89	29.62	31.84	34.05	36.21	38.30
Benzene	C_6H_6	78.11	16.32	18.43	18.86	20.59	22.73	24.79	26.74	28.59
n-Hexane	C_6H_{14}	86.18	30.10	32.49	33.00	35.10	37.79	40.47	43.10	45.62
n-Heptane	C_7H_{16}	100.21	34.84	37.64	38.24	40.69	43.83	46.96	50.00	52.93
Ammonia	NH_3	17.03	8.24	8.40	8.43	8.58	8.79	9.00	9.23	9.45
Air		28.96	6.94	6.95	6.95	6.96	6.97	6.99	7.01	7.03
Water	H_2O	18.02	7.99	8.01	8.01	8.03	8.07	8.12	8.17	8.23
Oxygen	O_2	32.00	6.97	6.99	7.00	7.02	7.07	7.11	7.17	7.23
Nitrogen	N_2	28.01	6.95	6.96	6.96	6.96	6.96	6.97	6.98	7.00
Hydrogen	H_2	2.02	6.80	6.85	6.86	6.89	6.93	6.95	6.97	6.99
Hydrogen sulfide	H_2S	34.08	8.06	8.13	8.14	8.20	8.29	8.38	8.48	8.59
Carbon monoxide	CO	28.01	6.95	6.96	6.96	6.96	6.97	6.98	7.00	7.03
Carbon dioxide	CO_2	44.01	8.34	8.71	8.78	9.05	9.36	9.65	9.90	10.13
Methanol	CH_3OH	32.04	9.94	10.29	10.37	10.71	11.19	11.71	12.24	12.78

Pure fluid values computed by application of coefficients given in Table B.6. Values for air taken from Engineering Data Book (2004b).

TABLE B.14
Molar Ideal Gas Heat Capacities for Various Gases [kJ/(kmol-K)]

Gas	Formula	Mass	Temperature (°C)								
			-25	0	10	25	50	75	100	125	150
Methane	CH_4	16.043	34.20	34.84	35.15	35.68	36.70	37.87	39.15	40.53	41.96
Ethylene	C_2H_4	28.054	38.41	40.57	41.50	42.95	45.44	47.97	50.50	53.00	55.43
Ethane	C_2H_6	30.07	46.58	49.34	50.53	52.39	55.60	58.88	62.16	65.40	68.58
Propane	C_3H_8	44.097	63.57	68.34	70.36	73.47	78.75	84.02	89.20	94.21	99.04
Isobutane	C_4H_{10}	58.124	82.63	89.36	92.18	96.51	103.79	111.00	118.02	124.77	131.25
n-Butane	C_4H_{10}	58.124	85.80	91.83	94.39	98.35	105.08	111.84	118.49	124.94	131.17
Isopentane	C_5H_{12}	72.151	100.31	109.35	113.06	118.65	127.89	136.86	145.47	153.68	161.51
n-Pentane	C_5H_{12}	72.151	104.66	111.88	114.98	119.80	128.07	136.45	144.74	152.84	160.69
Benzene	C_6H_6	78.114	66.13	73.92	77.14	82.03	90.17	98.12	105.78	113.07	119.99
n-Hexane	C_6H_{14}	86.178	123.66	132.30	136.04	141.84	151.87	162.04	172.12	181.94	191.41
n-Heptane	C_7H_{16}	100.205	143.07	153.23	157.61	164.41	176.12	187.97	199.69	211.09	222.08
Ammonia	NH_3	17.031	34.36	34.90	35.15	35.55	36.29	37.09	37.92	38.76	39.61
Air		28.964	29.11	29.12	29.12	29.13	29.14	29.17	29.20	29.25	29.31
Water	H_2O	18.015	33.43	33.49	33.52	33.58	33.70	33.85	34.03	34.23	34.45
Oxygen	O_2	31.999	29.18	29.25	29.28	29.35	29.48	29.65	29.84	30.06	30.29
Nitrogen	N_2	28.013	29.11	29.12	29.12	29.13	29.14	29.17	29.20	29.25	29.31
Hydrogen	H_2	2.016	28.41	28.60	28.68	28.78	28.92	29.04	29.13	29.20	29.25
Hydrogen sulfide	H_2S	34.076	33.69	33.92	34.02	34.19	34.50	34.84	35.20	35.58	35.97
Carbon monoxide	CO	28.01	29.11	29.12	29.13	29.14	29.16	29.20	29.26	29.33	29.42
Carbon dioxide	CO_2	44.01	34.51	35.91	36.45	37.24	38.48	39.62	40.65	41.58	42.44
Methanol	CH_3OH	32.04	41.29	42.51	43.08	44.01	45.72	47.58	49.55	51.57	53.61

Pure fluid values computed using coefficients given in Table B.6. Values for air taken from Engineering Data Book (2004b).

B.3.3 PHYSICAL PROPERTIES OF ALKANOLAMINES
AND ALKANOLAMINE-WATER SYSTEMS

TABLE B.15
Selected Physical Properties of Alkanolamines

Compound	MEA[a]	DEA[a]	TEA[a]	MDEA[b]
Molecular weight	61.08	15.14	149.19	119.16
Specific gravity at 20/20°C	1.017	1.092	1.126	1.041
ΔSp. Gr./Δt (°C⁻¹)	0.00080	0.00065	0.00059	0.00076
Boiling point, °C at 760 mm Hg	170.4	268	335	247.3
At 50 mm Hg, °C	101	182	245	163.5
At 10 mm Hg, °C	71	150	205	128.6
Vapor pressure at 20°C, mm Hg	<1	<0.01	<0.001	<0.01
Freezing point, °C (°F), (pour point)	10.5 (50.9)	28.0 (82.4)	21.6 (70.9)	−21 (−6)
Viscosity, cP				
At 20°C	24.1		921	101
At 40°C				33.8
Heat of combustion, Btu/lb (cal/g)				−12,200
at 25°C				(−6780)
Flash point, Tag closed cup (ASTM D56), °C (°F)	96 (205)	191 (375)	208 (407)	138 (280)[c]

[a] Dow Chemical (2003b). Ethanolamine brochure
[b] Dow Chemical (2003a).
[c] Pensky-Martens Closed Cup (ASTM D93).

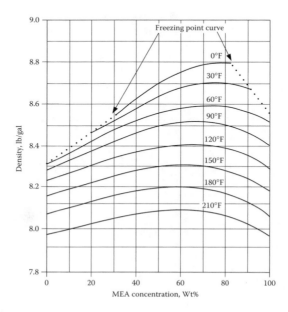

FIGURE B.1 Densities of monoethanolamine (MEA) – water mixtures as a function of composition and temperature. (Adapted from Dow Chemical, 2003b.)

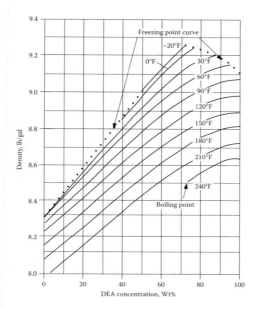

FIGURE B.2 Densities of diethanolamine (DEA) – water mixtures as a function of composition and temperature. (Adapted from Dow Chemical, 2003b.)

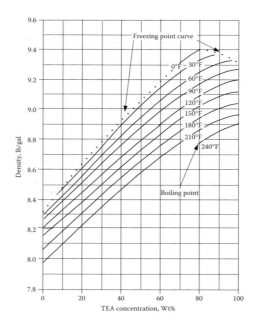

FIGURE B.3 Densities of triethanolamine (TEA) – water mixtures as a function of composition and temperature. (Adapted from Dow Chemical, 2003b.)

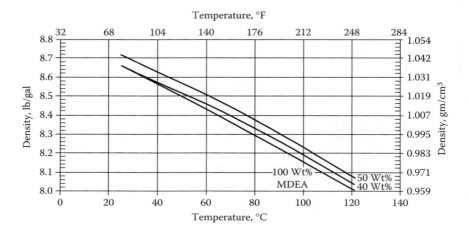

FIGURE B.4 Densities of methyl diethanolamine (MDEA) – water mixtures as a function of composition and temperature. (Adapted through the courtesy of INEOS Oxide).

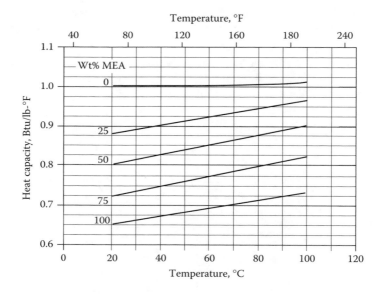

FIGURE B.5 Heat capacities of monoethanolamine (MEA) – water mixtures as a function of composition and temperature. (Adapted from Dow Chemical, 2003b.)

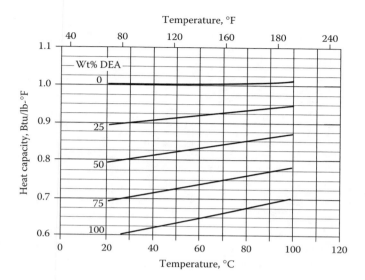

FIGURE B.6 Heat capacities of diethanolamine (DEA) – water mixtures as a function of composition and temperature. (Adapted from Dow Chemical, 2003b.)

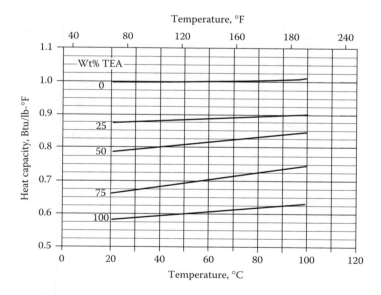

FIGURE B.7 Heat capacities of triethanolamine (TEA) – water mixtures as a function of composition and temperature. (Adapted from Dow Chemical, 2003b.)

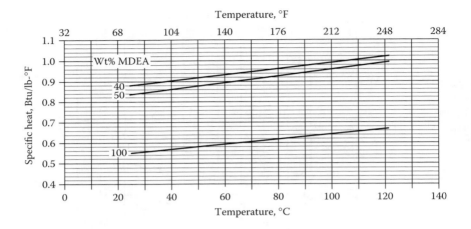

FIGURE B.8 Heat capacities of methyldiethanolamine (MDEA) – water mixtures as a function of composition and temperature. (Adapted through the courtesy of INEOS Oxide).

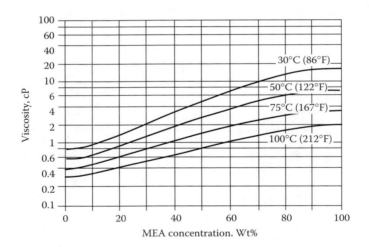

FIGURE B.9 Viscosities of monoethanolamine (MEA) – water mixtures as a function of composition and temperature. (Adapted from Dow Chemical, 2003b.)

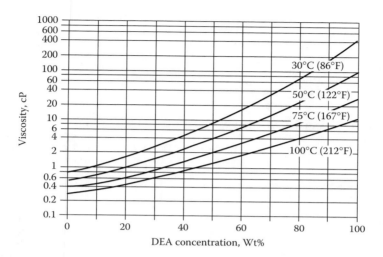

FIGURE B.10 Viscosities of diethanolamine (DEA) – water mixtures as a function of composition and temperature. (Adapted from Dow Chemical, 2003b.) Reference implies that DEA and TEA viscosities are sufficiently close that this figure can be used to estimate the viscosity of TEA–water mixtures.

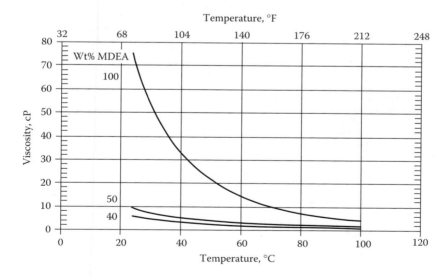

FIGURE B.11 Viscosities of methyldiethanolamine (MDEA) – water mixtures as a function of composition and temperature. (Adapted through the courtesy of INEOS Oxide).

B.3.4 Physical Properties of Glycols and Glycol-Water Systems

---

TABLE B.16
Selected Properties of Glycols Used for Dehydration

	Ethylene Glycol[a]	Diethylene Glycol[b]	Triethylene Glycol[c]	Tetraethylene Glycol[d]
Molar mass, lb/lb-mol, kg/kg-mol	62.07	106.12	150.17	194.23
Normal boiling point, °F (°C)	386.8 (197.1)	473.5 (245.3)	550.4 (288.0)	625.5 (329.7) (decomposes)
Vapor pressure at 20°C, psia (kPa)	0.00012 (0.0075)	$4\times10-5$ (0.0003)	$< 1\times10-5$ (<0.001)	$< 1\times10-5$ (<0.001)
Specific gravity (20/20°C)	1.1153	1.1182	1.1255	1.1247
Δ Specific gravity/ΔT (10 to 40°C), °F^{-1} (°C^{-1})	0.00039 (0.00070)	0.00040 (0.00073)	0.00043 (0.00078)	0.00044 (0.00080)
Viscosity at 20°C, cP (mPs)	~25 (~25)	35.7 (35.7)	49.0 (49.0)	58.3 (58.3)
Critical pressure, psia (bar)	1190 (82)	668 (46.05)	480 (33.1)	375 (25.9)
Critical specific volume, ft³/lb-mol (m³/kmol)	3.06 (0.191)	5.00 (0.312)	7.10 (0.443)	9.03 (0.564)
Critical temperature, °F (°C)	836.33 (446.85)	764.33 (406.85)	824(440)	971.6 (522)
Onset of initial decomposition, °F (°C)	464 (240)	464 (240)	464 (240)	464 (240)
Autoignition temperature, °F (°C)	801 (427)	687(364)	660 (349)	676 (358)
Flammable limits in air, lower, vol%		2.0	0.9	
Flammable limits in air, upper, vol%		12.3	9.2	

[a] MEGlobal (2005b).
[b] MEGlobal (2005a).
[c] Dow Chemical (2003d).
[d] Dow Chemical (2003c).

TABLE B.17
Selected Solubility for Compounds in Glycols at 25°C (g solute/100 ml of glycol)

Solute	Ethylene Glycol[a]	Diethylene Glycol[b]	Triethylene Glycol[c]	Tetraethylene Glycol[d]
Heptane	Slightly Soluble	0.03	Slightly Soluble	Slightly Soluble
Benzene	6.0	45.5	Completely miscible	Completely miscible
Toluene	3.1	20.7	33	89

[a] MEGlobal (2005b).
[b] MEGlobal (2005a).
[c] Dow Chemical (2003d).
[d] Dow Chemical (2003c).

The following solutes are reported to be completely miscible with all of the above glycols:

Water, Methanol, Monoethanolamine, Diethanolamine

TABLE B.18
Vapor Pressure Equations for Glycols

	Ethylene Glycol[a]	Diethylene Glycol[b]	Triethylene Glycol[c]	Tetraethylene Glycol[d]
A	8.21211	7.7007954	7.6302007	7.3101
B	2161.91	2019.2548	2156.4581	2076.0
C	208.43	173.66153	165.92442	139.0
Range of applicability, °C	20 to 198	10 to 250	150 to 300	190 to 330

The coefficients are for the equation $\log_{10}$ (P [mm Hg]) = A − B/(C + t [°C]).

[a] MEGlobal (2005b).
[b] MEGlobal (2005a).
[c] Dow Chemical (2003d).
[d] Dow Chemical (2003c).

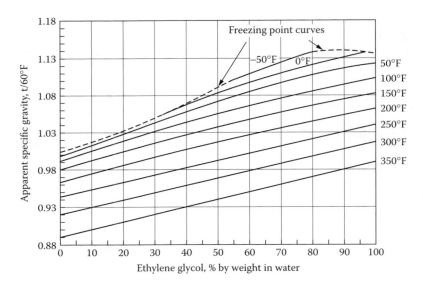

FIGURE B.12 Specific gravities of ethylene glycol–water mixtures as a function of concentration and temperature. (Courtesy of MEGlobal.)

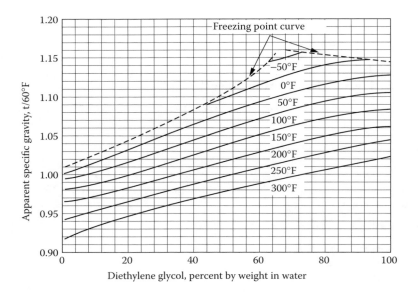

FIGURE B.13 Specific gravities of diethylene glycol–water mixtures as a function of concentration and temperature. (Courtesy of MEGlobal.)

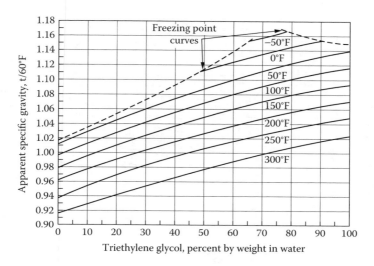

FIGURE B.14 Specific gravities of triethylene glycol–water mixtures as a function of concentration and temperature. (Courtesy of Dow Chemical.)

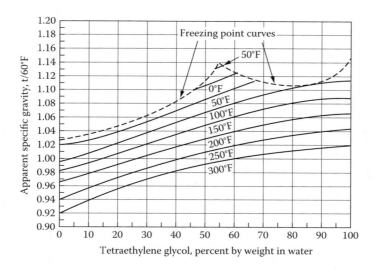

FIGURE B.15 Specific gravities of tetraethylene glycol–water mixtures as a function of concentration and temperature. (Courtesy of Dow Chemical.)

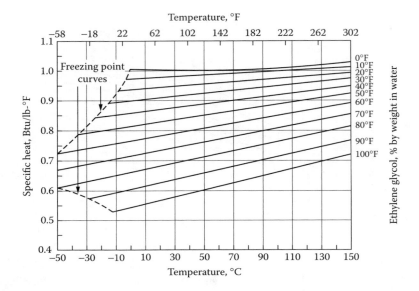

FIGURE B.16 Specific heats of ethylene glycol–water mixtures as a function of concentration and temperature. (Courtesy of MEGlobal.)

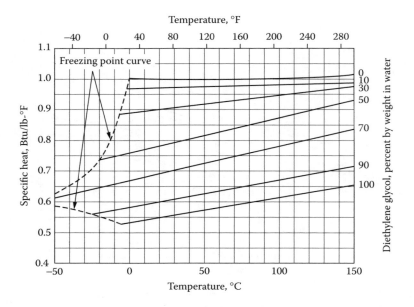

FIGURE B.17 Specific heats of diethylene glycol–water mixtures as a function of concentration and temperature. (Courtesy of MEGlobal.)

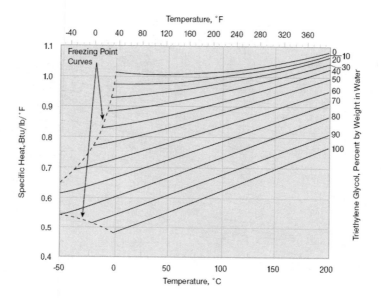

FIGURE B.18 Specific heats of triethylene glycol–water mixtures as a function of concentration and temperature. (Courtesy of Dow Chemical.)

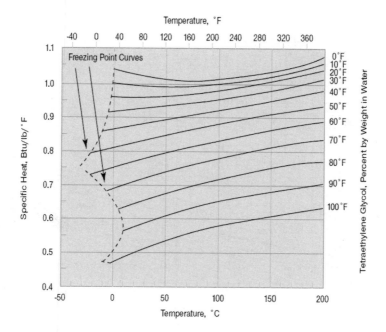

FIGURE B.19 Specific heats of tetraethylene glycol–water mixtures as a function of concentration and temperature. (Courtesy of Dow Chemical.)

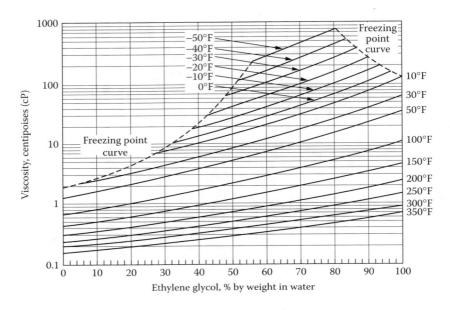

FIGURE B.20 Viscosities of ethylene glycol–water mixtures as a function of concentration and temperature. (Courtesy of MEGlobal.)

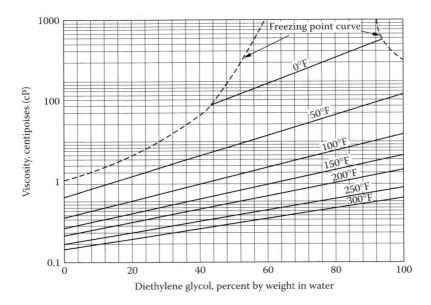

FIGURE B.21 Viscosities of diethylene glycol–water mixtures as a function of concentration and temperature. (Courtesy of MEGlobal.)

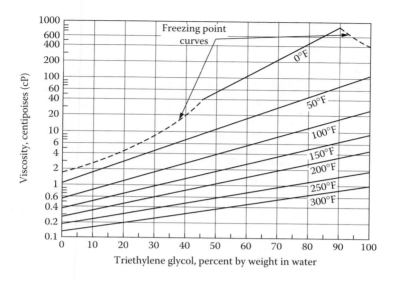

FIGURE B.22 Viscosities of triethylene glycol–water mixtures as a function of concentration and temperature. (Courtesy of Dow Chemical.)

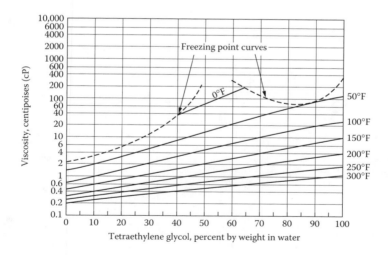

FIGURE B.23 Viscosities of tetraethylene glycol–water mixtures as a function of concentration and temperature. (Courtesy of Dow Chemical.)

B.3.5 PROPERTIES OF SATURATED STEAM

TABLE B.19
Properties of Saturated Steam in Engineering Units

Temp(°F)	Pressure (psia)	Liquid Density (lb/ft³)	Vapor Volume (ft³/lb)	Enthalpy (Btu/lb)		Entropy (Btu/lb-°F)		Temp (°F)
				Liquid	Vapor	Liquid	Vapor	
32.02	0.0887	62.42	3299.6	0.000	1075.900	0.00000	2.18820	32.02
35.00	0.1000	62.42	2945.4	3.006	1077.200	0.00609	2.17770	35.00
40.00	0.122	62.42	2443.3	8.038	1079.400	0.01622	2.16040	40.00
45.00	0.148	62.42	2035.5	13.061	1081.600	0.02622	2.14350	45.00
50.00	0.178	62.41	1702.8	18.078	1083.800	0.03611	2.12710	50.00
55.00	0.214	62.39	1430.2	23.089	1086.000	0.04590	2.11110	55.00
60.00	0.256	62.36	1206.0	28.096	1088.200	0.05558	2.09540	60.00
65.00	0.306	62.33	1020.7	33.100	1090.300	0.06516	2.08020	65.00
70.00	0.363	62.30	867.11	38.101	1092.500	0.07464	2.06530	70.00
75.00	0.430	62.26	739.23	43.100	1094.700	0.08404	2.05080	75.00
80.00	0.507	62.21	632.38	48.097	1096.800	0.09334	2.03660	80.00
85.00	0.597	62.16	542.79	53.093	1099.000	0.10255	2.02280	85.00
90.00	0.699	62.11	467.40	58.088	1101.100	0.11168	2.00930	90.00
95.00	0.816	62.05	403.75	63.084	1103.300	0.12073	1.99610	95.00
100.00	0.951	61.99	349.83	68.078	1105.400	0.12969	1.98320	100.00
110.00	1.277	61.86	264.97	78.069	1109.700	0.14738	1.95830	110.00
120.00	1.695	61.71	202.95	88.062	1114.000	0.16477	1.93460	120.00
130.00	2.226	61.55	157.09	98.059	1118.200	0.18187	1.91180	130.00
140.00	2.893	61.38	122.82	108.060	1122.300	0.19868	1.89010	140.00
150.00	3.723	61.19	96.930	118.070	1126.500	0.21523	1.86930	150.00
160.00	4.747	61.00	77.183	128.080	1130.600	0.23152	1.84930	160.00
170.00	6.000	60.79	61.980	138.110	1134.600	0.24756	1.83020	170.00
180.00	7.520	60.58	50.169	148.140	1138.600	0.26336	1.81180	180.00
190.00	9.350	60.35	40.916	158.190	1142.600	0.27893	1.79420	190.00
200.00	11.538	60.12	33.609	168.240	1146.500	0.29429	1.77720	200.00
210.00	14.136	59.88	27.794	178.310	1150.300	0.30943	1.76090	210.00
212.00	14.709	59.83	26.779	180.330	1151.100	0.31243	1.75770	212.00
220.00	17.201	59.63	23.133	188.400	1154.100	0.32436	1.74510	220.00
230.00	20.795	59.37	19.371	198.510	1157.700	0.33911	1.72990	230.00
240.00	24.986	59.10	16.314	208.630	1161.300	0.35366	1.71530	240.00
250.00	29.844	58.82	13.815	218.780	1164.800	0.36804	1.70110	250.00
260.00	35.447	58.54	11.759	228.950	1168.200	0.38225	1.68740	260.00
270.00	41.878	58.24	10.058	239.140	1171.500	0.39629	1.67410	270.00
280.00	49.222	57.94	8.6431	249.370	1174.700	0.41017	1.66120	280.00
290.00	57.574	57.63	7.4600	259.620	1177.800	0.42391	1.64870	290.00
300.00	67.029	57.31	6.4658	269.910	1180.800	0.43750	1.63650	300.00
310.00	77.691	56.99	5.6263	280.230	1183.600	0.45096	1.62460	310.00
320.00	89.667	56.65	4.9142	290.600	1186.300	0.46428	1.61310	320.00
330.00	103.070	56.31	4.3075	301.000	1188.800	0.47749	1.60180	330.00
340.00	118.020	55.96	3.7884	311.450	1191.300	0.49058	1.59080	340.00
350.00	134.630	55.59	3.3425	321.950	1193.500	0.50355	1.58000	350.00
360.00	153.030	55.23	2.9580	332.500	1195.600	0.51643	1.56940	360.00
370.00	173.360	54.85	2.6252	343.110	1197.500	0.52921	1.55910	370.00

TABLE B.19 (Continued)
Properties of Saturated Steam in Engineering Units

Temp (°F)	Pressure (psia)	Liquid Density (lb/ft³)	Vapor Volume (ft³/lb)	Enthalpy (Btu/lb)		Entropy (Btu/lb-°F)		Temp (°F)
				Liquid	Vapor	Liquid	Vapor	
380.00	195.740	54.46	2.3361	353.770	1199.300	0.54190	1.54890	380.00
390.00	220.330	54.06	2.0841	364.500	1200.900	0.55450	1.53880	390.00
400.00	247.260	53.65	1.8638	375.300	1202.200	0.56703	1.52890	400.00
410.00	276.680	53.23	1.6706	386.170	1203.400	0.57948	1.51920	410.00
420.00	308.760	52.80	1.5006	397.120	1204.400	0.59187	1.50960	420.00
430.00	343.640	52.36	1.3505	408.150	1205.100	0.60420	1.50000	430.00
440.00	381.480	51.91	1.2177	419.270	1205.600	0.61648	1.49050	440.00
450.00	422.460	51.45	1.0999	430.490	1205.900	0.62872	1.48110	450.00
460.00	466.750	50.97	0.99506	441.810	1205.900	0.64092	1.47180	460.00
470.00	514.520	50.48	0.90154	453.240	1205.700	0.65309	1.46240	470.00
480.00	565.950	49.98	0.81791	464.780	1205.100	0.66524	1.45310	480.00
490.00	621.230	49.46	0.74293	476.460	1204.300	0.67738	1.44380	490.00
500.00	680.550	48.92	0.67555	488.270	1203.100	0.68952	1.43440	500.00
520.00	812.100	47.80	0.56007	512.350	1199.800	0.71384	1.41550	520.00
540.00	962.240	46.59	0.46553	537.140	1194.800	0.73829	1.39620	540.00
560.00	1132.700	45.29	0.38742	562.770	1188.000	0.76299	1.37620	560.00
580.00	1325.500	43.88	0.32227	589.450	1179.000	0.78812	1.35520	580.00
600.00	1542.500	42.32	0.26739	617.420	1167.400	0.81388	1.33290	600.00
620.00	1786.200	40.57	0.22062	647.110	1152.300	0.84062	1.30850	620.00
640.00	2059.200	38.57	0.18017	679.190	1132.600	0.86888	1.28120	640.00
660.00	2364.900	36.15	0.14439	714.960	1106.100	0.89974	1.24910	660.00
680.00	2707.300	32.94	0.11132	757.890	1067.600	0.93611	1.20780	680.00
700.00	3093.000	27.28	0.07480	823.000	991.660	0.99065	1.13610	700.00
705.10	3200.100	20.10	0.04975	896.670	896.670	1.05330	1.05330	705.10

Source: Lemmon (2005).

TABLE B.20
Properties of Saturated Water in SI Units

Temp (°C)	Pressure (bar)	Liquid Density (kg/m³)	Vapor Volume (m³/kg)	Enthalpy (kJ/kg)		Entropy (kJ/kg-°C)		Temp (°C)
				Liquid	Vapor	Liquid	Vapor	
0.01	0.00612	999.79	205.99	0.001	2500.900	0.00000	9.15550	0.01
2.00	0.00706	999.89	179.76	8.392	2504.600	0.03061	9.10270	2.00
4.00	0.00814	999.93	157.12	16.813	2508.200	0.06110	9.05050	4.00
6.00	0.00935	999.89	137.63	25.224	2511.900	0.09134	8.99930	6.00
8.00	0.0107	999.80	120.83	33.627	2515.600	0.12133	8.94910	8.00
10.00	0.0123	999.65	106.30	42.021	2519.200	0.15109	8.89980	10.00
12.00	0.0140	999.45	93.719	50.409	2522.900	0.18061	8.85130	12.00
14.00	0.0160	999.20	82.793	58.792	2526.500	0.20990	8.80370	14.00
16.00	0.0182	998.90	73.286	67.170	2530.200	0.23897	8.75700	16.00
18.00	0.0206	998.55	64.998	75.544	2533.800	0.26783	8.71110	18.00
20.00	0.0234	998.16	57.757	83.914	2537.400	0.29648	8.66600	20.00
22.00	0.0265	997.73	51.418	92.282	2541.100	0.32493	8.62170	22.00
24.00	0.0299	997.25	45.858	100.650	2544.700	0.35318	8.57810	24.00
26.00	0.0336	996.74	40.973	109.010	2548.300	0.38123	8.53530	26.00
28.00	0.0378	996.19	36.672	117.370	2551.900	0.40908	8.49330	28.00
30.00	0.0425	995.61	32.878	125.730	2555.500	0.43675	8.45200	30.00
32.00	0.0476	994.99	29.526	134.090	2559.200	0.46424	8.41130	32.00
34.00	0.0533	994.33	26.560	142.450	2562.800	0.49155	8.37140	34.00
36.00	0.0595	993.64	23.929	150.810	2566.300	0.51867	8.33210	36.00
38.00	0.0663	992.92	21.593	159.170	2569.900	0.54562	8.29350	38.00
40.00	0.0738	992.18	19.515	167.530	2573.500	0.57240	8.25550	40.00
42.00	0.0821	991.40	17.664	175.890	2577.100	0.59901	8.21820	42.00
44.00	0.0911	990.59	16.011	184.250	2580.600	0.62545	8.18150	44.00
46.00	0.101	989.75	14.534	192.620	2584.200	0.65173	8.14530	46.00
48.00	0.112	988.89	13.212	200.980	2587.800	0.67785	8.10980	48.00
50.00	0.124	988.00	12.027	209.340	2591.300	0.70381	8.07480	50.00
55.00	0.158	985.66	9.5643	230.260	2600.100	0.76802	7.98980	55.00
60.00	0.199	983.16	7.6672	251.180	2608.800	0.83129	7.90810	60.00
65.00	0.250	980.52	6.1935	272.120	2617.500	0.89365	7.82960	65.00
70.00	0.312	977.73	5.0395	293.070	2626.100	0.95513	7.75400	70.00
75.00	0.386	974.81	4.1289	314.030	2634.600	1.01580	7.68120	75.00
80.00	0.474	971.77	3.4052	335.010	2643.000	1.07560	7.61110	80.00
85.00	0.579	968.59	2.8258	356.010	2651.300	1.13460	7.54340	85.00
90.00	0.702	965.30	2.3591	377.040	2659.500	1.19290	7.47810	90.00
95.00	0.846	961.88	1.9806	398.090	2667.600	1.25040	7.41510	95.00
100.00	1.014	958.35	1.6718	419.170	2675.600	1.30720	7.35410	100.00
110.00	1.434	950.95	1.2093	461.420	2691.100	1.41880	7.23810	110.00
120.00	1.987	943.11	0.89121	503.810	2705.900	1.52790	7.12910	120.00
130.00	2.703	934.83	0.66800	546.380	2720.100	1.63460	7.02640	130.00
140.00	3.615	926.13	0.50845	589.160	2733.400	1.73920	6.92930	140.00
150.00	4.762	917.01	0.39245	632.180	2745.900	1.84180	6.83710	150.00
160.00	6.182	907.45	0.30678	675.470	2757.400	1.94260	6.74910	160.00
170.00	7.922	897.45	0.24259	719.080	2767.900	2.04170	6.66500	170.00

TABLE B.20 (Continued)
Properties of Saturated Water in SI Units

Temp (°C)	Pressure (bar)	Liquid Density (kg/m³)	Vapor Volume (m³/kg)	Enthalpy (kJ/kg)		Entropy (kJ/kg-°C)		Temp (°C)
				Liquid	Vapor	Liquid	Vapor	
180.00	10.028	887.00	0.19384	763.050	2777.200	2.13920	6.58400	180.00
190.00	12.552	876.08	0.15636	807.430	2785.300	2.23550	6.50590	190.00
200.00	15.549	864.66	0.12721	852.270	2792.000	2.33050	6.43020	200.00
210.00	19.077	852.72	0.10429	897.630	2797.300	2.42450	6.35630	210.00
220.00	23.196	840.22	0.08609	943.580	2800.900	2.51770	6.28400	220.00
230.00	27.971	827.12	0.07150	990.190	2802.900	2.61010	6.21280	230.00
240.00	33.469	813.37	0.05971	1037.600	2803.000	2.70200	6.14230	240.00
250.00	39.762	798.89	0.05008	1085.800	2800.900	2.79350	6.07210	250.00
260.00	46.923	783.63	0.04217	1135.000	2796.600	2.88490	6.00160	260.00
270.00	55.030	767.46	0.03562	1185.300	2789.700	2.97650	5.93040	270.00
280.00	64.166	750.28	0.03015	1236.900	2779.900	3.06850	5.85790	280.00
290.00	74.418	731.91	0.02556	1290.000	2766.700	3.16120	5.78340	290.00
300.00	85.879	712.14	0.02166	1345.000	2749.600	3.25520	5.70590	300.00
310.00	98.651	690.67	0.01834	1402.200	2727.900	3.35100	5.62440	310.00
320.00	112.840	667.09	0.01547	1462.200	2700.600	3.44940	5.53720	320.00
330.00	128.580	640.77	0.01298	1525.900	2666.000	3.55180	5.44220	330.00
340.00	146.010	610.67	0.01078	1594.500	2621.800	3.66010	5.33560	340.00
350.00	165.290	574.71	0.00880	1670.900	2563.600	3.77840	5.21100	350.00
360.00	186.660	527.59	0.00695	1761.700	2481.500	3.91670	5.05360	360.00
370.00	210.440	451.43	0.00495	1890.700	2334.500	4.11120	4.80120	370.00
373.95	220.640	322.00	0.00311	2084.300	2084.300	4.40700	4.40700	373.95

Source: Lemmon (2005).

B.3.6 PRESSURE-ENTHALPY DIAGRAMS AND SATURATION TABLES OF PURE HYDROCARBONS

TABLE B.21
Properties of Saturated Methane in Engineering Units

Temp (°F)	Pressure (psia)	Liquid Density (lb/ft³)	Vapor Volume (ft³/lb)	Enthalpy (Btu/lb)		Entropy (Btu/lb-°F)		Temp (°F)
				Liquid	Vapor	Liquid	Vapor	
−296.42	1.696	28.19	63.881	−30.897	203.250	−0.16968	1.26460	−296.42
−295.00	1.876	28.12	58.235	−29.753	203.910	−0.16271	1.25630	−295.00
−290.00	2.635	27.89	42.619	−25.716	206.230	−0.13858	1.22840	−290.00
−285.00	3.625	27.65	31.809	−21.663	208.510	−0.11508	1.20270	−285.00
−280.00	4.891	27.41	24.167	−17.593	210.750	−0.09216	1.17870	−280.00
−275.00	6.487	27.17	18.659	−13.505	212.950	−0.06977	1.15650	−275.00
−270.00	8.467	26.93	14.619	−9.396	215.100	−0.04789	1.13570	−270.00
−265.00	10.890	26.68	11.608	−5.265	217.190	−0.02648	1.11630	−265.00
−260.00	13.822	26.43	9.3292	−1.111	219.230	−0.00551	1.09800	−260.00
−258.67	14.695	26.37	8.8191	−0.001	219.760	0.00000	1.09330	−258.67
−255.00	17.327	26.18	7.5811	3.070	221.210	0.01505	1.08090	−255.00
−250.00	21.476	25.92	6.2230	7.279	223.120	0.03522	1.06470	−250.00
−245.00	26.341	25.66	5.1555	11.517	224.960	0.05503	1.04930	−245.00
−240.00	31.996	25.39	4.3071	15.789	226.720	0.07451	1.03470	−240.00
−235.00	38.517	25.12	3.6261	20.095	228.410	0.09368	1.02090	−235.00
−230.00	45.984	24.84	3.0742	24.439	230.000	0.11256	1.00760	−230.00
−225.00	54.476	24.56	2.6230	28.824	231.510	0.13117	0.99486	−225.00
−220.00	64.073	24.27	2.2512	33.255	232.910	0.14955	0.98258	−220.00
−215.00	74.859	23.98	1.9423	37.734	234.210	0.16770	0.97071	−215.00
−210.00	86.918	23.67	1.6839	42.267	235.390	0.18566	0.95918	−210.00
−205.00	100.330	23.36	1.4662	46.858	236.460	0.20345	0.94793	−205.00
−200.00	115.190	23.04	1.2817	51.514	237.390	0.22109	0.93691	−200.00
−195.00	131.580	22.71	1.1243	56.241	238.190	0.23862	0.92605	−195.00
−190.00	149.590	22.37	0.98929	61.047	238.830	0.25605	0.91531	−190.00
−185.00	169.300	22.01	0.87280	65.940	239.310	0.27343	0.90461	−185.00
−180.00	190.820	21.64	0.77175	70.932	239.610	0.29078	0.89390	−180.00
−175.00	214.220	21.26	0.68364	76.035	239.710	0.30815	0.88311	−175.00
−170.00	239.620	20.86	0.60640	81.263	239.590	0.32557	0.87215	−170.00
−165.00	267.110	20.44	0.53835	86.636	239.220	0.34312	0.86095	−165.00
−160.00	296.800	19.99	0.47805	92.177	238.580	0.36085	0.84938	−160.00
−155.00	328.790	19.52	0.42432	97.918	237.600	0.37885	0.83733	−155.00
−150.00	363.200	19.01	0.37614	103.900	236.240	0.39725	0.82460	−150.00
−145.00	400.170	18.45	0.33261	110.180	234.400	0.41618	0.81096	−145.00
−140.00	439.830	17.84	0.29290	116.840	231.970	0.43590	0.79606	−140.00
−135.00	482.350	17.15	0.25622	124.010	228.740	0.45678	0.77935	−135.00
−130.00	527.920	16.34	0.22164	131.950	224.360	0.47950	0.75982	−130.00
−125.00	576.810	15.32	0.18779	141.170	218.090	0.50552	0.73537	−125.00
−120.00	629.390	13.83	0.15113	153.360	207.530	0.53971	0.69917	−120.00
−116.66	667.060	10.23	0.09779	178.280	178.280	0.61095	0.61095	−116.66

Source: Lemmon (2005).

TABLE B.22
Properties of Saturated Methane in SI Units

Temp (°C)	Pressure (bar)	Liquid Density (kg/m³)	Vapor Volume (m³/kg)	Enthalpy (kJ/kg)		Entropy (kJ/kg-°C)		Temp (°C)
				Liquid	Vapor	Liquid	Vapor	
−182.46	0.117	451.47	3.9877	−71.816	472.441	−0.70990	5.29111	−182.46
−180.00	0.159	448.20	3.0071	−63.531	477.221	−0.61986	5.18531	−180.00
−175.00	0.282	441.43	1.7755	−46.574	486.761	−0.44284	4.99101	−175.00
−170.00	0.472	434.51	1.1081	−29.485	495.991	−0.27345	4.82081	−170.00
−165.00	0.751	427.44	0.72466	−12.238	504.851	−0.11079	4.67041	−165.00
−161.48	1.013	422.36	0.55054	0.000	510.831	0.00000	4.57461	−161.48
−160.00	1.143	420.18	0.49291	5.189	513.281	0.04589	4.53631	−160.00
−155.00	1.676	412.72	0.34665	22.823	521.221	0.19728	4.41561	−155.00
−150.00	2.378	405.02	0.25077	40.694	528.601	0.34399	4.30581	−150.00
−145.00	3.282	397.06	0.18583	58.838	535.341	0.48661	4.20491	−145.00
−140.00	4.418	388.79	0.14054	77.299	541.381	0.62571	4.11111	−140.00
−135.00	5.819	380.16	0.10813	96.130	546.631	0.76185	4.02281	−135.00
−130.00	7.520	371.11	0.08440	115.391	550.971	0.89560	3.93841	−130.00
−125.00	9.555	361.57	0.06667	135.171	554.291	1.02761	3.85661	−125.00
−120.00	11.959	351.44	0.05315	155.581	556.431	1.15851	3.77591	−120.00
−115.00	14.770	340.56	0.04268	176.741	557.191	1.28931	3.69491	−115.00
−110.00	18.026	328.76	0.03442	198.851	556.281	1.42091	3.61171	−110.00
−105.00	21.767	315.72	0.02779	222.211	553.291	1.55481	3.52381	−105.00
−100.00	26.040	300.98	0.02236	247.251	547.561	1.69341	3.42781	−100.00
−95.00	30.895	283.64	0.01781	274.801	537.861	1.84081	3.31741	−95.00
−90.00	36.399	261.66	0.01384	306.711	521.521	2.00621	3.17911	−90.00
−85.00	42.648	227.46	0.00993	349.751	488.811	2.22411	2.96321	−85.00
−82.59	45.992	162.66	0.00615	415.591	415.591	2.56241	2.56241	−82.59

Source: Lemmon, 2005.

TABLE B.23
Properties of Saturated Ethane in Engineering Units

Temp (°F)	Pressure (psia)	Liquid Density (lb/ft³)	Vapor Volume (ft³/lb)	Enthalpy (Btu/lb) Liquid	Enthalpy (Btu/lb) Vapor	Entropy (Btu/lb-°F) Liquid	Entropy (Btu/lb-°F) Vapor	Temp (°F)
−250.00	0.0308	38.91	2432.6	−68.958	174.902	−0.25843	0.90462	−250.00
−245.00	0.0463	38.71	1655.0	−66.152	176.372	−0.24520	0.88456	−245.00
−240.00	0.0682	38.52	1148.9	−63.350	177.852	−0.23230	0.86574	−240.00
−235.00	0.0987	38.33	812.62	−60.553	179.342	−0.21971	0.84807	−235.00
−230.00	0.140	38.14	584.86	−57.760	180.842	−0.20742	0.83146	−230.00
−225.00	0.196	37.94	427.81	−54.970	182.342	−0.19541	0.81585	−225.00
−220.00	0.269	37.75	317.69	−52.184	183.842	−0.18366	0.80115	−220.00
−215.00	0.365	37.55	239.25	−49.400	185.352	−0.17216	0.78729	−215.00
−210.00	0.488	37.36	182.56	−46.616	186.852	−0.16090	0.77422	−210.00
−205.00	0.643	37.16	141.02	−43.833	188.362	−0.14987	0.76188	−205.00
−200.00	0.839	36.96	110.19	−41.049	189.862	−0.13905	0.75021	−200.00
−195.00	1.082	36.76	87.029	−38.263	191.362	−0.12843	0.73917	−195.00
−190.00	1.381	36.56	69.431	−35.475	192.852	−0.11799	0.72871	−190.00
−185.00	1.744	36.36	55.916	−32.683	194.342	−0.10774	0.71879	−185.00
−180.00	2.183	36.16	45.432	−29.886	195.822	−0.09766	0.70938	−180.00
−175.00	2.708	35.96	37.222	−27.083	197.292	−0.08774	0.70044	−175.00
−170.00	3.331	35.75	30.735	−24.274	198.742	−0.07797	0.69194	−170.00
−165.00	4.066	35.55	25.565	−21.458	200.192	−0.06834	0.68385	−165.00
−160.00	4.927	35.34	21.413	−18.633	201.622	−0.05885	0.67615	−160.00
−155.00	5.928	35.13	18.052	−15.799	203.042	−0.04949	0.66880	−155.00
−150.00	7.085	34.92	15.312	−12.956	204.452	−0.04025	0.66180	−150.00
−145.00	8.414	34.71	13.063	−10.101	205.842	−0.03113	0.65511	−145.00
−140.00	9.934	34.50	11.206	−7.234	207.212	−0.02212	0.64872	−140.00
−135.00	11.661	34.29	9.6620	−4.355	208.572	−0.01321	0.64260	−135.00
−130.00	13.617	34.07	8.3715	−1.463	209.902	−0.00440	0.63675	−130.00
−127.48	14.694	33.96	7.8015	0.000	210.572	0.00000	0.63390	−127.48
−125.00	15.819	33.85	7.2866	1.443	211.232	0.00431	0.63115	−125.00
−120.00	18.288	33.63	6.3699	4.365	212.532	0.01293	0.62578	−120.00
−115.00	21.046	33.41	5.5913	7.302	213.812	0.02147	0.62063	−115.00
−110.00	24.113	33.18	4.9269	10.256	215.072	0.02993	0.61568	−110.00
−105.00	27.513	32.96	4.3574	13.227	216.312	0.03832	0.61093	−105.00
−100.00	31.267	32.73	3.8670	16.217	217.532	0.04663	0.60635	−100.00
−95.00	35.400	32.49	3.4431	19.226	218.722	0.05487	0.60195	−95.00
−90.00	39.933	32.26	3.0750	22.256	219.902	0.06305	0.59770	−90.00
−85.00	44.893	32.02	2.7544	25.307	221.042	0.07118	0.59360	−85.00
−80.00	50.302	31.78	2.4740	28.380	222.162	0.07924	0.58964	−80.00
−75.00	56.186	31.54	2.2279	31.477	223.262	0.08725	0.58581	−75.00
−70.00	62.570	31.29	2.0112	34.598	224.322	0.09522	0.58210	−70.00
−65.00	69.479	31.04	1.8197	37.744	225.362	0.10314	0.57851	−65.00
−60.00	76.940	30.78	1.6501	40.917	226.362	0.11102	0.57501	−60.00
−55.00	84.978	30.53	1.4993	44.119	227.332	0.11886	0.57160	−55.00
−50.00	93.620	30.26	1.3650	47.350	228.272	0.12666	0.56829	−50.00
−45.00	102.890	30.00	1.2449	50.612	229.172	0.13444	0.56504	−45.00

TABLE B.23 (Continued)
Properties of Saturated Ethane in Engineering Units

Temp (°F)	Pressure (psia)	Liquid Density (lb/ft³)	Vapor Volume (ft³/lb)	Enthalpy (Btu/lb) Liquid	Enthalpy (Btu/lb) Vapor	Entropy (Btu/lb-°F) Liquid	Entropy (Btu/lb-°F) Vapor	Temp (°F)
−40.00	112.820	29.73	1.1372	53.907	230.032	0.14219	0.56187	−40.00
−35.00	123.440	29.45	1.0405	57.236	230.852	0.14992	0.55875	−35.00
−30.00	134.770	29.17	0.95341	60.601	231.632	0.15763	0.55569	−30.00
−25.00	146.840	28.88	0.87474	64.004	232.372	0.16532	0.55266	−25.00
−20.00	159.680	28.59	0.80354	67.448	233.052	0.17301	0.54967	−20.00
−15.00	173.310	28.29	0.73894	70.935	233.692	0.18070	0.54671	−15.00
−10.00	187.780	27.98	0.68020	74.467	234.262	0.18838	0.54375	−10.00
−5.00	203.100	27.67	0.62667	78.049	234.782	0.19608	0.54080	−5.00
0.00	219.310	27.34	0.57777	81.682	235.232	0.20379	0.53784	0.00
5.00	236.450	27.01	0.53301	85.372	235.622	0.21152	0.53486	5.00
10.00	254.530	26.67	0.49194	89.123	235.922	0.21928	0.53184	10.00
15.00	273.600	26.31	0.45418	92.940	236.142	0.22708	0.52877	15.00
20.00	293.690	25.95	0.41938	96.829	236.272	0.23493	0.52563	20.00
25.00	314.830	25.57	0.38722	100.802	236.292	0.24285	0.52240	25.00
30.00	337.070	25.17	0.35745	104.862	236.192	0.25085	0.51906	30.00
35.00	360.440	24.76	0.32979	109.012	235.972	0.25895	0.51558	35.00
40.00	384.970	24.32	0.30403	113.292	235.592	0.26717	0.51193	40.00
45.00	410.720	23.86	0.27996	117.692	235.032	0.27555	0.50806	45.00
50.00	437.740	23.36	0.25738	122.252	234.272	0.28412	0.50391	50.00
55.00	466.060	22.83	0.23610	127.002	233.272	0.29294	0.49942	55.00
60.00	495.740	22.25	0.21593	131.982	231.962	0.30209	0.49449	60.00
65.00	526.840	21.61	0.19666	137.242	230.262	0.31168	0.48897	65.00
70.00	559.430	20.88	0.17804	142.892	228.052	0.32185	0.48263	70.00
75.00	593.590	20.03	0.15974	149.052	225.122	0.33285	0.47511	75.00
80.00	629.430	19.01	0.14116	156.012	221.022	0.34517	0.46563	80.00
85.00	667.140	17.63	0.12082	164.492	214.602	0.36010	0.45210	85.00
89.92	706.560	13.19	0.07579	187.042	187.042	0.40042	0.40042	89.92

Source: Lemmon et al (2005).

TABLE B.24
Properties of Saturated Ethane in SI Units

Temp (°C)	Pressure (bar)	Liquid Density (kg/m³)	Vapor Volume (m³/kg)	Enthalpy (kJ/kg)		Entropy (kJ/kg-°C)		Temp (°C)
				Liquid	Vapor	Liquid	Vapor	
−155.00	0.00272	621.36	120.22	−156.375	408.605	−1.04787	3.73393	−155.00
−152.50	0.00389	618.59	85.874	−150.505	411.695	−0.99879	3.66093	−152.50
−150.00	0.00547	615.81	62.310	−144.655	414.805	−0.95075	3.59213	−150.00
−147.50	0.00757	613.03	45.883	−138.805	417.925	−0.90374	3.52703	−147.50
−145.00	0.0103	610.24	34.255	−132.965	421.055	−0.85771	3.46543	−145.00
−142.50	0.0139	607.44	25.907	−127.125	424.195	−0.81263	3.40713	−142.50
−140.00	0.0186	604.64	19.833	−121.295	427.335	−0.76844	3.35203	−140.00
−137.50	0.0244	601.82	15.356	−115.475	430.485	−0.72510	3.29963	−137.50
−135.00	0.0318	599.00	12.018	−109.655	433.635	−0.68258	3.25003	−135.00
−132.50	0.0409	596.17	9.5008	−103.825	436.785	−0.64082	3.20283	−132.50
−130.00	0.0521	593.32	7.5823	−98.005	439.935	−0.59980	3.15803	−130.00
−127.50	0.0658	590.47	6.1055	−92.179	443.075	−0.55947	3.11543	−127.50
−125.00	0.0823	587.60	4.9578	−86.349	446.205	−0.51980	3.07493	−125.00
−122.50	0.102	584.72	4.0580	−80.513	449.325	−0.48076	3.03623	−122.50
−120.00	0.126	581.83	3.3464	−74.670	452.425	−0.44231	2.99943	−120.00
−117.50	0.154	578.92	2.7793	−68.817	455.515	−0.40444	2.96423	−117.50
−115.00	0.187	575.99	2.3237	−62.954	458.585	−0.36710	2.93063	−115.00
−112.50	0.225	573.05	1.9552	−57.078	461.635	−0.33028	2.89853	−112.50
−110.00	0.270	570.10	1.6549	−51.188	464.665	−0.29395	2.86783	−110.00
−107.50	0.321	567.12	1.4088	−45.284	467.665	−0.25809	2.83853	−107.50
−105.00	0.380	564.13	1.2057	−39.362	470.645	−0.22267	2.81043	−105.00
−102.50	0.447	561.12	1.0371	−33.422	473.595	−0.18768	2.78343	−102.50
−100.00	0.524	558.08	0.89645	−27.463	476.525	−0.15309	2.75763	−100.00
−97.50	0.610	555.03	0.77841	−21.482	479.415	−0.11889	2.73283	−97.50
−95.00	0.708	551.95	0.67885	−15.480	482.285	−0.08506	2.70903	−95.00
−92.50	0.817	548.85	0.59447	−9.453	485.115	−0.05157	2.68613	−92.50
−90.00	0.939	545.73	0.52261	−3.401	487.915	−0.01842	2.66413	−90.00
−88.60	1.013	543.97	0.48703	0.000	489.465	0.00000	2.65223	−88.60
−87.50	1.075	542.58	0.46115	2.678	490.685	0.01441	2.64303	−87.50
−85.00	1.225	539.41	0.40836	8.785	493.405	0.04693	2.62263	−85.00
−82.50	1.392	536.20	0.36283	14.921	496.105	0.07917	2.60303	−82.50
−80.00	1.575	532.97	0.32340	21.088	498.755	0.11113	2.58413	−80.00
−77.50	1.777	529.71	0.28913	27.288	501.375	0.14282	2.56593	−77.50
−75.00	1.998	526.42	0.25923	33.521	503.945	0.17427	2.54833	−75.00
−72.50	2.239	523.10	0.23307	39.790	506.475	0.20548	2.53133	−72.50
−70.00	2.501	519.74	0.21009	46.096	508.965	0.23646	2.51493	−70.00
−67.50	2.786	516.34	0.18984	52.440	511.405	0.26723	2.49903	−67.50
−65.00	3.095	512.91	0.17195	58.825	513.805	0.29780	2.48363	−65.00
−62.50	3.429	509.45	0.15609	65.251	516.155	0.32818	2.46873	−62.50
−60.00	3.790	505.94	0.14200	71.722	518.445	0.35838	2.45423	−60.00
−57.50	4.178	502.39	0.12943	78.238	520.685	0.38841	2.44013	−57.50
−55.00	4.595	498.79	0.11821	84.801	522.875	0.41829	2.42643	−55.00
−52.50	5.043	495.15	0.10815	91.414	525.005	0.44802	2.41313	−52.50

TABLE B.24 (Continued)
Properties of Saturated Ethane in SI Units

Temp (°C)	Pressure (bar)	Liquid Density (kg/m³)	Vapor Volume (m³/kg)	Enthalpy (kJ/kg)		Entropy (kJ/kg·°C)		Temp (°C)
				Liquid	Vapor	Liquid	Vapor	
−50.00	5.522	491.46	0.09912	98.079	527.075	0.47762	2.40003	−50.00
−47.50	6.033	487.71	0.09098	104.795	529.085	0.50709	2.38733	−47.50
−45.00	6.579	483.91	0.08365	111.575	531.025	0.53646	2.37493	−45.00
−42.50	7.161	480.06	0.07701	118.405	532.895	0.56572	2.36283	−42.50
−40.00	7.779	476.14	0.07100	125.305	534.705	0.59490	2.35083	−40.00
−37.50	8.435	472.16	0.06553	132.265	536.425	0.62401	2.33913	−37.50
−35.00	9.132	468.12	0.06056	139.295	538.075	0.65305	2.32753	−35.00
−32.50	9.869	464.00	0.05603	146.395	539.635	0.68204	2.31613	−32.50
−30.00	10.649	459.80	0.05189	153.565	541.105	0.71100	2.30483	−30.00
−27.50	11.472	455.52	0.04810	160.815	542.475	0.73994	2.29363	−27.50
−25.00	12.341	451.15	0.04462	168.155	543.745	0.76888	2.28243	−25.00
−22.50	13.258	446.69	0.04143	175.585	544.915	0.79784	2.27133	−22.50
−20.00	14.222	442.12	0.03849	183.095	545.965	0.82682	2.26023	−20.00
−17.50	15.236	437.45	0.03578	190.715	546.885	0.85586	2.24903	−17.50
−15.00	16.302	432.65	0.03328	198.445	547.675	0.88498	2.23783	−15.00
−12.50	17.422	427.72	0.03096	206.285	548.325	0.91420	2.22643	−12.50
−10.00	18.596	422.65	0.02881	214.245	548.815	0.94355	2.21493	−10.00
−7.50	19.826	417.42	0.02681	222.345	549.125	0.97307	2.20323	−7.50
−5.00	21.115	412.01	0.02496	230.585	549.255	1.00283	2.19113	−5.00
−2.50	22.464	406.40	0.02323	238.995	549.165	1.03273	2.17883	−2.50
0.00	23.875	400.57	0.02161	247.575	548.845	1.06303	2.16603	0.00
2.50	25.350	394.48	0.02009	256.355	548.265	1.09363	2.15273	2.50
5.00	26.891	388.09	0.01867	265.355	547.385	1.12473	2.13873	5.00
7.50	28.501	381.36	0.01733	274.615	546.175	1.15643	2.12403	7.50
10.00	30.181	374.22	0.01607	284.175	544.565	1.18873	2.10833	10.00
12.50	31.934	366.58	0.01487	294.085	542.485	1.22193	2.09153	12.50
15.00	33.763	358.33	0.01373	304.415	539.845	1.25613	2.07323	15.00
17.50	35.671	349.31	0.01263	315.265	536.525	1.29183	2.05303	17.50
20.00	37.660	339.30	0.01158	326.765	532.315	1.32923	2.03043	20.00
22.50	39.735	327.96	0.01054	339.125	526.945	1.36913	2.00433	22.50
25.00	41.901	314.76	0.00951	352.665	519.875	1.41243	1.97323	25.00
27.50	44.163	298.66	0.00845	368.045	510.075	1.46133	1.93373	27.50
30.00	46.535	276.65	0.00725	387.165	494.495	1.52193	1.87603	30.00
32.18	48.718	206.58	0.00484	438.265	438.265	1.68683	1.68683	32.18

Source: Lemmon et al. (2005).

TABLE B.25
Properties of Saturated Ethylene in Engineering Units

Temp (°F)	Pressure (psia)	Liquid Density (lb/ft³)	Vapor Volume (ft³/lb)	Enthalpy (Btu/lb)		Entropy (Btu/lb-°F)		Temp (°F)
				Liquid	Vapor	Liquid	Vapor	
−272.50	0.0177	40.87	4046.9	−68.031	176.123	−0.28183	1.02254	−272.50
−270.00	0.0225	40.76	3219.5	−66.582	176.823	−0.27414	1.00914	−270.00
−265.00	0.0358	40.54	2077.4	−63.676	178.243	−0.25902	0.98368	−265.00
−260.00	0.0555	40.32	1374.6	−60.769	179.653	−0.24427	0.95981	−260.00
−255.00	0.0840	40.09	930.98	−57.863	181.063	−0.22990	0.93745	−255.00
−250.00	0.124	39.87	644.22	−54.958	182.463	−0.21588	0.91648	−250.00
−245.00	0.180	39.65	454.73	−52.055	183.863	−0.20220	0.89680	−245.00
−240.00	0.257	39.43	326.95	−49.155	185.263	−0.18884	0.87830	−240.00
−235.00	0.359	39.20	239.12	−46.258	186.653	−0.17581	0.86089	−235.00
−230.00	0.493	38.98	177.68	−43.365	188.043	−0.16307	0.84450	−230.00
−225.00	0.668	38.75	133.99	−40.474	189.423	−0.15062	0.82905	−225.00
−220.00	0.891	38.52	102.44	−37.587	190.803	−0.13846	0.81446	−220.00
−215.00	1.174	38.29	79.326	−34.703	192.163	−0.12655	0.80068	−215.00
−210.00	1.527	38.07	62.164	−31.822	193.523	−0.11490	0.78765	−210.00
−205.00	1.963	37.83	49.260	−28.943	194.863	−0.10349	0.77532	−205.00
−200.00	2.496	37.60	39.441	−26.065	196.193	−0.09231	0.76363	−200.00
−195.00	3.141	37.37	31.888	−23.189	197.513	−0.08135	0.75254	−195.00
−190.00	3.915	37.14	26.016	−20.314	198.823	−0.07060	0.74200	−190.00
−185.00	4.835	36.90	21.407	−17.438	200.113	−0.06005	0.73199	−185.00
−180.00	5.922	36.66	17.754	−14.562	201.393	−0.04970	0.72246	−180.00
−175.00	7.195	36.42	14.835	−11.684	202.643	−0.03952	0.71338	−175.00
−170.00	8.675	36.18	12.483	−8.804	203.883	−0.02951	0.70472	−170.00
−165.00	10.386	35.94	10.572	−5.920	205.103	−0.01967	0.69645	−165.00
−160.00	12.351	35.70	9.0094	−3.032	206.303	−0.00999	0.68854	−160.00
−155.00	14.594	35.45	7.7219	−0.139	207.473	−0.00045	0.68097	−155.00
−154.76	14.710	35.44	7.6660	0.000	207.533	0.00000	0.68062	−154.76
−150.00	17.143	35.20	6.6543	2.760	208.623	0.00894	0.67372	−150.00
−145.00	20.022	34.95	5.7635	5.667	209.753	0.01820	0.66676	−145.00
−140.00	23.259	34.69	5.0159	8.581	210.853	0.02734	0.66007	−140.00
−135.00	26.883	34.44	4.3848	11.505	211.923	0.03635	0.65364	−135.00
−130.00	30.922	34.18	3.8493	14.440	212.963	0.04526	0.64744	−130.00
−125.00	35.406	33.92	3.3926	17.387	213.973	0.05405	0.64147	−125.00
−120.00	40.365	33.65	3.0012	20.346	214.963	0.06275	0.63569	−120.00
−115.00	45.829	33.38	2.6643	23.320	215.903	0.07136	0.63010	−115.00
−110.00	51.829	33.11	2.3730	26.310	216.823	0.07987	0.62469	−110.00
−105.00	58.398	32.84	2.1200	29.317	217.693	0.08831	0.61944	−105.00
−100.00	65.567	32.56	1.8995	32.343	218.533	0.09666	0.61433	−100.00
−95.00	73.369	32.27	1.7065	35.390	219.333	0.10495	0.60936	−95.00
−90.00	81.836	31.99	1.5370	38.459	220.093	0.11318	0.60451	−90.00

TABLE B.25 (Continued)
Properties of Saturated Ethylene in Engineering Units

Temp (°F)	Pressure (psia)	Liquid Density (lb/ft³)	Vapor Volume (ft³/lb)	Enthalpy (Btu/lb)		Entropy (Btu/lb-°F)		Temp (°F)
				Liquid	Vapor	Liquid	Vapor	
−85.00	91.001	31.69	1.3876	41.552	220.803	0.12134	0.59976	−85.00
−80.00	100.900	31.40	1.2555	44.671	221.473	0.12946	0.59512	−80.00
−75.00	111.560	31.09	1.1383	47.818	222.083	0.13753	0.59056	−75.00
−70.00	123.030	30.78	1.0340	50.995	222.653	0.14556	0.58608	−70.00
−65.00	135.330	30.47	0.94091	54.205	223.163	0.15355	0.58166	−65.00
−60.00	148.500	30.14	0.85756	57.451	223.623	0.16152	0.57729	−60.00
−55.00	162.580	29.82	0.78274	60.735	224.013	0.16947	0.57295	−55.00
−50.00	177.600	29.48	0.71538	64.060	224.343	0.17741	0.56865	−50.00
−45.00	193.590	29.13	0.65458	67.430	224.603	0.18534	0.56436	−45.00
−40.00	210.610	28.77	0.59955	70.850	224.783	0.19327	0.56007	−40.00
−35.00	228.680	28.41	0.54963	74.322	224.883	0.20122	0.55576	−35.00
−30.00	247.840	28.03	0.50421	77.853	224.903	0.20919	0.55142	−30.00
−25.00	268.140	27.64	0.46278	81.448	224.813	0.21720	0.54703	−25.00
−20.00	289.610	27.24	0.42490	85.114	224.633	0.22525	0.54257	−20.00
−15.00	312.300	26.81	0.39016	88.858	224.323	0.23337	0.53802	−15.00
−10.00	336.250	26.38	0.35822	92.691	223.893	0.24157	0.53334	−10.00
−5.00	361.500	25.92	0.32876	96.622	223.313	0.24987	0.52851	−5.00
0.00	388.120	25.43	0.30150	100.663	222.563	0.25829	0.52348	0.00
5.00	416.150	24.92	0.27618	104.843	221.623	0.26688	0.51821	5.00
10.00	445.640	24.38	0.25258	109.173	220.463	0.27568	0.51262	10.00
15.00	476.650	23.79	0.23045	113.693	219.023	0.28474	0.50664	15.00
20.00	509.260	23.15	0.20957	118.433	217.243	0.29414	0.50014	20.00
25.00	543.540	22.45	0.18970	123.473	215.043	0.30401	0.49294	25.00
30.00	579.580	21.65	0.17056	128.903	212.253	0.31454	0.48476	30.00
35.00	617.490	20.72	0.15172	134.913	208.623	0.32607	0.47508	35.00
40.00	657.420	19.56	0.13244	141.883	203.573	0.33936	0.46282	40.00
45.00	699.580	17.86	0.11056	151.073	195.343	0.35682	0.44453	45.00
48.56	731.240	13.38	0.07477	171.823	171.823	0.39703	0.39703	48.56

Source: Lemmon et al. (2005).

TABLE B.26
Properties of Saturated Ethylene in SI Units

Temp (°C)	Pressure (bar)	Liquid Density (kg/m³)	Vapor Volume (m³/kg)	Enthalpy (kJ/kg)		Entropy (kJ/kg-°C)		Temp (°C)
				Liquid	Vapor	Liquid	Vapor	
−169.16	0.00122	654.60	252.64	−158.095	409.425	−1.17893	4.27867	−169.16
−167.50	0.00163	652.50	192.14	−154.055	411.385	−1.14033	4.21167	−167.50
−165.00	0.00247	649.33	129.69	−147.975	414.345	−1.08353	4.11597	−165.00
−162.50	0.00367	646.15	89.335	−141.895	417.295	−1.02793	4.02577	−162.50
−160.00	0.00534	642.96	62.719	−135.815	420.235	−0.97358	3.94077	−160.00
−157.50	0.00764	639.76	44.817	−129.735	423.185	−0.92045	3.86047	−157.50
−155.00	0.0107	636.55	32.557	−123.665	426.115	−0.86849	3.78477	−155.00
−152.50	0.0149	633.33	24.018	−117.595	429.045	−0.81766	3.71317	−152.50
−150.00	0.0203	630.10	17.974	−111.525	431.965	−0.76792	3.64537	−150.00
−147.50	0.0273	626.86	13.633	−105.475	434.875	−0.71923	3.58117	−147.50
−145.00	0.0362	623.61	10.472	−99.420	437.775	−0.67156	3.52037	−145.00
−142.50	0.0474	620.35	8.1383	−93.374	440.665	−0.62485	3.46267	−142.50
−140.00	0.0614	617.07	6.3951	−87.335	443.535	−0.57908	3.40797	−140.00
−137.50	0.0788	613.78	5.0777	−81.301	446.395	−0.53421	3.35587	−137.50
−135.00	0.1000	610.48	4.0710	−75.273	449.235	−0.49019	3.30647	−135.00
−132.50	0.126	607.16	3.2938	−69.249	452.045	−0.44701	3.25937	−132.50
−130.00	0.156	603.82	2.6880	−63.228	454.845	−0.40461	3.21447	−130.00
−127.50	0.193	600.47	2.2114	−57.210	457.625	−0.36298	3.17177	−127.50
−125.00	0.237	597.10	1.8331	−51.194	460.365	−0.32207	3.13097	−125.00
−122.50	0.288	593.72	1.5305	−45.179	463.095	−0.28187	3.09197	−122.50
−120.00	0.347	590.31	1.2865	−39.163	465.785	−0.24233	3.05477	−120.00
−117.50	0.416	586.88	1.0883	−33.145	468.445	−0.20343	3.01907	−117.50
−115.00	0.496	583.44	0.92612	−27.124	471.075	−0.16514	2.98497	−115.00
−112.50	0.587	579.97	0.79260	−21.098	473.665	−0.12744	2.95227	−112.50
−110.00	0.691	576.47	0.68196	−15.067	476.215	−0.09029	2.92097	−110.00
−107.50	0.809	572.95	0.58972	−9.028	478.735	−0.05368	2.89087	−107.50
−105.00	0.942	569.41	0.51239	−2.980	481.205	−0.01758	2.86187	−105.00
−103.77	1.013	567.65	0.47895	0.000	482.405	0.00000	2.84807	−103.77
−102.50	1.091	565.83	0.44721	3.079	483.635	0.01803	2.83407	−102.50
−100.00	1.259	562.23	0.39198	9.151	486.025	0.05318	2.80727	−100.00
−97.50	1.445	558.60	0.34495	15.237	488.355	0.08789	2.78147	−97.50
−95.00	1.651	554.93	0.30471	21.339	490.645	0.12217	2.75647	−95.00
−92.50	1.880	551.23	0.27014	27.459	492.875	0.15606	2.73237	−92.50
−90.00	2.132	547.50	0.24030	33.600	495.055	0.18956	2.70917	−90.00
−87.50	2.409	543.72	0.21445	39.762	497.175	0.22271	2.68657	−87.50
−85.00	2.712	539.91	0.19196	45.950	499.245	0.25551	2.66477	−85.00
−82.50	3.043	536.06	0.17232	52.163	501.245	0.28800	2.64347	−82.50
−80.00	3.403	532.16	0.15510	58.406	503.175	0.32018	2.62287	−80.00
−77.50	3.795	528.21	0.13997	64.681	505.045	0.35207	2.60287	−77.50

TABLE B.26 (Continued)
Properties of Saturated Ethylene in SI Units

Temp (°C)	Pressure (bar)	Liquid Density (kg/m³)	Vapor Volume (m³/kg)	Enthalpy (kJ/kg)		Entropy (kJ/kg-°C)		Temp (°C)
				Liquid	Vapor	Liquid	Vapor	
−75.00	4.219	524.21	0.12662	70.989	506.845	0.38370	2.58337	−75.00
−72.50	4.677	520.16	0.11480	77.334	508.575	0.41508	2.56427	−72.50
−70.00	5.172	516.06	0.10431	83.719	510.225	0.44623	2.54567	−70.00
−67.50	5.703	511.90	0.09497	90.146	511.785	0.47717	2.52747	−67.50
−65.00	6.274	507.67	0.08663	96.619	513.275	0.50792	2.50967	−65.00
−62.50	6.886	503.38	0.07916	103.145	514.675	0.53848	2.49217	−62.50
−60.00	7.541	499.02	0.07246	109.715	515.985	0.56889	2.47487	−60.00
−57.50	8.239	494.58	0.06643	116.345	517.195	0.59917	2.45797	−57.50
−55.00	8.984	490.07	0.06099	123.035	518.305	0.62932	2.44127	−55.00
−52.50	9.777	485.46	0.05607	129.795	519.315	0.65937	2.42467	−52.50
−50.00	10.619	480.77	0.05161	136.615	520.205	0.68934	2.40827	−50.00
−47.50	11.513	475.98	0.04756	143.515	520.985	0.71925	2.39207	−47.50
−45.00	12.460	471.08	0.04387	150.495	521.635	0.74913	2.37587	−45.00
−42.50	13.462	466.06	0.04051	157.565	522.155	0.77900	2.35967	−42.50
−40.00	14.521	460.92	0.03743	164.725	522.525	0.80888	2.34357	−40.00
−37.50	15.639	455.65	0.03461	171.975	522.745	0.83881	2.32737	−37.50
−35.00	16.818	450.23	0.03202	179.345	522.805	0.86881	2.31097	−35.00
−32.50	18.059	444.64	0.02964	186.835	522.685	0.89892	2.29457	−32.50
−30.00	19.366	438.88	0.02745	194.445	522.375	0.92918	2.27787	−30.00
−27.50	20.739	432.93	0.02542	202.205	521.855	0.95963	2.26087	−27.50
−25.00	22.182	426.75	0.02354	210.115	521.105	0.99032	2.24357	−25.00
−22.50	23.696	420.33	0.02180	218.205	520.095	1.02127	2.22577	−22.50
−20.00	25.284	413.62	0.02017	226.485	518.795	1.05267	2.20737	−20.00
−17.50	26.949	406.60	0.01866	234.985	517.175	1.08447	2.18827	−17.50
−15.00	28.692	399.21	0.01724	243.735	515.195	1.11687	2.16837	−15.00
−12.50	30.517	391.39	0.01591	252.775	512.775	1.14997	2.14747	−12.50
−10.00	32.428	383.04	0.01466	262.165	509.865	1.18387	2.12517	−10.00
−7.50	34.426	374.06	0.01347	271.955	506.345	1.21897	2.10127	−7.50
−5.00	36.516	364.28	0.01233	282.255	502.075	1.25537	2.07517	−5.00
−2.50	38.703	353.46	0.01124	293.215	496.845	1.29377	2.04617	−2.50
0.00	40.990	341.21	0.01018	305.055	490.315	1.33497	2.01317	0.00
2.50	43.385	326.84	0.00912	318.185	481.885	1.38017	1.97407	2.50
5.00	45.896	308.84	0.00802	333.515	470.245	1.43267	1.92427	5.00
7.50	48.535	282.24	0.00674	353.995	451.255	1.50277	1.84927	7.50
9.20	50.417	214.24	0.00467	399.425	399.425	1.66137	1.66137	9.20

Source: Lemmon et al. (2005).

TABLE B.27
Properties of Saturated Propane in Engineering Units

Temp (°F)	Pressure (psia)	Liquid Density (lb/ft³)	Vapor Volume (ft³/lb)	Enthalpy (Btu/lb)		Entropy (Btu/lb-°F)		Temp (°F)
				Liquid	Vapor	Liquid	Vapor	
−200.00	0.0202	42.05	3133.5	−80.216	137.530	−0.23924	0.59932	−200.00
−190.00	0.0401	41.70	1637.0	−75.457	140.150	−0.22126	0.57825	−190.00
−180.00	0.0754	41.35	902.43	−70.677	142.800	−0.20386	0.55945	−180.00
−170.00	0.135	40.99	521.84	−65.873	145.480	−0.18698	0.54266	−170.00
−160.00	0.231	40.64	314.91	−61.043	148.200	−0.17059	0.52765	−160.00
−150.00	0.381	40.28	197.42	−56.184	150.950	−0.15464	0.51424	−150.00
−145.00	0.482	40.11	158.36	−53.743	152.330	−0.14683	0.50807	−145.00
−140.00	0.605	39.93	128.06	−51.294	153.720	−0.13911	0.50224	−140.00
−135.00	0.754	39.75	104.36	−48.836	155.120	−0.13148	0.49672	−135.00
−130.00	0.932	39.57	85.666	−46.370	156.520	−0.12394	0.49150	−130.00
−125.00	1.143	39.38	70.808	−43.894	157.930	−0.11649	0.48657	−125.00
−120.00	1.394	39.20	58.912	−41.409	159.350	−0.10912	0.48190	−120.00
−115.00	1.687	39.02	49.321	−38.913	160.760	−0.10184	0.47749	−115.00
−110.00	2.030	38.84	41.537	−36.407	162.180	−0.09462	0.47332	−110.00
−105.00	2.428	38.65	35.179	−33.891	163.610	−0.08748	0.46937	−105.00
−100.00	2.888	38.47	29.954	−31.363	165.040	−0.08041	0.46564	−100.00
−95.00	3.416	38.28	25.636	−28.824	166.470	−0.07341	0.46212	−95.00
−90.00	4.020	38.09	22.047	−26.272	167.900	−0.06647	0.45879	−90.00
−85.00	4.707	37.90	19.048	−23.708	169.330	−0.05959	0.45565	−85.00
−80.00	5.485	37.71	16.530	−21.132	170.770	−0.05276	0.45268	−80.00
−75.00	6.364	37.52	14.404	−18.542	172.210	−0.04600	0.44988	−75.00
−70.00	7.351	37.33	12.603	−15.938	173.640	−0.03929	0.44723	−70.00
−65.00	8.456	37.14	11.068	−13.320	175.080	−0.03262	0.44473	−65.00
−60.00	9.689	36.94	9.7563	−10.687	176.510	−0.02601	0.44238	−60.00
−55.00	11.059	36.75	8.6295	−8.039	177.950	−0.01944	0.44016	−55.00
−50.00	12.577	36.55	7.6580	−5.376	179.380	−0.01292	0.43807	−50.00
−45.00	14.254	36.35	6.8174	−2.696	180.810	−0.00644	0.43610	−45.00
−43.75	14.700	36.30	6.6252	−2.024	181.170	−0.00483	0.43562	−43.75
−40.00	16.101	36.15	6.0872	0.000	182.240	0.00000	0.43424	−40.00
−35.00	18.128	35.94	5.4509	2.713	183.660	0.00640	0.43250	−35.00
−30.00	20.348	35.74	4.8944	5.444	185.080	0.01277	0.43085	−30.00
−25.00	22.772	35.53	4.4062	8.193	186.500	0.01910	0.42931	−25.00
−20.00	25.412	35.32	3.9766	10.960	187.910	0.02540	0.42786	−20.00
−15.00	28.281	35.11	3.5974	13.747	189.320	0.03167	0.42650	−15.00
−10.00	31.390	34.90	3.2618	16.553	190.720	0.03791	0.42522	−10.00
−5.00	34.754	34.68	2.9638	19.380	192.110	0.04412	0.42403	−5.00
0.00	38.385	34.46	2.6987	22.227	193.500	0.05030	0.42290	0.00
5.00	42.296	34.24	2.4620	25.096	194.880	0.05646	0.42185	5.00
10.00	46.500	34.02	2.2503	27.987	196.250	0.06260	0.42087	10.00

TABLE B.27 (Continued)
Properties of Saturated Propane in Engineering Units

Temp (°F)	Pressure (psia)	Liquid Density (lb/ft³)	Vapor Volume (ft³/lb)	Enthalpy (Btu/lb)		Entropy (Btu/lb-°F)		Temp (°F)
				Liquid	Vapor	Liquid	Vapor	
15.00	51.012	33.80	2.0605	30.900	197.620	0.06872	0.41994	15.00
20.00	55.844	33.57	1.8899	33.836	198.970	0.07482	0.41908	20.00
25.00	61.011	33.34	1.7362	36.796	200.310	0.08090	0.41827	25.00
30.00	66.527	33.10	1.5975	39.781	201.640	0.08696	0.41751	30.00
35.00	72.406	32.86	1.4719	42.791	202.960	0.09301	0.41680	35.00
40.00	78.662	32.62	1.3581	45.827	204.270	0.09905	0.41613	40.00
45.00	85.310	32.38	1.2548	48.889	205.560	0.10507	0.41550	45.00
50.00	92.365	32.13	1.1607	51.979	206.830	0.11108	0.41491	50.00
55.00	99.841	31.88	1.0749	55.097	208.090	0.11708	0.41435	55.00
60.00	107.750	31.62	0.99652	58.245	209.330	0.12308	0.41382	60.00
65.00	116.120	31.36	0.92479	61.422	210.560	0.12907	0.41332	65.00
70.00	124.950	31.10	0.85903	64.630	211.760	0.13506	0.41283	70.00
75.00	134.260	30.83	0.79865	67.870	212.940	0.14104	0.41237	75.00
80.00	144.080	30.56	0.74310	71.144	214.090	0.14703	0.41191	80.00
85.00	154.410	30.28	0.69192	74.452	215.220	0.15301	0.41146	85.00
90.00	165.270	29.99	0.64468	77.797	216.320	0.15900	0.41102	90.00
95.00	176.680	29.70	0.60102	81.178	217.390	0.16500	0.41058	95.00
100.00	188.650	29.40	0.56059	84.600	218.430	0.17100	0.41013	100.00
105.00	201.200	29.10	0.52310	88.062	219.430	0.17702	0.40966	105.00
110.00	214.360	28.79	0.48828	91.568	220.390	0.18305	0.40918	110.00
115.00	228.140	28.47	0.45588	95.120	221.300	0.18911	0.40868	115.00
120.00	242.550	28.14	0.42569	98.721	222.170	0.19518	0.40814	120.00
125.00	257.620	27.80	0.39751	102.370	222.980	0.20129	0.40756	125.00
130.00	273.370	27.45	0.37116	106.080	223.730	0.20743	0.40694	130.00
135.00	289.820	27.08	0.34648	109.860	224.420	0.21361	0.40625	135.00
140.00	306.980	26.71	0.32331	113.690	225.030	0.21984	0.40549	140.00
145.00	324.900	26.32	0.30152	117.610	225.550	0.22613	0.40464	145.00
150.00	343.580	25.91	0.28098	121.600	225.980	0.23249	0.40369	150.00

Source: Lemmon et al. (2005).

TABLE B.28
Properties of Saturated Propane in SI Units

Temp (°C)	Pressure (bar)	Liquid Density (kg/m³)	Vapor Volume (m³/kg)	Enthalpy (kJ/kg)		Entropy (kJ/kg-°C)		Temp (°C)
				Liquid	Vapor	Liquid	Vapor	
−130.00	0.00120	674.66	224.32	−83.924	423.230	−0.39033	3.15250	−130.00
−125.00	0.00226	669.60	123.43	−73.980	428.670	−0.32205	3.07080	−125.00
−120.00	0.00406	664.54	71.058	−63.996	434.190	−0.25578	2.99710	−120.00
−115.00	0.00699	659.46	42.600	−53.969	439.770	−0.19136	2.93060	−115.00
−110.00	0.0116	654.37	26.489	−43.895	445.420	−0.12865	2.87050	−110.00
−105.00	0.0186	649.27	17.024	−33.770	451.130	−0.06753	2.81620	−105.00
−100.00	0.0289	644.13	11.273	−23.590	456.900	−0.00788	2.76710	−100.00
−97.50	0.0356	641.56	9.2683	−18.477	459.800	0.02143	2.74430	−97.50
−95.00	0.0436	638.97	7.6695	−13.349	462.710	0.05041	2.72270	−95.00
−92.50	0.0531	636.38	6.3856	−8.205	465.630	0.07908	2.70210	−92.50
−90.00	0.0642	633.78	5.3479	−3.044	468.570	0.10744	2.68250	−90.00
−87.50	0.0773	631.17	4.5040	2.134	471.520	0.13551	2.66380	−87.50
−85.00	0.0924	628.55	3.8136	7.330	474.470	0.16330	2.64610	−85.00
−82.50	0.110	625.92	3.2455	12.545	477.430	0.19082	2.62930	−82.50
−80.00	0.130	623.28	2.7756	17.780	480.400	0.21808	2.61320	−80.00
−77.50	0.153	620.62	2.3847	23.034	483.380	0.24509	2.59800	−77.50
−75.00	0.180	617.96	2.0581	28.309	486.370	0.27186	2.58350	−75.00
−72.50	0.210	615.28	1.7837	33.606	489.360	0.29840	2.56980	−72.50
−70.00	0.243	612.59	1.5523	38.924	492.350	0.32471	2.55670	−70.00
−67.50	0.282	609.89	1.3561	44.266	495.350	0.35082	2.54430	−67.50
−65.00	0.325	607.17	1.1891	49.631	498.350	0.37671	2.53250	−65.00
−62.50	0.373	604.43	1.0464	55.020	501.350	0.40241	2.52120	−62.50
−60.00	0.426	601.68	0.92404	60.434	504.360	0.42792	2.51060	−60.00
−57.50	0.486	598.91	0.81861	65.874	507.360	0.45325	2.50050	−57.50
−55.00	0.552	596.13	0.72749	71.340	510.360	0.47840	2.49090	−55.00
−52.50	0.624	593.32	0.64844	76.834	513.370	0.50338	2.48180	−52.50
−50.00	0.705	590.50	0.57965	82.356	516.370	0.52820	2.47310	−50.00
−47.50	0.793	587.66	0.51958	87.906	519.370	0.55287	2.46500	−47.50
−45.00	0.889	584.80	0.46697	93.487	522.360	0.57739	2.45720	−45.00
−42.50	0.995	581.91	0.42074	99.097	525.360	0.60177	2.44980	−42.50
−42.08	1.014	581.43	0.41359	100.040	525.850	0.60583	2.44870	−42.08
−40.00	1.110	579.01	0.38001	104.740	528.340	0.62602	2.44290	−40.00
−37.50	1.235	576.08	0.34402	110.410	531.320	0.65013	2.43630	−37.50
−35.00	1.371	573.13	0.31213	116.120	534.300	0.67413	2.43010	−35.00
−32.50	1.518	570.15	0.28381	121.860	537.260	0.69800	2.42420	−32.50
−30.00	1.677	567.15	0.25858	127.640	540.220	0.72176	2.41860	−30.00
−27.50	1.849	564.12	0.23605	133.450	543.170	0.74542	2.41330	−27.50
−25.00	2.034	561.06	0.21590	139.300	546.100	0.76897	2.40830	−25.00
−22.50	2.232	557.98	0.19781	145.180	549.030	0.79242	2.40360	−22.50

TABLE B.28 (Continued)
Properties of Saturated Propane in SI Units

Temp (°C)	Pressure (bar)	Liquid Density (kg/m³)	Vapor Volume (m³/kg)	Enthalpy (kJ/kg)		Entropy (kJ/kg-°C)		Temp (°C)
				Liquid	Vapor	Liquid	Vapor	
−20.00	2.445	554.86	0.18156	151.110	551.940	0.81579	2.39920	−20.00
−17.50	2.673	551.71	0.16692	157.070	554.840	0.83907	2.39500	−17.50
−15.00	2.916	548.54	0.15370	163.070	557.730	0.86227	2.39100	−15.00
−12.50	3.176	545.32	0.14174	169.120	560.600	0.88539	2.38730	−12.50
−10.00	3.453	542.08	0.13090	175.210	563.450	0.90843	2.38380	−10.00
−7.50	3.748	538.79	0.12106	181.340	566.290	0.93141	2.38050	−7.50
−5.00	4.061	535.47	0.11211	187.510	569.100	0.95433	2.37740	−5.00
−2.50	4.394	532.11	0.10395	193.730	571.900	0.97719	2.37440	−2.50
0.00	4.746	528.71	0.09650	200.000	574.670	1.00000	2.37170	0.00
2.50	5.119	525.27	0.08969	206.320	577.420	1.02280	2.36910	2.50
5.00	5.513	521.79	0.08345	212.680	580.150	1.04550	2.36660	5.00
7.50	5.929	518.26	0.07772	219.100	582.840	1.06810	2.36420	7.50
10.00	6.368	514.68	0.07246	225.560	585.510	1.09080	2.36200	10.00
12.50	6.831	511.05	0.06762	232.080	588.150	1.11340	2.35990	12.50
15.00	7.318	507.37	0.06315	238.660	590.750	1.13600	2.35790	15.00
17.50	7.830	503.64	0.05904	245.290	593.320	1.15850	2.35600	17.50
20.00	8.368	499.85	0.05523	251.980	595.850	1.18110	2.35410	20.00
22.50	8.932	495.99	0.05170	258.720	598.340	1.20360	2.35230	22.50
25.00	9.524	492.08	0.04844	265.530	600.780	1.22620	2.35060	25.00
27.50	10.144	488.10	0.04540	272.410	603.180	1.24870	2.34890	27.50
30.00	10.793	484.05	0.04259	279.350	605.530	1.27120	2.34720	30.00
32.50	11.472	479.93	0.03996	286.360	607.820	1.29380	2.34550	32.50
35.00	12.181	475.73	0.03752	293.430	610.060	1.31640	2.34390	35.00
37.50	12.923	471.45	0.03524	300.590	612.230	1.33900	2.34220	37.50
40.00	13.696	467.07	0.03311	307.820	614.330	1.36160	2.34040	40.00
42.50	14.503	462.60	0.03112	315.130	616.360	1.38430	2.33870	42.50
45.00	15.344	458.03	0.02925	322.520	618.300	1.40710	2.33680	45.00
47.50	16.221	453.34	0.02750	330.010	620.160	1.42990	2.33480	47.50
50.00	17.133	448.53	0.02586	337.590	621.920	1.45290	2.33270	50.00
52.50	18.083	443.59	0.02431	345.280	623.580	1.47590	2.33050	52.50
55.00	19.071	438.50	0.02286	353.070	625.120	1.49900	2.32810	55.00
57.50	20.098	433.25	0.02148	360.980	626.530	1.52230	2.32550	57.50
60.00	21.166	427.81	0.02018	369.020	627.800	1.54580	2.32260	60.00
62.50	22.275	422.17	0.01896	377.190	628.910	1.56950	2.31940	62.50
65.00	23.427	416.30	0.01779	385.520	629.840	1.59340	2.31590	65.00

Source: Lemmon et al. (2005).

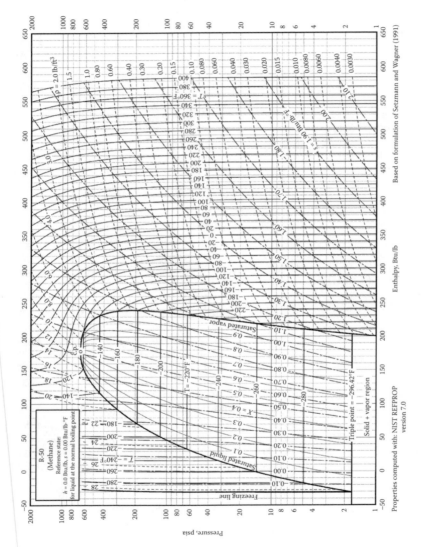

FIGURE B.24 Pressure–enthalpy diagram for methane in engineering units (ASHRAE, 2005).

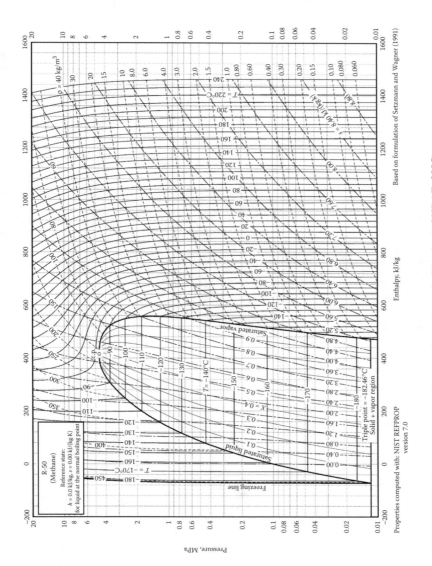

FIGURE B.25 Pressure–enthalpy diagram for methane in SI units (ASHRAE, 2005).

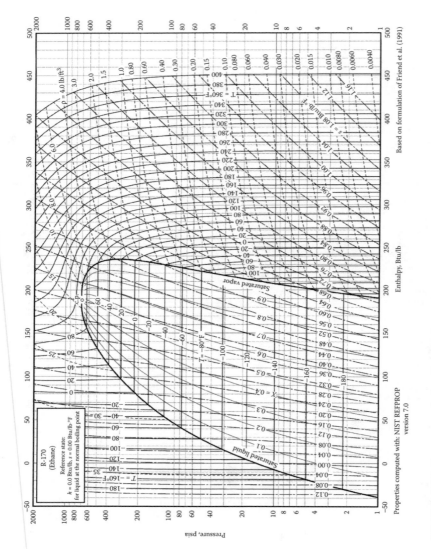

FIGURE B.26 Pressure–enthalpy diagram for ethane in engineering units (ASHRAE, 2005).

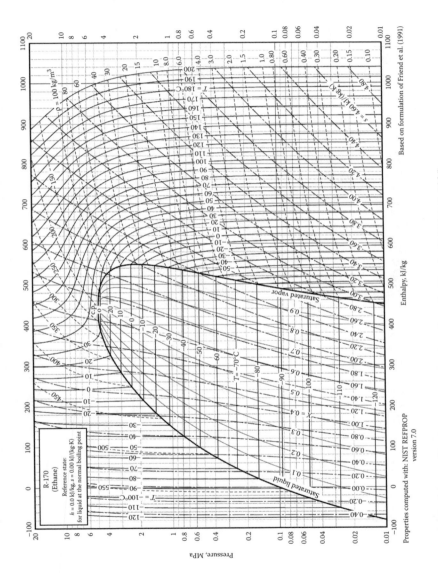

FIGURE B.27 Pressure–enthalpy diagram for ethane in SI units (ASHRAE, 2005).

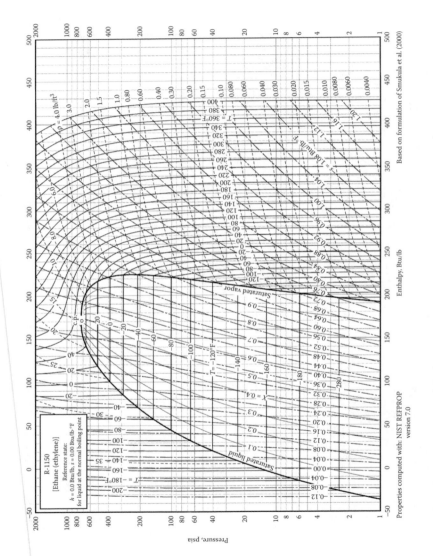

FIGURE B.28 Pressure–enthalpy diagram for ethylene in engineering units (ASHRAE, 2005).

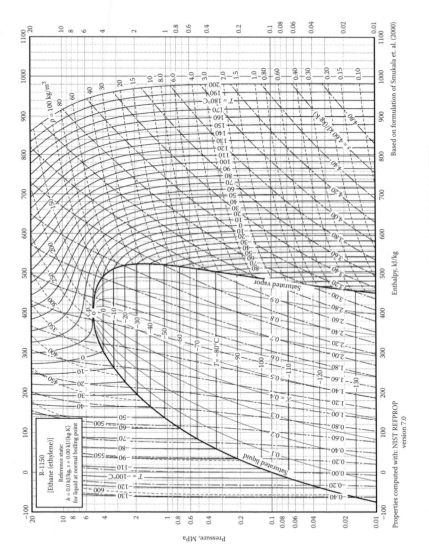

FIGURE B.29 Pressure–enthalpy diagram for ethylene in SI units (ASHRAE, 2005).

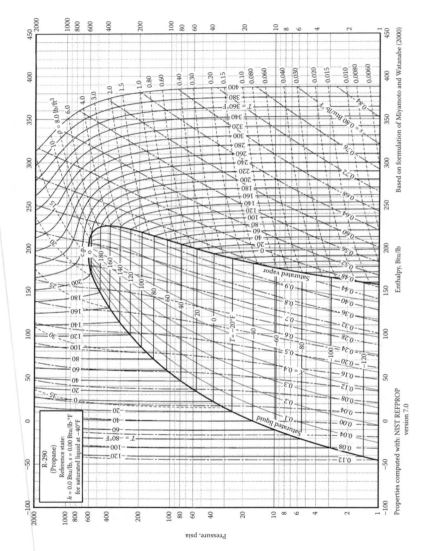

FIGURE B.30 Pressure–enthalpy diagram for propane in engineering units (ASHRAE, 2005).

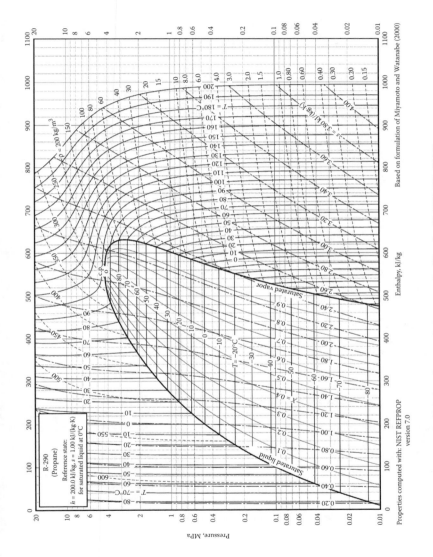

FIGURE B.31 Pressure–enthalpy diagram for propane in SI units (ASHRAE, 2005).

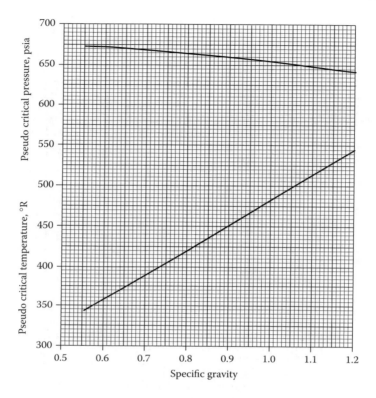

FIGURE B.32 Pseudocritical temperature and pressure as a function of gas specific gravity. Applicable to natural gas mixtures that contain less than 5 mol% N_2, 2 mol% CO_2, and 2 mol% H_2S. (Adapted from Engineering Data Book, 1987.)

B.4 HYDROCARBON COMPRESSIBILITY FACTORS

The compressibility factor z for natural gases can be estimated by use of Figures B.32 and B.33 with the following procedure:

Use the gas specific gravity and Figure B.32 to determine the pseudocritical temperature, T_c, and pressure, P_c. If only gas composition is known, compute molar mass of gas and divide by molar mass of air (28.959) to obtain specific gravity.

Compute pseudoreduced temperature T_R ($= (t[°F] +460)/T_c$) and pressure P_R $= P(psia)/P_c$.

Read compressibility factor from Figure B.33.

As noted in the figure captions, the correlations are not valid at high acid gas concentrations. The Engineering Data Book (2004c) provides a method applicable to high hydrogen sulfide concentrations. For best results, computer programs with equations of state that properly predict phase behavior of sour gases should be used.

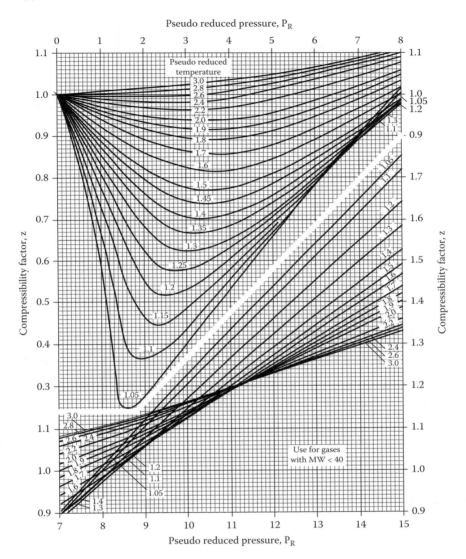

FIGURE B.33 Compressibility factor chart as a function of pseudocritical temperature and pressure. Chart applies to gases with a molar mass of less than 40. (Adapted from Engineering Data Book, 2004c.)

REFERENCES

American Society of Heating, Refrigerating and Air-Conditioning Engineers, Inc. (www.ashrae.org). Reprinted by permission from 2005 ASHRAE Handbook—Fundamentals, © 2005. This material may not be copied nor distributed in either paper or digital form without ASHRAE's permission.

Design Institute for Physical Properties, DIPPR Evaluated Pure Component Database, American Institute of Chemical Engineers, New York, 2005.

Dow, Alkyl alkanolamines, Dow Chemical Company, Midland, MI, 2003a, http://www.dow.com/PublishedLiterature/dh_03bf/09002f13803bfe14.pdf, Retrieved November 2005.

Dow, Ethanolamines, Dow Chemical Company, Midland, MI, 2003b. http://www.dow.com/PublishedLiterature/dh_03bf/09002f13803bfe11.pdf, Retrieved November 2005.

Dow, Tetraethylene Glycol, Dow Chemical Company, Midland MI, 2003c, http://www.dow.com/PublishedLiterature/dh_0451/09002f13804518f4.pdf, Retrieved November 2005.

Dow, Triethylene Glycol, Dow Chemical Company, Midland MI, 2003d. http://www.dow.com/PublishedLiterature/dh_0451/09002f13804518f1.pdf, Retrieved November 2005.

Engineering Data Book, 12th ed. Sec. 1, General Information, Gas Processors Supply Association., Tulsa, OK, 2004a.

Engineering Data Book, 12th ed. Sec. 13, Compressors and Expanders, Gas Processors Supply Association, Tulsa, OK, 2004b.

Engineering Data Book, Physical Properties, 12th ed., Sec. 23, Gas Processors Supply Association, Tulsa, OK, 2004c.

GPA Publication 1167-77, Glossary, Definition of Words and Terms Used in the Gas Processing Industry, Gas Processors Association, Tulsa, OK, 1977.

GPA Standard 2145-03 Rev. 1, Table of Physical Constants for Hydrocarbons and Other Compounds of Interest to the Natural Gas Industry, Gas Processors Association, Tulsa, OK, 2005.

INEOS, GAS/SPEC MDEA Specialty Amine, INEOS Oxide, Houston, TX, undated http://www.coastalchem.com/PDFs/GAS_SPEC/MDEA.pdf, Retrieved November 2005.

MEGlobal, Diethylene Glycol, MEGlobal Group of Companies, London, UK, 2005a, http://www.meglobal.biz/literature/ product_ guides/ MEGlobal_DEG.pdf, Retrieved November 2005.

MEGlobal, Ethylene Glycol, MEGlobal Group of Companies, London, UK, 2005b, http://www.meglobal.biz/literature/product_ guides/ MEGlobal_MEG.pdf, Retrieved November 2005.

Lemmon, E.W., McLinden, M.O., and Friend, D.G., Thermophysical properties of fluid systems, in *NIST Chemistry WebBook, NIST Standard Reference Database Number 69*, Linstrom P.J. and Mallard, W.G.,Eds., 2005, National Institute of Standards and Technology, Gaithersburg MD,. http://webbook.nist.gov/chemistry/fluid/,Retrieved November 2005.

Author Index

Subject Index

425